Aggregates in Concrete

Modern concrete technology series

A series of books presenting the state of the art in concrete technology
Series Editors

Arnon Bentur
National Building Research Institute
Faculty of Civil and Environmental Engineering
Technion-Israel Institute of Technology
Technion City
Haifa 32 000
Israel

Sidney Mindess
Department of Civil Engineering
University of British Columbia
6250 Applied Science Lane
Vancouver, B.C. V6T 1Z4
Canada

Aggregates in Concrete

Mark Alexander

and

Sidney Mindess

CRC Press
Taylor & Francis Group
Boca Raton London New York

CRC Press is an imprint of the
Taylor & Francis Group, an **informa** business
A TAYLOR & FRANCIS BOOK

CRC Press
Taylor & Francis Group
6000 Broken Sound Parkway NW, Suite 300
Boca Raton, FL 33487-2742

First issued in paperback 2019

© 2005 by Taylor & Francis Group, LLC
CRC Press is an imprint of Taylor & Francis Group, an Informa business

No claim to original U.S. Government works

ISBN-13: 978-0-415-25839-5 (hbk)
ISBN-13: 978-0-367-86459-0 (pbk)

Typeset in Sabon by
Integra Software Services Pvt. Ltd, Pondicherry, India

British Library Cataloguing in Publication Data
A catalogue record for this book is available from the British Library

Library of Congress Cataloging in Publication Data
A catalog record for this book has been requested

Visit the Taylor & Francis Web site at
http://www.taylorandfrancis.com

and the CRC Press Web site at
http://www.crcpress.com

Contents

Preface

Cementitious materials are certainly the oldest *manufactured* materials of construction, their use going back at least 9000 years. Today, portland cement concrete is the most widely used construction material worldwide, its production far outstripping that of asphalt, timber, steel or other building materials. Indeed, it is second only to water as the most widely used material of any type. Since concrete aggregates typically make up about 70 per cent of the mass of concrete, they are clearly a vitally important ingredient for two main reasons:

1 Their properties must affect to a considerable degree the properties of the concrete; and
2 The vast quantities of aggregates used in concrete production have a significant environmental impact.

In spite of their importance, however, aggregates tend very much to play 'second fiddle' to the other principal ingredient of modern concrete, namely portland cement. Most concrete research over the past decades has focused on the *binder phase* (i.e. the portland cement and the supplementary cementitious materials and chemical admixtures that are commonly combined with it). This is, perhaps, understandable, since it is largely through the intelligent manipulation (or 'engineering') of the binder that we can now 'tailor-make' concretes with such a wide range of properties, such as ultra-high strength concretes or self-compacting concretes. However, it must be remembered that even these very high performance concretes would not be possible without an intelligent selection of their aggregates as well.

What aggregate research there is has tended to focus either on aggregates which are chemically reactive with portland cement, such as those involved in alkali-aggregate reactions, or on aggregates for special concretes (low density concretes, concretes for radiation shielding, etc.). These aggregates, while important, make up only a small fraction of modern concrete production. Indeed, it is still commonly assumed that aggregates are essentially an inert component of concrete, used primarily as an economical filler and to give the concrete some volume stability. At least part of the purpose of this book is to show the falsity of this belief.

The aim of this book is to describe and explain the role of the aggregate phase *in concrete*. That is, it is not intended to be merely a bald description of the origins of various rock types and their chemical and physical properties (although some of that material will of necessity be included). Rather it is intended to show the relevance of these materials and their properties to the behaviour of concrete. While much of the material presented will, perforce, be of an empirical nature, the underlying science will be dealt with whenever possible. The intent is to present a unified view of the role of concrete aggregates in the light of this science, rather than simply as a series of more or less unrelated facts.

The book comprises eight chapters. After a general introduction, Chapter 2 describes the origin, classification and production methods for naturally occurring aggregates, while Chapter 3 discusses the physical, mechanical, and chemical properties of these aggregates as they relate to the properties of concrete. The role of aggregates in fresh concrete is covered in Chapter 4, with particular emphasis on particle packing and concrete rheology. Chapter 5 deals with how aggregates affect the physical and mechanical properties of hardened concrete. Chapter 6 is concerned primarily with the effects of aggregates on transport of substances in concrete, and on concrete deterioration and durability. Special aggregates (such as lightweight aggregates, synthetic aggregates, etc.) and the more stringent requirements of aggregates for special concretes are discussed in Chapter 7. Finally, a description of standards for aggregates is given in Chapter 8. Various standards (ASTM, BS, CSA, and SANS) are referred to throughout the text. The relevant standards are gathered in lists in the Appendices, and therefore are not referenced on each occasion when they appear. Canadian (CSA) Standards are only quoted where they differ from ASTM standards, or where no comparable ASTM standard exists.

This book is not written specifically as a textbook; however, we hope that it will be of use to students as well as to practising engineers. It includes an extensive and up-to-date reference list, and numerous graphs and tables.

We wish to thank the following people who assisted in different ways: Gill Owens of the Cement and Concrete Institute (C&CI), who sourced various photographs and diagrammatic material, as did Prof. Yunus Ballim; the wonderfully helpful librarians at the C&CI – Hanlie Turner, Ansie Martinek, and Grace Legoale – who dug out much material that was obscure and not so obscure; Lynette Alexander who typed the manuscript so diligently; students at the University of Cape Town, in particular Tom Gardner and Mafanelo Sibuyi, who helped with drawing many of the diagrams; several people who provided useful criticism and input on various sections, notably Prof. Geoff Blight, Dr Ian Sims, and Dr Bertie Oberholster for assistance with the section on alkali-aggregate reaction in Chapter 6,

Dr Clive Stowe for critical review of the section on aggregate origins, petrography and chemical properties, and Dr Graham Grieve for a critical reading of Chapter 8; Dr V.R. Kulkarni of the Associated Cement Companies Limited of India, also Editor of the *Indian Concrete Journal*, for providing data on cement and aggregate production in India; Mrs Elly Yelverton for assisting hugely with getting the necessary permissions for use of copyright material; others who provided photographic or diagrammatic material – Profs J. Newman and T. Bremner, Dr E. Garboczi, Dr H. Frimmel, Mr D. Labuschagne, Mr P. Evans, Mr N. Hassen, Mr H. Hale, Mr J. Cokart or Mr C. Casalena.

Without doubt, this book would not have been written were it not for the keen interest and enquiring mind of Dr Derek Davis, previous Director of the Portland Cement Institute (now the Cement and Concrete Institute), who initiated several studies on the effects of aggregates in concrete, and provided inspiration for much of the work done by Dr Alexander.

Mark Alexander

Sidney Mindess

Acknowledgements

The authors gratefully acknowledge the provision of material in various figures and tables in the book. Sources are given below.

American Ceramic Society
Chapter 3: Figure 3.17a, b; Table 3.12
Chapter 5: Figures 5.20, 5.28, 5.29, 5.30, 5.31, and 5.36; Table 5.3

American Concrete Institute
Chapter 5: Figures 5.5, 5.6, 5.13, 5.18, 5.19, 5.25, and 5.32
Chapter 7: Figure 7.1

American Society for Testing and Materials
Chapter 3: Figures 3.23 and 3.28; Tables 3.3 and 3.14
Chapter 5: Figure 5.4

British Standards Institute
Chapter 2: Figure 2.11; Tables 2.3, 2.4, 2.8, 2.9, 2.10, and 2.11
Chapter 3: Tables 3.4 and 3.5
Chapter 7: Table 7.11

Cement Association of Canada
Chapter 3: Figures 3.3a, b and 3.4
Chapter 4: Figure 4.12

Cement and Concrete Institute
Chapter 1: Figure 1.2
Chapter 5: Figures 5.22, 5.23, 5.33, 5.34, 5.35, and 5.44; Tables 5.4, 5.5, and 5.7
Chapter 6: Figures 6.8, 6.11, and 6.18; Table 6.2 and 6.5

Concrete Society (London)
Chapter 4: Figure 4.13
Chapter 7: Tables 7.10, 7.11, and 7.12

Concrete Society of Southern Africa
Chapter 5: Figures 5.7, 5.9, and 5.16; Table 5.6

E&F N Spon
Chapter 3: Figure 3.2

Elsevier
Chapter 2: Figures 2.3 and 2.8
Chapter 3: Figures 3.5 and 3.27
Chapter 5: Figures 5.11
Chapter 6: Figures 6.2 and 6.3

Geological Society of London
Chapter 2: Table 2.10
Chapter 5: Figure 5.43

Her Majesty's Stationery Office
Chapter 4: Figure 4.6
Chapter 5: Figure 5.21

National Sand, Gravel and Stone Association
Chapter 2: Figure 2.10

Natural Resources, Canada
Chapter 6: Figures 6.13 and 6.14

Pearson Education
Chapter 4: Table 4.3

Prentice Hall
Chapter 3: Figures 3.9 and 3.15
Chapter 5: Figure 5.1

Quarry Management
Chapter 5: Figure 5.10

RILEM
Chapter 5: Table 5.1
Chapter 3: Figures 3.18 and 3.31

South African Institution of Civil Engineering
Chapter 5: Figures 5.8 and 5.42
Chapter 6: Figure 6.16

Dr M. R. Smith (Institute of Quarrying)
Chapter 2: Table 2.10
Chapter 5: Figure 5.43

Thomas Telford
Chapter 5: Figures 5.39 and 5.41

Chapter 1

Introduction

This book concerns aggregates *in concrete*. It deals not primarily with aggregates *per se*, but with the role that aggregates play in producing this construction material. Its theme is the importance of aggregates in modern concrete science and engineering. The book argues for a better understanding and appreciation of the role of aggregates, and illustrates how aggregates crucially influence the properties of the composite material. It departs from the outdated view of aggregates as being simply inert fillers, little more than a bulk constituent required for mass and economy. A materials science view is taken in which each constituent of the concrete is important in its own right, with interaction between the constituents governing the overall properties.

Concrete is used for engineering purposes only because it contains aggregates – cement alone is unsuitable for most purposes except a few special applications. In one form or another, aggregates have been used in bound materials of construction for millennia. Roman concrete incorporated aggregates in the form of 'rubble', probably broken stone or broken brick (Vitruvius, trans. Morgan, 1960). Modern concretes use aggregates of various types, with ever-increasing sophistication as concrete science and technology advance. Today, highly complex mixtures, which may comprise several binders, admixtures, and aggregates of different types and sizes, are used in modern construction. These mixtures are a far distance from the crude mixtures produced by the simple volumetric proportions of even half a century ago. They rely critically on the engineering design of grading, particle packing, rheological properties, and internal chemistry for their success.

The greater part of this book deals with conventional concrete mixtures which make up the vast bulk of concrete produced worldwide. Natural aggregates occupy the majority of the discussion, since they remain the ubiquitous constituent of modern concretes. However, newer aggregates in the form of synthetic and recycled materials are covered, although they really require a book in their own right. Special aggregates, actually aggregates for special concretes or special applications, are also dealt with. These include lightweight aggregates, high density aggregates, aggregates for thermal insulation, radiation shielding, and so on.

Concrete engineers and technologists must take the role of aggregates more seriously, since there are increasing demands of modern concrete mixtures in terms of technological properties and greater economy. Also, sources of natural aggregates are becoming increasingly scarce. The focus of concrete science and technology in the last half century has mostly been on the binder component; increasing focus on the aggregate component will now also be required in the coming decades.

This book has the flavour of concrete and aggregate practice in those parts of the world related to the experience of the authors. Thus, Canadian practice, which reflects general North American practice mostly, is covered. South African practice is also covered, and since standards and practice used in South Africa often relate to British standards and practice, the book refers to UK sources. The different approaches are compared and contrasted in a way that helps to elucidate the underlying principles of concrete engineering. The book should therefore be useful not only to students, practitioners and researchers in these fields who are active in the geographical areas covered, but also to those involved in concrete engineering worldwide.

Some definitions

Concrete is a mixture of water, cement or binder, and aggregates. Chemical admixtures are also incorporated in most modern concrete mixtures. Here, the binder phase for concrete is assumed to be based on portland cement, although several additional components – extenders or supplementary cementitious materials – may also be present. Aggregates are defined as mineral constituents of concrete in granular or particulate form, usually comprising both coarse and fine fractions. The definition of aggregates in ASTM C125 is 'a granular material such as sand, gravel, crushed stone or iron blast-furnace slag, used with a cementing medium to form hydraulic-cement concrete or mortar'. Our concern is mainly with natural aggregates, composed of rock fragments which are used in their natural state except for operations such as crushing, washing, and sizing. They are generally hard, non-cohesive granular materials of varying sizes. Of course, 'aggregate' does not need to be restricted to use with a mineral binder. Aggregates are extensively employed as ingredients of asphalt mixtures, ballast and fill materials, for road bases and formations, for decorative purposes, as filter and drainage media, and for various manufacturing processes such as metal casting, fluxing, and so on.

Concrete is essentially an artificial conglomerate comprising fragments (aggregates) that are held together by a cementing phase. In cross section, it appears much like coarse-grained natural geological conglomerate. This

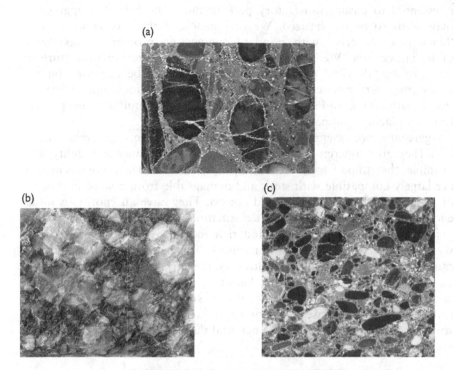

Figure 1.1 Sections of geological conglomerates and modern concrete: (a) Polymictic conglomerate with detrital pyrite, Steyn Reef, Welkom Goldfield, South Africa; max. size of inclusions *c*.20 mm; (b) Quartz-pebble conglomerate, Ventersdorp Contact Reef, Klerksdorp Goldfield, South Africa; max. size of pebble inclusions *c*.30–50 mm; (c) Concrete; max. size of aggregate *c*.20 mm (Figures 1.1a and b courtesy of Dr H. Frimmel, Department of Geological Sciences, University of Cape Town).

is illustrated in Figure 1.1, which shows a natural conglomerate and a modern concrete.

The use of correct terminology is very important in science and engineering. Therefore, a glossary and list of notations is provided elsewhere in the book.

Purpose and role of aggregates

Between 70 and 80 per cent of the volume of concrete is occupied by aggregates. Consequently, we should expect them to exercise profound influences on concrete properties and performance. Despite this expectation, many engineers continue to view aggregates as merely inert fillers. This notion must be dispelled finally. The proper 'engineering' of all the constituents of modern concrete mixtures, including the aggregates, is essential to ensure satisfactory performance. The role that aggregates play needs to be emphasized. We can no longer think of concrete as a 'black box' material, with standard engineering properties irrespective of its ingredients. We must begin to think of concrete manufacture in terms of an 'alloying' process whereby control can be exercised on the engineering performance by intelligent use of the various input variables (the constituents), and by giving attention to the manufacturing process (mixing, placing, compacting, and curing).

Aggregates are essential in making concrete into an engineering material. They give concrete its necessary property of volumetric stability. For example, they impart to the concrete properties of thermal movement that are largely compatible with steel, and manageable from a structural point of view – which pure paste would not do. They have an enormous influence on reducing moisture-related deformations (e.g. shrinkage) of concrete, a fact that renders pure paste and rich mortars very difficult to work with. Figure 1.2 illustrates this in terms of drying shrinkage; in the normal range of aggregate volume concentration, the shrinkage of concrete is only 10–15 per cent that of pure paste. Also, they impart wear resistance to concrete making it suitable for use as trafficked surfaces and in hydraulic structures. Aggregates exercise an important influence on concrete strength and stiffness, providing rigidity to the material that is necessary for engineering

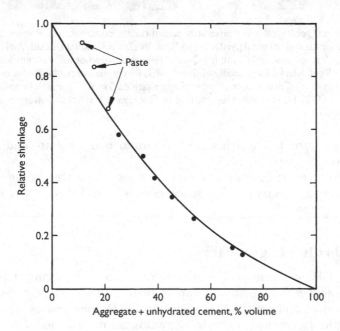

Figure 1.2 Effect of aggregate on reducing shrinkage of pure paste (Powers, 1971) (from Addis and Owens, 2001).

use. They restrain creep of the paste, giving acceptable long-term deformation properties. Lastly, aggregates are generally the more durable and stable of the materials incorporated into concrete mixtures, and thus provide durability. Occasionally, aggregates themselves may exhibit a lack of durability, but this tends to be the exception.

Aggregates thus help to produce an engineering material in two important ways. They reduce the cost, being generally the cheaper constituent, at the same time transforming the binder phase into a useful material. They fulfil the basic requirements of successful engineering: to produce a product that is fit-for-purpose in the most economical way. To quote from Legg:

> The conclusion becomes inescapable that aggregates are not simply fillers used to dilute the expensive water-cement paste and thus make a cheaper product. Economics are important, but significant improvements in the workability of the fresh concrete are contributed by proper choice of aggregates. Such choice influences highly important properties of the hardened concrete as well, such as volume stability, unit weight, resistance to destructive environments, strength, thermal properties, and pavement slipperiness.
>
> (Legg, 1974)

Consumption of aggregates

Besides water and soil, aggregates are the most abundantly used materials worldwide. Aggregate production and consumption vary widely, depending on the development of a country or region, its economic activity, the nature of construction carried out, and so on. An estimate by the US Bureau of Mines put aggregate annual demand in that country at nearly 2.5 billion short tons of natural aggregate in 2000. This estimate presumably covers all uses of natural aggregates. The UK production of aggregates from all sources in 1995 was approximately 250 million tons. Regarding aggregate usage *in concrete*, a conservative estimate is that at least 4.5 billion tons of concrete aggregates per year are consumed worldwide. Thus, the sheer bulk of global aggregate usage is staggering.

Turning to measures of per capita consumption of aggregates, this also varies widely worldwide. For example, aggregate (sand and gravel) consumption for British Columbia, Canada, in 2003 was estimated at about 15 tons per year for each resident of the province. On the other hand, estimates for annual UK per capita consumption are 4 tons, USA 8 tons, France 7 tons, and Italy and Japan 6 tons. In Europe, the figures vary from a high of about 14 tons per person in Norway, to a low of 3 tons in the Netherlands, with an average figure of about 7 tons. These figures are assumed to represent total aggregate production, including usage in road bases and wearing courses, fills, and so on. Aggregate usage in concrete constitutes perhaps between 25 and 35 per cent of total aggregate production. Thus, if these figures are converted to rough estimates of *concrete aggregate* usage, they become British Columbia 4.5 tons, UK 1.2 tons, USA 2.4 tons, France 2.1 tons, and Italy and Japan 1.8 tons, the figures being per capita per annum. By contrast, concrete aggregate usage in South Africa, a developing country, was estimated in 2001 at about 0.65 tons per capita per annum. For India, a rapidly developing country, it is presently about 0.35 tons per capita per annum, while the corresponding figure for China is 1.5 tons.

The implications of these figures for the responsible and proper use of such immense amounts of natural resources are clear. The statistics also indicate why aggregate usage is increasingly being subject to environmental restrictions and controls, a point further elaborated later.

Challenging issues for concrete aggregates

Several challenging issues present themselves for the next decades. These issues will all require sustained attention to research, development, and appropriate usage of concrete aggregates.

Environmental concerns

Possibly the most pressing issue is the environmental impact of aggregate production. In many parts of the world there are legislative restrictions on the operation of pits and quarries. In parts of southeastern England for example, aggregates are in such short supply that they are imported from the Midlands, Ireland, and across the English Channel. Large tonnages of marine aggregates are also used to make up the shortfall in local aggregate sources. The UK Government levies a tax of £1.60 per ton (2004 figure) on land-based aggregates but not on marine aggregates. This is likely to increase in the future as environmental pressures grow. Since aggregates are generally high-volume low-value materials, the economic implications for concrete production of transporting aggregates over large distances are obvious.

The environmental effects of aggregate quarries include impacts on

1   *Atmospheric environment* – mainly continuous background noise from plant and machinery, as well as intermittent noise from blasting; dust from drilling, moving vehicles, and crushing and screening.
2   *Water environment* – this includes ground waters where quarries cause modification of the ground water flow and the water table which can affect local water exploitation and water quality, and surface waters due to alteration of water courses, change in runoff quantity and quality and pattern, and increase in sediment load.

3 *Landscape* – this concerns the visual impact of a quarry, which can often be 'shrouded' by judicious planting of trees and wooded banks.
4 *Natural environment* – such as flora and fauna, and ultimate rehabilitation to produce a safe site with appropriate use, such as recreation (water sport), nature reserves, aquaculture, agriculture, and so on. Figure 1.3 shows a large regional quarry in South Africa, in which some of these aspects are illustrated.

During the operation of a quarry or pit, these impacts can be mitigated and reduced by judicious controls. These may include controls on hours of operation (particularly if adjacent to built-up areas) and on movement of heavy vehicles, blasting restrictions, noise and dust abatement measures, and strict safety requirements. An environmental plan for ultimate closure of the facility and proposals for reclamation and rehabilitation will also usually be required, and the development of the facility must conform to the environmental plan. Increasingly, it is becoming very difficult to open new production sources in urban or near-urban localities due to opposition from resident groups and concerned environmentalists, who hold influence with authorities. Therefore, it should be expected that restrictions on winning aggregate materials in

Figure 1.3 Large regional quarry showing various environmental aspects: remote from built-up areas; dam in upper LH corner; large visual impact; containment of operation. Long-term plan to return site to agriculture and recreation uses (photograph courtesy of ASPASA, via Gill Owens of C&CI).

developed areas will become ever stricter. At the same time, it is uneconomical to extract aggregates from sources remote from the areas of demand. Alternative sources will certainly need to be found in the future.

Aggregate usage in the developing world

The focus of new construction has shifted inexorably from the developed to the developing world in recent decades, and this trend will continue well into the present century. While restrictions on aggregate usage for minimizing environmental damage and other issues may become increasingly strict in the developed world, restrictions of the same severity may not always apply to the developing world, at least not initially. This reflects the urgent need in the majority of the world for adequate housing, transportation, health and education facilities, and so on. While such needs remain paramount, aggregate production will be required to meet those needs. The population of the developing world is increasing and will continue to do so into the near future. Consequently, the major demand for aggregates globally will be from the developing world. In the vast majority of cases, these aggregates will be used in conventional applications and will comprise natural aggregates. However, as local industries develop and become more sophisticated, trends established in the developed world will probably be reflected in the developing world.

Mention was made earlier of aggregate usage in various parts of the world. To illustrate the trend of increasing cement and aggregate usage in the developing world, Figure 1.4 shows that usage and production of cement in India is growing very rapidly. Whereas Indian cement consumption on a per capita basis amounted to only 26.2 per cent of average world consumption in 1990, it grew to 38.7 per cent in 2003. Over the same period, Indian cement consumption grew by 86 per cent, while for the world it grew by 26.1 per cent. Aggregate consumption in India for concrete, as well as for other cement-based building materials such as plasters and mortars, would have grown at rates similar to those indicated in Figure 1.4.

Shortages and alternative sources

A major challenge for the aggregate and construction industries is finding alternative aggregate sources to overcome shortages. There are numerous publications dealing with existing or anticipated shortages of aggregates, particularly in developed cities. It is not that suitable aggregate sources are always unavailable – in fact, the opposite is often true. Other development on areas that contain good aggregate sources often occurs. This locks up the sources, leaving them unable to be exploited for construction. Supplies are then brought in from further afield, increasing the costs. By way of example, there are plentiful supplies of sand and gravel close to Vancouver,

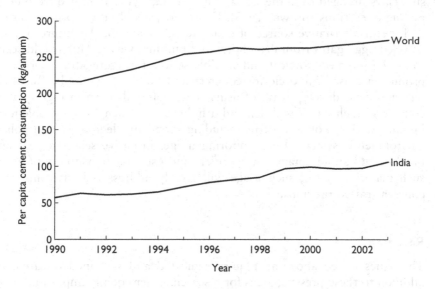

Figure 1.4 Per capita consumption of cement for India and the world as a whole
over the period 1990–2003 (data supplied by Dr Kulkarni, Associated
Cement Companies Ltd, India).

BC, but the relevant areas are intended for other development in the future.
Consequently, aggregate supplies in that area are rapidly running out, and
alternative sources are now being brought in by barge – at greater cost.

Marine, and to a lesser extent river sources, may be exploited increasingly
in the future as alternatives to land-based deposits. While this may be one
way of alleviating local shortages, even this form of production will come
under increasing criticism on environmental grounds, and indeed this is
already occurring. Oceans and rivers are important to natural resources
such as fish breeding and maintenance of biological life cycles and these
sources are not inexhaustible. In countries such as Japan, there has been
a steady swing for several decades from river sources to crushed quarry
sources, due to shortages of natural materials and environmental pressures.
Yet a further option in the future may be underground sources, although
the costs of extraction and haulage to the surface may be rather high.

As pressures against virgin aggregate extraction grow, conventional mineral aggregates will likely become a lesser proportion of total aggregate
usage in concrete, though this may take some time. Alternative sources
will need to be developed, researched, and successfully applied. These
will include industrial waste products from the metallurgical industry (i.e.
slags), which are already in regular use in concrete in many areas, marginal
materials previously considered unacceptable (e.g. porous and low strength
rocks), use of artificial materials such as sintered domestic refuse or sintered

silty clays dredged from the sea, and use of aggregates obtained from deep mining operations (Brown, 1993). However, possibly the most promising and useful alternative source of concrete aggregates in the future will be recycled aggregates from demolition and building waste. This would seem to be the most economical and sensible source for aggregates in order to produce a closed life cycle for concrete and building materials. Research will doubtless develop ways of using these materials even in high quality concretes. Such usage will demand a holistic approach to the problem of building and demolition waste, including supportive legislative and policy environments, systems for monitoring usage, incentive schemes (e.g. tax breaks), and an acceptance by specifiers and users of the viability of using such materials. Also, such usage will largely address the environmental concerns raised previously.

Research

The issues raised above all require dedicated and sustained research. In addition to these pressing areas for research, other equally important areas present themselves. The role of aggregates in concrete durability remains very important for research. The whole area of development of appropriate aggregate test methods is crucial: many current tests are time-consuming, require expensive equipment, and may take weeks or even months to obtain results which sometimes have questionable precision. To deal with increasing demand to use alternative and marginal aggregates, and to ensure structural aggregates are adequate, fast and functional inexpensive test methods combined with better understanding of how the results predict actual performance are needed. The development of a universal model for predicting the properties of concrete as a composite material, based in part on measured properties of the aggregates, remains an elusive goal but one worth pursuing. Newer developments in computational modelling of concrete, which have led to far greater understanding of the interaction of the constituents of concrete, are an exciting area of research. Thus there is a need for ongoing innovative and imaginative research into the use of aggregates in concrete.

Developments in standards and specifications

This subject is covered in Chapter 8 of this book, but a few comments are appropriate here. With the move towards alternative aggregate sources in the future, the need will grow for appropriate standards and specifications for aggregates. Standard test methods may need to be expanded to incorporate unusual properties of newer aggregates. Specifications, rather than being over prescriptive, should move towards identifying performance requirements and allowing the greatest possible flexibility within a performance framework. This will require an emphasis upon 'fitness for purpose', with aggregate quality and type being linked to concrete quality and use as far as possible (Brown, 1993). For example, lower grade recycled aggregates may become the 'stock in trade' for conventional concrete mixtures, with higher grade materials being reserved for high performance concretes or concretes for specialist applications. Modern concrete mixtures require to be engineered to produce specific properties, and consequently standards and tests will need to be developed to measure the important aggregate properties for use in these sophisticated approaches.

Texts on concrete aggregates

There are not many comprehensive texts dealing directly or exclusively with the subject of aggregates for concrete. This largely reflects the preoccupation of concrete scientists and engineers with binder systems rather than with aggregates. To be sure, engineers and technologists have studied aggregates, but mainly concerning their influence on water demand of concrete, which has an important influence on economics. There have also been the dramatic cases where aggregate durability has been at fault, such as alkali–aggregate reaction, which has focused attention on aggregates, albeit negatively.

Aggregates are covered in most concrete textbooks, although such coverage is sometimes fairly cursory. Nevertheless, there are a number of useful and comprehensive texts on concrete aggregates. These include ASTM STP 597 (1976) and STP 169C (1994), Dolar-Mantuani (1983), Murdoch, Brook and Dewar (1991), Orchard (1976), Popovics (1979), and Sims and Brown (1998). The standard work on concrete technology in South Africa, *Fulton's Concrete Technology* (Addis and Owens, 2001), contains comprehensive information on South African aggregates and their use in concrete.

Texts on aggregates tend to be written more from the point of view of geological sourcing and classification, or aggregate production and testing. A reputable text is *Aggregates – Sand Gravel and Crushed Rock Aggregates for Construction Purposes* (Smith and Collis, 2001). This was first published in 1985, with revisions in 1993 and 2001. Produced by a working party of the UK Geological Society, it covers aspects such as source occurrence, field investigation, extraction and processing of deposits, description and classification of aggregates, sampling and testing, and aggregates for concrete, mortar, unbound pavement construction, bituminous bound materials, railway track ballast and use in filter media. This text has been in demand in the UK quarrying industry. The book makes the point that increasing recognition is being given to the performance of rock aggregates in concrete structures, but an inhibiting factor has been the highly dispersed nature of the literature on aggregate materials. Consequently, non-material specialists find difficulty in identifying important literature giving information on aggregates. The book is aimed therefore at providing a knowledge of aggregates in their own right and of their production and testing to meet construction standards.

Another comprehensive UK reference is *Standards for Aggregates* (Pike, 1990). It comprises seven chapters contributed by eminent authors, and covers a similar field to the Geological Society publication, but with more of a view towards necessary standards for construction. It is also intended to provide authoritative guidance to commercial and technical specialists in the quarrying industry. Pertinent to UK construction, it covers six important aspects of aggregate technology: sampling of aggregates and precision tests, aggregates for concrete, sands for building mortars, aggregates for bituminous materials, unbound aggregates, and international and European standards. Significantly, the book was written because of the incidences of concrete deterioration that arose in the UK chiefly in the 1970s, relating to unsatisfactory aggregates. Examples of these are alkali–aggregate reaction and presence of chlorides in marine-dredged aggregates. The book makes the point that, despite the very large tonnages of aggregates produced annually in the UK, there is rather limited general literature on the use of aggregates. Much of the relevant research is published in papers not usually read by producers and users of aggregates, and many useful findings of practical work in the industry are not published at all.

The American Concrete Institute has also produced guides on aggregates for concrete, for example the publication by ACI Committee 221: 'Guide for use of normal weight and heavyweight aggregates in concrete' (ACI, 2001). This guide is divided into six major parts: (1) properties of hardened concrete influenced by aggregate properties, (2) properties of freshly mixed concrete influenced by aggregate properties, (3) aspects of processing and handling which have a bearing on concrete quality and uniformity, (4) quality control, (5) marginal and recycled aggregates, and (6) heavyweight aggregates. It presents information on selection and use of normal weight and heavyweight aggregates, which should be based on technical criteria as well as economic considerations and knowledge of types of aggregates available in the area of construction. It recognises that aggregates have an important influence on hardened concrete properties as well as on the plastic properties of freshly mixed concrete. This is reflected in a comprehensive table (Table 1.1 in ACI 221R) on properties of concrete influenced by aggregate properties. In many cases, the aggregate properties and test methods listed are not routinely used in specifications for aggregates. Their use may be needed only for research purposes, for investigation of new sources, or when aggregate sources are being investigated for a special application. Typical values in the table are listed only for guidance. Acceptable aggregates may have values outside the ranges shown and, conversely, not all aggregates within these limits may be acceptable for some uses. Therefore, service records are important in evaluating and specifying aggregate sources.

The ACI guide is intended primarily to assist designers in specifying aggregate properties, and it may also assist the aggregate producer and user in evaluating the influence of aggregate properties on concrete, including identifying aspects of processing and handling that have a bearing on concrete quality. It deals with natural aggregates, crushed stone, air-cooled blast-furnace slag, and heavyweight aggregates, but not with lightweight aggregates. The types of normal weight and heavyweight aggregates listed are those covered by ASTM C33, ASTM C63, and other standard specifications. In most cases, fine and coarse aggregates meeting ASTM C33 are regarded as adequate to ensure satisfactory performance. Experience and test results of these materials are the basis for discussion of effects on concrete properties in the guide. Other aggregate types such as slag, waste materials, and marginal or recycled materials may require special investigations for use as concrete aggregate.

The ACI has also published an education bulletin on aggregates for concrete (ACI, 1999). This bulletin deals with aspects such as classification, aggregate properties and test methods, sampling aggregates, and blast furnace slag and lightweight aggregates. It is aimed more at the practical user of aggregates.

A further American text is *The Aggregate Handbook* (Barksdale, 1991). Published by the National Stone Association, it is an industry handbook with extensive information on the American aggregate industry, basic properties of aggregates, geology, exploration and exploitation, and product manufacture and testing.

A book reflecting French practice has recently been translated from a text titled *Granulats* which appeared in 1990 (Primel and Tourenq, 2000). It is mainly concerned with the identification, exploitation, and manufacture of natural sources for construction. Topics covered include environmental issues concerning aggregate exploitation, geological prospecting methods, sand and gravel deposits, manufacturing equipment (crushing, screening, washing), planning and developing of new quarries, transport of materials, and quality control and safety. While reflecting French legislative requirements and practice, the book provides an interesting addition to relevant literature in this field.

It is perhaps worth noting that all these useful texts tend to regard aggregates in their own right rather than in terms of their important contributions to concrete performance – an important focus of this book.

Scope and objectives of this book

This book focuses on the *nature and performance of aggregates in concrete*. Its emphasis is on the interaction between the matrix or binder phase and the aggregate phase, and how this interaction governs the properties of concrete. Aggregates can be active components of concrete to a greater or lesser extent, governed by intelligent selection of the aggregates themselves, and by 'engineering' the interfacial transition zone (ITZ) between matrix and aggregate. The main objective of the book is therefore to create a better understanding of the interaction between aggregates and matrix, and a greater appreciation of the role that aggregates play in successfully producing fit-for-purpose concretes for modern construction.

The book deals mainly with natural aggregates, but Chapter 7 covers special concretes and special aggregates. Chapter 2 introduces aggregate sources and production, while Chapter 3 deals with natural aggregates and their engineering properties. Chapters 4, 5, and 6 deal with the influence of aggregates on plastic and hardened properties of concrete, with a separate chapter devoted to issues of deterioration and durability. Finally, Chapter 8 covers issues concerned with standards and specifications for aggregates, and looks at the future of specifications.

The study of aggregates, and concrete, is fascinating because it represents a highly interdisciplinary area. Important contributions are made by geologists, physicists, chemists, as well as engineers and technologists. Many seminal contributions have come from those without any formal training, and doubtless this will continue in the future in the light of the ubiquitous use of concrete. In addition, substantial knowledge resides in personnel who operate concrete production facilities, with much of this knowledge highly relevant to local conditions and unfortunately seldom published.

This book should be a valuable resource for students and researchers of concrete, as well as practitioners, specifiers, and users of aggregates and concrete. It is hoped that it will provide guidance to commercial and technical specialists in these areas as well as providing a useful resource. The comparison of practice and usage across several different countries should also provide interesting and useful contrasts for study.

References

Addis, B.J. and Owens, G. (eds) (2001) *Fulton's Concrete Technology*, 8th edn, Midrand: Cement and Concrete Institute.

American Concrete Institute Education Bulletin E1-99 (1999) *Aggregates for Concrete*, Farmington Hills, MI: American Concrete Institute.

American Concrete Institute Committee 221 (2001) 'Guide for use of normal weight and heavyweight aggregates in concrete', ACI 221R-96 (Re-approved 2001), *American Concrete Institute Manual of Concrete Practice*, Farmington Hills, MI: American Concrete Institute.

American Society for Testing and Materials (1976) *Living with Marginal Aggregates*, ASTM STP 597, West Conshohocken, PA: American Society for Testing and Materials.

American Society for Testing and Materials (1994) *Significance of Tests and Properties of Concrete and Concrete Making Materials*, ASTM STP 169C, West Conshohocken, PA: American Society for Testing and Materials.

Barksdale, R.D. (1991) *The Aggregate Handbook*, Washington, DC: National Stone Association.

Brown, B.V. (1993) 'Aggregates: The greater part of concrete', in R.K. Dhir and M.R. Jones (eds), *Concrete 2000*, London: E&FN Spon.

Dolar-Mantuani, L. (1983) *Handbook of Concrete Aggregates*, Park Ridge, NJ: Noyes Publications.

Legg, F.E. (1974) 'Aggregates', in J.J. Waddell (ed.) *Concrete Construction Handbook*, New York: McGraw-Hill, 2.1–2.67.

Murdoch, L.J., Brook, K.M. and Dewar, J.D. (1991) *Concrete Materials and Practice*, 6th edn, London: Edward Arnold.

Orchard, D.F. (1976) *Concrete Technology*, Volume 3, *Properties and Testing of Aggregates*, London: Applied Science.

Pike, D.C. (ed.) (1990) *Standards for Aggregates*, Chichester: Ellis Horwood.

Popovics, S. (1979) *Concrete-Making Materials*, Washington: Hemisphere Publications Corporation.

Powers, T.C. (1971) 'Fundamental aspects of concrete shrinkage', *Rev. Materiaux et Constructions*, 545: 79–85.

Primel, L. and Tourenq, C. (eds) (2000) *Aggregates* (English Translation), Rotterdam: A.A. Balkema.

Sims, I. and Brown, B. (1998) 'Concrete aggregates', in P.C. Hewlett (ed.) *Lea's Chemistry of Cement and Concrete*, 4th edn, London: Arnold.

Smith, M.R. and Collis, L. (2001) *Aggregates: Sand, Gravel and Crushed Rock Aggregates for Construction Purposes*, Engineering Geology Special Publications, 17, London: Geological Society.

Vitruvius (1960) *The Ten Books on Architecture*, Book 2, Chapter 6, trans. M.H. Morgan, New York: Dover Publications Inc.

Chapter 2

Natural aggregate sources and production

This book focuses on the nature and performance of aggregates *in concrete*. It emphasizes the interaction between matrix and aggregate phases, and how this interaction governs properties of the composite material. Before discussing the properties of aggregates themselves and their performance in concrete, we need to consider the origin and sources of aggregates, and how they become useful engineering materials. The subject matter in this and the subsequent four chapters will cover mainly conventional (normal strength) concretes and *natural aggregates*. Such aggregates may be defined as materials composed of rock fragments which are used in their natural state except for such operations as crushing, washing, and sizing. Natural aggregates are derived from naturally occurring geological sources, which are processed and beneficiated to a greater or lesser extent to produce hard, non-cohesive granular materials of varying sizes that can be incorporated into concrete. Natural aggregates are processed by crushing, screening, and washing to render them useful for engineering purposes. They exclude synthetic or artificial aggregates or marginal aggregates which are covered in Chapter 7. Also included in this chapter are aggregates that require minimal processing, and may be usable virtually from source, such as certain sands and gravels.

Natural aggregates may be derived from land- or sea-based deposits, from gravel pits or hard-rock quarries, from sand dunes or river courses. They thus comprise *gravels*, either crushed or uncrushed, *crushed stone*, produced from the artificial crushing of rock, boulders, or large cobble stones, and *sand*, which is fine granular material passing the 4.75-mm sieve. This chapter discusses the wide variety of natural aggregates that are available to the concrete engineer, indicating briefly their geological origins and nature, which is important in order to set the scene for an understanding of aggregate properties and performance in concrete, covered in subsequent chapters. It avoids unnecessary technical detail on aggregate sources and production. The interested reader can obtain such information from other publications (Barksdale, 1991; Sims and Brown, 1998; Grieve, 2001; Smith and Collis, 2001).

Origin and classification of aggregates

Natural concrete aggregates are derived from rocks of the earth's crust. Their properties are governed firstly by the chemical and physical properties of the parent rocks. Rocks undergo various processes of alteration, including natural geothermal and/or weathering processes which occur over long periods of geological time. Such processes may produce granular materials in the form of natural gravels and sands that can be used in concrete with a minimum of further processing or beneficiation. On the other hand, production of granular materials may require processes encompassing human-related activities in the form of rock breaking, crushing, and so on. These processes, which convert the rock in a very short period of time into useful engineering materials, must be linked to the nature of the parent rock and the required properties of the aggregates, in order to produce acceptable materials. Accordingly, this section will deal with the origin and classification of aggregates; subsequent sections will cover sources and production processes necessary to produce aggregate materials and the sampling techniques required to obtain representative results.

Origin

Many properties of an aggregate derive from its parent rock: physical and mechanical properties such as relative density, strength, stiffness, hardness, permeability and pore structure, and chemical and mineral composition. Thus the origin of the parent rock is very important. Rocks themselves are comprised of various minerals, defined as 'naturally occurring inorganic substances of more or less definite chemical composition and usually of a specific crystalline structure' (ASTM C294). Rocks of the earth's crust are generally classified as igneous, sedimentary, or metamorphic, relating to their origin. Table 2.1 gives a description of these primary rock types. (Further useful information can be found in ASTM C294 which contains a descriptive nomenclature for constituents of concrete aggregates.)

Igneous rocks

All rocks originate as igneous rocks, derived from solidification of the molten material underlying the crustal zone of the earth. Igneous rocks form either as intrusive rocks (coarse- to medium-grained) which solidify slowly beneath the crust, or as extrusive rocks (fine grained) which force their way to the surface and crystallize much more rapidly. Their mineralogical composition is important to engineers, since these minerals form not only the primary igneous rocks but also may be physically disintegrated or chemically decomposed to form the basis of sedimentary and metamorphic

Table 2.1 Igneous, sedimentary, and metamorphic rocks as origins for concrete aggregates

	Processes of formation	Distinguishing characteristics and features	Rock types for concrete aggregates
Igneous rocks	Formed by solidification of molten rock (magma), the structure being massive. Characteristics are largely governed by rate and condition of cooling. Igneous rocks include porphyries characterized by the presence of large mineral grains in a fine grained or glassy ground mass. A magma may be supersaturated, saturated, or undersaturated with respect to silica (SiO_2) present. Supersaturation gives rise to formation of free quartz, while in saturated or undersaturated rocks, no free quartz is present. *Acid* igneous rocks contain >65% SiO_2, *intermediate* rocks have 55–65% SiO_2 and *basic* igneous rocks contain <55% SiO_2.		
a Extruded volcanic rocks (fine grained)	Extruded on the earth's surface, thus cooling very rapidly, resulting in only partial crystallization of component minerals.	Partial crystallization gives rocks which are mixtures of crystalline components and glassy material. Extremely rapid cooling may produce an entirely glassy or vitreous rock. The glassy portion usually has higher silica content than the crystalline portion, and therefore may be alkali-reactive leading to deleterious reactions with the products of cement hydration in concrete. Texture is generally compact or glassy, and individual crystals cannot be distinguished with the naked eye. Generally contain vesicles (cavities remnant from escaping steam and vapours) which may be elongated by lava flow. If the quantity of vesicles is excessive, the rock can appear as frothy glass. Igneous rocks that may contain high silica glass are obsidian, pumice, trachyte, rhyolite, basalt, andesite, and perlite.	Rhyolite Felsite Trachyte Andesite Basalt Glassy Volcanics Obsidian Perlite Pumice Scoria
b Shallow intruded igneous rocks (Hypabyssal) (medium-grained)	Intruded into overlying rock masses to form sills, dykes, and laccoliths. Slower cooling occurs, sometimes under great pressure.	Crystallization of rock minerals is generally complete, grain size depending on length of cooling and pressure. At the upper end, these rocks grade into the volcanic rocks; at the lower end they grade into the plutonic rocks. Texture is therefore variable, from finely crystalline to medium-grained (1–5-mm grain size).	Microgranite Granophyre Diabase Dolerite

c Plutonic rocks (coarse-grained)	Formed at great depth beneath the earth's surface to form 'plutons' which cool extremely slowly and usually under great pressure.	Rock minerals are fully crystallized, with little or no glassy material. Consequently, these rocks are usually highly stable chemically, with little or no tendency to react detrimentally with cement hydration products. Texture can vary from medium to very coarse-grained (>5-mm grain size). Rocks in the plutonic class generally have chemical equivalents in the volcanic class, although their textures will differ.	Granite Pegmatite Syenite Diorite Norite Gabbro Peridotite

Sedimentary rocks	Sedimentary rocks are derived from physical or chemical degradation of other rocks, resulting in fragments which accumulate as deposits, usually under water. These are physically or chemically deposited and consolidated, and frequently bound by siliceous, argillaceous, ferruginous, or calcareous cementing agents, often by precipitation. Process of formation leads to stratified deposits (except for glacial deposits).	Composed of other rock fragments of varying sizes: large in the case of conglomerates, very small in the case of shale or limestone. Strength of the rock mass depends on strength of cementing agent; likewise durability depends largely on cementing agent. Shapes of fragments may be angular, but more commonly are rounded due to transport to the deposit. Sedimentary rocks may be less liable to subsequent degradation than igneous or metamorphic rocks. Successive layers of a deposit can be vastly different, resulting in interbedded layers of different composition. Limestones and chert are generally chemically deposited and consolidated.	Conglomerate
			Sandstone
			  Orthoquartzite
			  Greywacke
			  Arkose
			Chert
			Claystone, Siltstone, Argillite, and Shale
			Carbonates
			  Limestone and Dolomite
			  Marl and Chalk
			Tillite

	A further group of sedimentary materials comprise unconsolidated sediments, representing gravels, sands, silts, and clays. These are important materials for aggregate sources.	These materials carry the distinguishing characteristics and features of their parent rocks. They often represent transported and weathered materials, giving well-shaped and stable concrete aggregates. However, in other cases, the weathered minerals may comprise dimensionally and chemically unstable materials, such as shrinking aggregates.	Various types of gravels and sands derived from parent rocks.

Table 2.1 (Continued)

	Processes of formation	Distinguishing characteristics and features	Rock types for concrete aggregates
Metamorphic rocks	Metamorphic rocks derive from pre-existing igneous, sedimentary, or other metamorphic rocks which are altered texturally, structurally, or mineralogically in situ by the actions of high temperature and/or pressure, often together with percolating waters or vapours. There is therefore a wide variety of metamorphic rocks which differ greatly in structure and texture. For example, degrees of foliation and crystallinity vary widely.	Metamorphic rocks formed under both high temperature and pressure show massive internal structures and equigranular texture, often resulting in great strength and toughness, for example quartzite and hornfels. Metamorphosed igneous rocks: properties are not usually greatly improved by metamorphism. Metamorphosed sedimentary rocks: metamorphism often improves properties such as strength, toughness, hardness, durability, etc. Re-crystallization under conditions of high shear stress produces rocks that are foliated and schistose.	Gneiss Hornfels Marble Quartzite (Metaquartzite) Schist, Slate, and Phyllite (preferably not used as concrete aggregates)

Primary sources: ASTM C294; Grieve, 2001; Kosmatka et al., 2002.

Notes
1 For the minerals that occur in various rocks, see Table 2.2. They can be summarized as:
  a Free silica in the form of quartz.
  b Feldspathic silicates in the form of feldspars and feldspathoids.
  c Ferromagnesian silicates in the form of pyroxenes, amphiboles, biotites, and olivines.
  d Accessory minerals such as magnetite, haematite, ilmenite, apatite, rutile, zircon, etc.
2 With the exception of the ultrabasic rocks, feldspathic silicates are usually associated with one or more ferromagnesian silicates together with minor quantities of accessory minerals. Quartz may or may not be present.
3 See Table 2.4 for a description of rock types and their characteristics for use as concrete aggregates.
4 ASTM C294 has further useful information on rock types used as concrete aggregates.

Table 2.2 Minerals found in concrete aggregates

Mineral	Examples
Silica	**Quartz**: A very common hard mineral composed of silica (SiO_2). Pure quartz is colourless and glassy, without visible cleavage. Highly resistant to weathering. Abundant in sands, gravels, many sandstones, and many light-coloured igneous and metamorphic rocks. Some strained or microcrystalline quartz may be alkali-reactive. **Opal**: A hydrous form of silica, generally without crystal structure. It is usually highly alkali-reactive. **Chalcedony**: A fibrous form of quartz with submicroscopic porosity. Frequently occurs with chert and is usually alkali-reactive. **Tridymite and Cristobalite**: High temperature crystalline forms of silica associated with volcanic rocks, and alkali-reactive.
Feldspars	These alumino-silicate minerals are the most abundant rock forming minerals in the earth's crust, and are important constituents of most major rock groups. Feldspar minerals are differentiated by chemical composition and crystallographic properties. **Potassium Feldspars: orthoclase and microcline** **Sodium Feldspars: albite, plagioclase** } Intermediate feldspars are oligoclase, andesine, labradorite, and bytownite. **Calcium Feldspars: anorthite** Potassium and sodium feldspars occur typically in igneous rocks such as granites and rhyolites; calcium feldspars occur in igneous rocks of lower silica content such as andesite, basalt, and gabbro.
Ferromagnesian minerals	These are constituents of many rocks, comprising dark minerals, generally silicates of iron or magnesium or both. **Amphiboles**: e.g. hornblende **Pyroxenes**: e.g. augite **Olivines**: e.g. forsterite. Found only in dark igneous rocks without quartz. **Dark micas**: e.g. biotite, phlogopite. Easily cleave into thin flakes and plates. See also below.
Micaceous minerals	These have perfect cleavage in one plane, splitting into thin flakes. Micas are common in all rock types, and occur as minor or trace constituents in many sands and gravels. **Muscovites**: Colourless to light green **Biotites**: Dark coloured to black **Lepidolites**: Light coloured to white **Chlorites**: Dark green coloured **Vermiculites**: Formed by alteration of other micas, brown coloured.
Clay minerals	These are layered silicate minerals, the size range being less than 1 μm, e.g. hydrous aluminium, magnesium, and iron silicates with variable cations such as calcium, magnesium, potassium, sodium, etc. Formed by alteration of other silicates and volcanic glass.

Table 2.2 (Continued)

Mineral	Examples
	They are major constituents of clays and shales, also found in altered and weathered igneous and metamorphic rocks, as seams and lenses in carbonate rocks, and as matrix or cementing material in sandstones and other sedimentary rocks. Rocks containing large amounts of clay minerals are generally soft and unsuitable for use as aggregates. Different types of clay minerals are frequently interlayered. **Kaolinites, illites, and chlorites**: These are relatively stable clay minerals, but are absorptive. **Smectites and montmorillonites**: These comprise the swelling clays, and are highly unstable volumetrically. If included in concrete they give rise to high volume changes on wetting and drying.
Zeolites	Zeolites are a large group of hydrated alkali–aluminium silicates, soft and light coloured. Usually formed from hydrothermal alteration of feldspars. Can contribute releasable alkalis to concrete through cation exchange. Some varieties give substantial volume change with wetting and drying. These minerals are therefore not favoured in concrete aggregates. They are rare except in basalt cavities.
Carbonate minerals	**Calcite**: Calcium carbonate, $CaCO_3$ **Dolomite**: Calcium and magnesium carbonate, $CaCO_3 \cdot MgCO_3$ Both are relatively soft minerals, soluble in acid.
Sulphate minerals	**Gypsum**: Hydrous calcium sulphate, $CaSO_4 \cdot 2H_2O$, or anhydrite, $CaSO_4$. Typically forms a whitish coating on sand and gravel, and is slightly soluble in water. Other sulphates, e.g. sodium and magnesium, can also be present. All sulphates can attack concrete and mortar.
Iron sulphide minerals	**Pyrite, marcasite, and pyrrhotite**: Frequently found in natural aggregates. They may oxidize to sulphuric acid, and form iron oxides and hydroxides, which can attack or stain the concrete. See also Table 3.15.
Iron oxides	**Magnetite**: Black-coloured common mineral, Fe_3O_4 **Haematite**: Red-coloured common mineral, Fe_2O_3 **Ilmenite**: Black, weakly magnetic, less common mineral, $FeTiO_3$ **Limonite**: Brown weathering product of iron-bearing mineral. These minerals give colour to rocks and also colour concrete. They are frequently found as accessory minerals in igneous rocks and sediments. Magnetite, ilmenite, and haematite ores are used as heavy aggregates.

Source: Based on information in ASTM C294.

Note
Minerals are defined as naturally occurring inorganic substances of more or less definite chemical composition and usually of a specific crystalline structure (ASTM C294 definition).

rocks. The nature of the primary mineral constituents, given in Table 2.2, varies greatly from highly stable quartz to sometimes rapidly decomposing ferromagnesians. The chemical nature of these materials is also discussed in Chapter 3.

Sedimentary rocks

Sedimentary rocks are formed from the physico/mechanical and chemical breakdown of other pre-existing rocks, and depend for their properties largely on the nature of the binding or cementing phases. They are highly variable as might be expected, and grade from weakly cemented sandstones to very strongly cemented limestones. Since their minerals have endured processes of degradation, they tend to be more stable than those in igneous (or metamorphic) rocks. Their sometimes-layered structure can result in poorly shaped and flaky particles in the case of aggregates crushed from such rocks. The sandstone group, provided they have adequate strength, represent the most abundant source of concrete aggregates due to their chemical stability. Calcareous rocks are also widely used as concrete aggregates.

Metamorphic rocks

These are comprised of pre-existing igneous or sedimentary rocks, altered under conditions of high temperature and pressure usually at great depth. Consequently they are even more variable in composition and structure than the other rock types. Metamorphic rocks generally exist either as foliated rocks with parallel arrangement of platy minerals often appearing with distinct bedding or banding, such as gneiss or schist, or non-foliated rocks which are massive, such as quartzite. Hydrothermal metamorphism can result in the minerals in these rocks being re-formed and re-crystallized to render them less durable and less stable in some cases, for example the formation of alkali-reactive silicates. On the other hand, similar processes can greatly improve the basic properties of the parent rock, for example when weak sandstones are converted into strong, massive quartzite (metaquartzite). Metamorphic rocks make important contributions to the production of concrete aggregates such as quartzite and hornfels. Examples of the surfaces of metamorphic and igneous rocks are shown in Figure 2.1.

Any of the three primary rock types can and frequently do undergo subsequent alteration which can change their original properties. This gives rise to wide variability of rock properties, and one cannot generalize too much about these properties. In doubtful cases, or for new or untried sources, it is important for a thorough examination to be done on the source, involving physical, mechanical, chemical, and petrographic tests and techniques.

Geological processes that alter rocks subsequent to their initial formation include those that tend to improve the rock (from a concrete-aggregate perspective), that is constructive processes, or those that cause deterioration of the rock and its properties, that is destructive processes. These latter processes tend to predominate and are the more frequent.

(a)

(b)

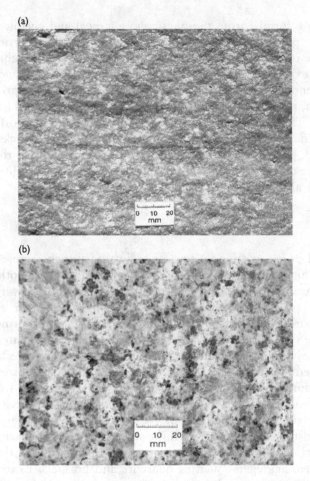

Figure 2.1 Typical rocks used as aggregates: (a) Fine-grained quartzitic sandstone (metamorphic rock); (b) Coarse-grained granite (igneous rock). Fair faces produced by stonemasons.

CONSTRUCTIVE GEOLOGICAL PROCESSES

These include:

a   Re-crystallization and re-formation of rock masses, by great heat and pressure, such as occurs during metamorphism.
b   Replacement of leached constituents by other cementing compounds, giving improved properties; for example the alteration of granular sand or sandstone to quartzite by deposition of silica in solution from percolating water.

DESTRUCTIVE GEOLOGICAL PROCESSES

These are too numerous to mention in this context; some of the more important from a concrete-aggregate perspective are:

a   Physico/mechanical destructive processes such as freezing and thawing, heating and cooling, and mechanical erosion and attrition. Tectonic stresses and earth movements can cause fracturing and jointing of rock masses, which may influence quarrying and crushing.
b   Hydrochemically destructive processes such as wetting and drying, oxidation and hydration, leaching due to percolating waters, and direct chemical attack from groundwaters containing organic or inorganic compounds. Examples are conversion of feldspars to clay minerals, and dissolution of limestone and dolomite by acidic groundwaters and acid rain.
c   Rock deformations and straining that can result in silica becoming strained or re-crystallized, thus changing from stable and chemically resistant minerals to materials that are susceptible to attack by cement alkalis. This is an important process from a concrete aggregate perspective, and the alkali–aggregate problem is dealt with in Chapter 6. For example, while quartzite is generally a highly stable rock type for use in aggregate, its crystal structure may be strained by tectonic forces, rendering it susceptible to alkali attack.

These processes can all broadly be described as weathering processes, and all depend on the climatic and weathering conditions pertaining in a given locality. In cold or temperate climates, ice action and freezing/thawing will be important, while in warm, moist conditions chemical breakdown will predominate. Dry and hot areas will experience a greater degree of physical breakdown from thermal expansion and contraction. The rocks resulting from these processes will display different characteristics and properties, which is a good reason for the need to thoroughly characterize a concrete aggregate prior to use by petrographic techniques (see Chapter 3). Nevertheless, provided the rock type has an adequate inherent strength, and is reasonably stable both physically and chemically, it is likely to be suitable for use as a concrete aggregate, but possibly not in all situations.

Weathering of rocks produces fine granular materials such as sands, clays, and silts. In this sense weathering can be thought of as a process that produces concrete aggregates, sands in particular. Sands tend to be dominated by quartz, with feldspar, mica, and iron oxide as less abundant phases. Quartz is a hard, chemically stable mineral and is a desirable component of concrete sand. Feldspars are less stable and may experience decomposition resulting in the formation of clay minerals. Some sands may contain minerals of high relative density such as garnet, zircon, rutile and ilmenite which do not usually affect the behaviour of sand in concrete, provided they are present only in small amounts.

Classification of natural aggregates

Classification is an attempt to group similar materials together so that knowledge and experience with one material can be transferred to another similar material. However, even similar materials vary, and thus the engineering performance of an untried source of aggregate, though it may be readily classified, cannot be accepted uncritically before trials and tests have been carried out. Ultimately, it is the performance of the aggregate *in concrete* that is important, and for this reason existing service records are invaluable when judging an aggregate source.

Concrete aggregates are classified in different ways:

a   In terms of their origin. In this system, rocks are classified into the three broad groups already discussed: igneous, sedimentary, or metamorphic.
b   In terms of physical characteristics such as particle size or bulk density. *Coarse aggregate* is used to describe particles larger than 4.75 mm (usually), while *fine aggregates* are particles equal to or smaller than 4.75 mm. Regarding density, most natural aggregates have granular bulk densities (density of a confined mass of particles) in the region of $1500–1700\,kg/m^3$, referred to as *normal weight* aggregates, while those with bulk densities less than about $1120\,kg/m^3$ and greater than about $2000\,kg/m^3$ are termed *lightweight* and *heavyweight* aggregates respectively (alternative terms are *low density* and *high density* respectively).
c   Petrologically, that is, in terms of the types and relative proportions of the minerals present. For example, a dolerite will be composed of mainly plagioclase and pyroxene, with smaller amounts of olivine or quartz and minor amounts of other minerals. Table 2.2 gives a list of the most common minerals found in aggregates. This classification scheme is not directly useful for concrete aggregates, but can serve as a starting point. Minerals alone cannot be used as a basis for predicting aggregate performance in concrete, and in any event aggregate particles are normally composed of several minerals. The techniques of petrological and mineralogical characterization of aggregates are, however, very useful in arriving at an informed assessment of likely performance of an aggregate in concrete, and these are discussed in Chapter 3.

When a sample of granular aggregate is examined, it is necessary to determine whether all the particles are of the same petrological type, or whether they vary. For example, mixed gravels will contain various rock types and a proper description of the aggregate sample would have to include the relative proportions of the rock types.

In the UK, the petrological basis of classification is used (Sims and Brown, 1998). A review of the development of standardized aggregate classification schemes is given in Smith and Collis (2001), who discuss the so-called CADAM scheme (Classification and Description of Aggregate Material). This scheme attempted to provide a simple classification system which was independent of physical properties, except for those which could be inferred from mineral composition. The scheme also encouraged the use of a correct petrological name and additional information on geological age, colour, grain size, and fissility (i.e. the ability to be split) of the aggregate. However, the CADAM system has not been adopted in practice, partly because it is at variance with current British standards, and due to technical criticisms relating to the assumption that aggregates classified within one of the five simplified Classes would have similar characteristics, which is not always true.

Aggregates are classified according to BS 812: Part 1: 1975 in petrological terms, shown in Table 2.3. This system is more suited to crushed rock aggregates than to natural sands and gravels. It is necessary to expand the terms in this table to include, for example, the mineral 'quartz' when describing certain natural sands and gravels. The information in Table 2.3 gives no indication as to the abundance of usage of rock types as aggregates, and might imply that aggregates in any given class would have similar characteristics although, as pointed out, these characteristics are likely to vary widely.

Table 2.3 Classification of natural aggregates according to BS 812: Part 1: 1975

Basalt group	Andesite; Basalt; Basic porphyrites; Diabase; Dolerites of all kinds including theralite and teschenite; Epidiorite, Lamprophyre; Quartz-dolerite, Spilite
Flint group	Chert; Flint
Gabbro group	Basic diorite; Basic gneiss; Gabbro; Hornblende-rock; Norite; Peridotite, Picrite; Serpentinite
Granite group	Gneiss; Granite; Granodiorite; Granulite; Pegmatite; Quartz-diorite; Syenite
Gritstone group (including fragmental volcanic rocks)	Arkose; Greywacke; Grit; Sandstone; Tuff
Hornfels group	Contact-altered rocks of all kinds except Marble
Limestone group	Dolomite; Limestone; Marble
Porphyry group	Aplite; Dacite; Felsite; Granophyre; Keratophyre; Microgranite; Porphyry; Quartz-porphyrite; Rhyolite; Trachyte
Quartzite group	Ganister; Quartzitic sandstones, Re-crystallized quartzite
Schist group	Phyllite; Schist; Slate; All severely sheared rocks

Note
Both suitable and unsuitable materials for aggregates can be found in any group above.

The classification shown in Table 2.3 was withdrawn from BS 812 in 1976. The reasons reflected concerns that the scheme gave no reliable guide as to the engineering characteristics of the rock types, confusion arising from the mixture of terms within groups, and the fact that the system was biased towards igneous and away from carbonate rocks while carbonate rocks make up in excess of 60 per cent of crushed rock aggregate in the UK (Smith and Collis, 2001). Further, the system had no classification for sands and gravels although consumption of these materials is similar to that of crushed rock aggregates. The later version of BS 812: Part 102: 1989 refers to the 'nominal description' of the aggregate which should include the following: (a) type of aggregate, that is, crushed rock, sand, or gravel (whether crushed or partially crushed) or artificial, for example slag, broken rubble, and so on, (b) nominal particle size, and (c) other, for example deleterious materials, clay lumps, and so on. The Standard requires an aggregate to be described in petrological terms when necessary, preferably taken from the list reproduced here in Table 2.4. For mixed gravels, the composition should be indicated, for example flint/quartzite. In addition the Standard notes that a competent person or authority should provide the petrological description, and that such description does not account for suitability for any particular purpose.

In the USA and Canada, aggregates are also classified petrologically in terms of rock types (ASTM C294; Kosmatka *et al.*, 2002). However, the sources are grouped under the three main groups of igneous, sedimentary, and metamorphic rocks. The essentials of the North American approach are provided in the fourth column of Table 2.1.

In South Africa, aggregates are likewise classified in petrological terms, but more emphasis is given to the major sources with regard to abundance of use. Thus, quartzites and sandstones are used the most, followed by basic igneous rocks, and so on.

A summary of the characteristics of some of the more common rock types used as aggregates is provided in Table 2.5.

Table 2.4 Rock types commonly used for aggregates

Petrological term	Description
Andesite*	Fine grained, usually volcanic, variety of diorite
Arkose	Type of sandstone or gritstone containing over 25% feldspar
Basalt	Fine-grained basic rock, similar in composition to gabbro, usually volcanic
Breccia[†]	Rock consisting of angular, unworn rock fragments, bonded by natural cement
Chalk	Very fine-grained Cretaceous limestone, usually white
Chert	Cryptocrystalline[‡] silica

Table 2.4 (Continued)

Conglomerate[†]	Rock consisting of rounded pebbles bonded by natural cement
Diorite	Intermediate plutonic rock, consisting mainly of plagioclase, with hornblende, augite or biotite
Dolerite	Basic rock, with grain size intermediate between that of gabbro and basalt
Dolomite	Rock or mineral composed of calcium magnesium carbonate
Flint	Cryptocrystalline[‡] silica originating as nodules or layers in chalk
Gabbro	Coarse grained, basic, plutonic rock, consisting essentially of calcic plagioclase and pyroxene, sometimes with olivine
Gneiss	Banded rock, produced by intense metamorphic conditions
Granite	Acidic, plutonic rock, consisting essentially of alkali feldspars and quartz
Granulite	Metamorphic rock with granular texture and no preferred orientation of the minerals
Greywacke	Impure type of sandstone or gritstone, composed of poorly sorted fragments of quartz, other minerals and rock; the coarser grains are usually strongly cemented in a fine matrix
Gritstone	Sandstone, with coarse and usually angular grains
Hornfels	Thermally metamorphosed rock containing substantial amount of rock-forming silicate minerals
Limestone	Sedimentary rock, consisting predominantly of calcium carbonate
Marble	Metamorphosed limestone
Microgranite*	Acidic rock with grain size intermediate between that of granite and rhyolite
Quartzite	Metamorphic rock or sedimentary rock, composed almost entirely of quartz grains
Rhyolite*	Fine grained or glassy acidic rock, usually volcanic
Sandstone	Sedimentary rock, composed of sand grains naturally cemented together
Schist	Metamorphic rock in which the minerals are arranged in nearly parallel bands or layers
	Platy or elongate minerals such as mica or hornblende cause fissility in the rock which distinguishes it from a gneiss
Slate	Rock derived from argillaceous sediments or volcanic ash by metamorphism, characterized by cleavage planes independent of the original stratification
Syenite	Intermediate plutonic rock, consisting mainly of alkali feldspar with plagioclase, hornblende, biotite, or augite
Trachyte*	Fine grained, usually volcanic, variety of syenite
Tuff	Consolidated volcanic ash

Source: From BS 812: Part 102: 1989.

Notes

* The terms microgranite, rhyolite, andesite, or trachyte, as appropriate, are preferred for rocks alternatively described as porphyry or felsite.

† Some terms refer to structure or texture only, e.g. breccia or conglomerate, and these terms cannot be used alone to provide a full description.

‡ Composed of crystals so fine that they can be resolved only with the aid of a high power microscope.

Table 2.5 Characteristics of common rock types used as concrete aggregates

Rock type and primary minerals	Characteristics in terms of use as concrete aggregate
Quartzites and sandstones Quartzites: composed largely of quartz. Sandstones: depends on pre-existing rock particles, but quartz is often abundant; also cementing minerals (see alongside). Opal, which is alkali-reactive, sometimes present.	These rock types are abundantly used as concrete aggregates because they contain mainly the mineral quartz which is highly resistant to weathering. Quartzites are in many instances metamorphosed (metaquartzites) in which processes of re-crystallization of the original particles result in an interlocking mosaic of tightly bound grains. These metamorphic processes involve both very high pressure and temperature. Alternatively, quartzites may be formed by processes of grain cementation by silica. Quartzites generally make excellent concrete aggregates. Sandstones are formed by pressure alone, usually together with additional cementation by silica, iron oxide, or carbonates. Sandstones tend to be of lower strength, of greater friability and porosity, and fracture around the grains. They also frequently contain variable amounts of feldspars and other rock fragments in addition to quartz grains. The amount of non-quartz material governs the nature and type of the sandstone. Quartzites and sandstones can be thought of as lying in a continuous spectrum with their relative positions depending on the degree of metamorphism. Some quartzites and sandstones are alkali-reactive, and petrographic examination is required to identify these deleterious rocks.
Basic igneous rocks The most important mineral components are feldspars and ferromagnesian minerals such as amphiboles and pyroxenes. Numerous secondary minerals are also present such as olivine, quartz, magnetite, apatite, and biotite.	The earth's crust has an abundant amount of these rocks which include volcanic varieties (basalt, andesite) and hypabyssal or plutonic varieties (diabase, dolerite, diorite, gabbro). In their fresh form, these rocks generally produce excellent concrete aggregates. The texture varies depending on whether they were formed on the surface (e.g. basalt), at intermediate depth (e.g. dolerite), or at great depth (e.g. gabbro). The plutonic rocks exhibit coarser grain sizes than the volcanic varieties. The greater the proportion of ferromagnesian (mafic) minerals, the darker the colour. The primary minerals of basic igneous rocks are less inert and stable than those of sandstones and quartzites. Consequently, weathering and alteration are far more common in these rocks. Such alteration can produce undesirable clay minerals which frequently require petrographic examination to be positively identified.

Granites

Mainly comprise quartz and potassium feldspar with differing minor amounts of amphibole, mica, and iron oxides. Granites are generally light-coloured.

Granites are the basement rocks of the continental landmasses and consequently exist in vast quantities. Depending on their mode of formation and subsequent metamorphic history, granites vary in grain size from relatively fine-grained to excessively coarse-grained rocks such as pegmatite.

Fresh or slightly weathered granites generally make excellent concrete aggregates. Weathering results mainly in the conversion of the feldspars to clay minerals, but these are usually of a stable variety such as kaolinite and/or illite. Occasionally weathering of the mica minerals may cause some dimensional instability in concrete, and weathered granites must be carefully examined and tested before being used in concrete.

Granodiorite is a rock similar to granite, except the dominant feldspar is plagioclase.

Limestones and dolomites

Calcite (calcium carbonate)

Dolomite (calcium-magnesium carbonate)

The earth's surface contains vast quantities of calcareous rocks of sedimentary origin laid down in strata with thicknesses up to several kilometres in places. They are classified as limestones or dolomites depending on the relative proportions of calcium carbonate and magnesium carbonate, giving rise to rocks that can be classified as dolomitic limestones or calcitic dolomites. In general, they all produce excellent concrete aggregates, although the range of physical and chemical properties of individual carbonate rocks is large.

Arenaceous calcareous rocks contain 10–50 per cent sand, argillaceous calcareous rocks 10–50 per cent clay. Non-carbonate constituents are usually present, typically in lenses or bands in the calcareous layers. These comprise cherts, clays, quartz veins, and so on. Generally, these materials are not detrimental to the calcareous rocks as aggregates, but caution should be exercised.

Some dolomite carbonate rocks are alkali-reactive, and petrographic examination is required to identify these deleterious rocks.

Unconsolidated sediments

Cobbles (>75 mm)
Gravel (4.75–75 mm)
Sand (0.075–4.75 mm)
Silt (0.002 mm)
Shale (consolidated clay)

These materials can comprise all types of rocks and minerals, depending on their origin. Silt consists predominantly of silica and silicate minerals. Clays are largely composed of clay minerals.

These materials usually comprise hard and durable fragments with a rounded shape and smooth texture, due to their surviving various weathering processes.

Provided they are not contaminated with clay and silt, unconsolidated gravels and sands usually make strong and durable concrete aggregates. However, shales may disintegrate in water and are usually platy in appearance.

Table 2.5 (Continued)

Rock type and primary minerals	Characteristics in terms of use as concrete aggregate
Miscellaneous	Arkose is a coarse-grained sandstone derived from granite with large amounts of feldspar. Chert is the general term for a group of very fine-grained siliceous sedimentary rocks of microcrystalline or cryptocrystalline quartz, chalcedony, or opal, formed by chemical deposition and consolidation. They may be dense and very tough and of darker colour, or porous usually of lighter colour. Chert occurs frequently as lenses in calcareous deposits, or in extensive beds, and is often found in sands and gravels. Cherts are often alkali-reactive and should be treated with caution. Dense, black or grey chert is called flint. Felsites are taken to refer collectively to a group of usually light-coloured igneous rocks, including rhyolite, dacite, andesite, and trachyte, which are the equivalents of granite, quartz diorite, diorite, and syenite respectively. When microcrystalline or containing natural glass, they may be potentially alkali-reactive in concrete. When used as crushed materials, the shape may be characteristically flaky and elongated. Greywackes are grey to greenish-grey sedimentary or metamorphic rocks containing angular quartz, feldspar grains, and sand-sized fragments in a matrix of clay or shale. Schist is a highly layered rock generally containing quartz and/or feldspars which splits along parallel planes. Schists are seldom suitable rocks for concrete aggregates. Shales are composed of sedimentary clays with lesser amounts of silt-sized particles, and are of varying hardness. Shales are generally not very suitable for concrete aggregates, since they are sometimes soft or break into flaky elongated shapes. Tillite is a glacially deposited rock, generally metamorphosed in the case of concrete aggregates, and having highly variable properties.

Sources and production of natural aggregates

Natural aggregates are obtained from a variety of sources. Almost invariably, the sources of aggregate must be as close as possible to their demand locality, due to the large tonnage of aggregate used in concrete and the high cost of transportation. Use of special aggregates that require unusual sourcing or long transportation distances is rare, and technical or engineering requirements have to outweigh additional costs in these cases. In remote areas a local source may be exploited, even if only temporarily, to fulfil the aggregate needs of a construction job. With increasing environmental pressures on aggregate producers, there is a move away from small aggregate operations to larger regional facilities or 'mammoth quarries', where environmental controls and costs can be better handled.

Natural aggregates can be sourced from pits, river banks and beds, the seabed, gravelly or sandy terraces, beaches and dunes, or other deposits that provide granular materials that can be processed with minimal extra effort or cost. Sand and gravel, which are unconsolidated sedimentary materials, are important sources of natural aggregate. The occurrence of high quality natural sands and gravels within economic distance of major urban areas may be critical for viable concrete construction in those areas. In continental North America, glacial deposits are an important source of natural aggregates. Sand and gravel deposited from glaciers constitute most of the economic deposits found in previously glaciated areas. On the other hand, many natural aggregates are sourced from fresh hard rock quarries, and require blasting and crushing to produce engineering materials.

The various sources of natural aggregates and the production processes necessary to render the raw materials acceptable for use in concrete will be covered briefly. The interested reader should consult detailed texts for further information on aggregate production (Sims and Barksdale, 1998, and references therein; Brown, 1991; Smith and Collis, 2001). Aggregates requiring a minimum of extraction and processing will be dealt with first, followed by those requiring substantial effort in the form of blasting and/or crushing. Production of natural aggregates includes: (a) a reduction phase in which the material is reduced in size by crushing and rock-breaking techniques; this phase obviously not only applies to crushed rock sources, but may also apply to gravel sources that contain some oversized material or where there are technical advantages in crushing the material, (b) a processing phase and possibly a beneficiation phase, and (c) a sizing and sorting phase. Aggregates need to be handled, transported, and stockpiled in preparation for use, and these processes require careful control to ensure consistency of supply (Miller-Warden, 1967).

Pit or terrace sources

Physical and chemical weathering of rocks has produced vast quantities of granular materials, much of which are suitable for use as concrete aggregates. They may be present as residual materials from parent rocks, or as transported materials that have been moved over the earth's surface and then deposited by ice, water, or wind. They can be dug from pits or terraces and processed to produce aggregates. These materials can vary in particle size from boulders and cobbles down to very fine silt and clay. Some natural aggregate deposits, called bank gravel, contain both sand and gravel sizes that can be used for concrete with little need for processing. The properties and quality of these materials will vary enormously depending on their origin, weathering and transportation, and subsequent processing.

Granular sources are exploited in bulk by the use of heavy earth-moving equipment and excavators. Occasionally, they may be removed by high-pressure water jetting. Some processing possibilities are: screening out of oversize boulder fragments; partial crushing of large or oversize material; screening and washing to remove excessive quantities of clay or fine silt; and some screening and re-blending of appropriate sizes to produce acceptably graded materials.

These sources can produce either coarse or fine aggregates, or both. Whether they are residual or transported materials will govern properties such as particle shape and surface texture, as well as subsequent processing such as crushing. For example, sands from residual weathered granites may have less than ideal particle shapes and surface textures, and are rather absorptive due to the presence of decomposed feldspars. On the other hand, terrace gravels resulting from natural alluvial processes will tend to have rounded shapes and smooth textures. For granular sources, proper petrographic examination must be undertaken to identify the rock and mineral types present in a deposit, and to provide recommendations on their suitability and processing requirements for concrete aggregates. An example of a pit source of fine aggregate is shown in Figure 2.2.

Processing of pit and terrace sources usually requires the following:

a   *Washing and scrubbing.* These processes are carried out to remove unwanted materials such as excessive clay or silt fractions and/or soluble salts if present. Washing can occur by jetting during screening or by use of a washer barrel. More vigorous washing is carried out using 'scrubbers' which help to remove resistant clay lumps.

b   *Screening and sorting.* Consistent aggregate size gradation is very important for producing good quality concrete. For coarse aggregates, this will usually involve screening the material into different size

Figure 2.2 Pit source of fine aggregate (photo courtesy of J. Cokart, Holcim).

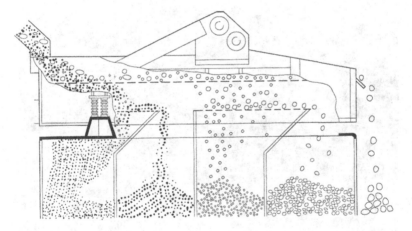

Figure 2.3 Schematic diagram of a horizontal vibrating screen plant (Reprinted from Sims and Brown, 1998, ©1997, with permission from Elsevier).

fractions. For fine aggregates, it may involve screening out excessive quantities of coarser or finer sizes. Alternatively, fine aggregates can be sorted by a water-settling process called 'Classification'. Figure 2.3 shows the design of a horizontal vibrating screen plant.

River sources

Material obtained from river sources will depend on the rocks present in the river's catchment area, and how readily they break down and weather to be transported in the river. During river transport, rock fragments will undergo further weathering, be slowly reduced in size, and be shaped by processes of attrition. Consequently, most river aggregates are reasonably well shaped, rounded, and smooth.

A feature of many rivers is erratic flow, and consequently different particle sizes tend to be transported and deposited during different regimes of flow. This can lead to marked stratification, making it necessary to blend or mix the materials to obtain consistency and uniformity and improve the grading. On the other hand, rivers with consistent flow will sort and deposit the material, with rapid velocities giving a deficiency of fine material and vice versa. Thus it may be necessary to blend the materials with other sources to extend the grading envelope. River aggregates will generally have lower water requirements in concrete due to their superior shape and surface texture, but the water requirement may be increased by poor grading and absorptive fines in some cases. Figure 2.4 shows a typical river source of

Figure 2.4 River source of aggregates (photo courtesy of H. Hale, Malans Quarries).
Note
The crushing and screening plant is shown in the foreground; the river channel is in the background.

aggregates. Processing and beneficiation of river sands and gravels are much the same as for pit or terrace sources.

Beach deposits

Beach materials fall within the influence of waves and tidal action. Being transported, they generally have reasonably smooth textures and good particle shapes, although in the larger sizes above 5–10 mm, particles tend to be discoidal rather than spherical. The major deficiency of beach sands is their poor grading, resulting from the sorting action of waves. Frequently, only one or two particle sizes are present, with shortage or absence of fine material. Consequently these materials, particularly sands, need to be blended with other aggregates to improve grading and provide adequate fines for cohesiveness of the concrete mix.

Beach sands are composed mainly of quartz grains, but varying amounts of shell fragments may also be present. This seldom presents a problem since these fragments are normally sound and non-fragile. Higher shell contents (in excess of say 30 per cent) may lead to increased mix-water requirements if the shell fragments are poorly shaped or partly hollow. Beach sands may also contain salts, but if they are washed so that the chloride content is no greater than 0.01 per cent by mass, there should be no problem with corrosion of steel embedded in concrete made with the sand. For permissible quantities of chlorides in aggregates, see Table 3.16. Figure 2.5 shows the dramatic consequences of how reinforced concrete can deteriorate if unwashed marine sand is used.

Figure 2.5 Corrosion of reinforcing bars at Alcatraz Prison, California, due to use of aggregates contaminated with sea salts.

Dune deposits

Dune sands are found either adjacent to the coast, or in dry desert-like areas where vegetation is sparse and sand can be heaped up by wind into hills or hillocks. Wind, like water, has a good sorting power, which means that particles in dunes tend to be similar in particle shape, relative density, and even mineral composition. Dune sands usually have well-rounded shapes, but suffer from poor grading with particles tending to be concentrated in the 0.1–1.0-mm size range. These sands are often characterized by severe deficiency of fines, resulting in concrete that lacks cohesion and tends to bleed excessively. For this reason, dune sands usually benefit by being blended with other sands to address the grading deficiencies.

Coastal dune sands should be checked for their chloride content before being used in reinforced concrete. Chlorides in dune sands are usually lower than in adjacent beach sands and may be well within the 0.01 per cent prescribed limit.

Dredged sources

Dredging is an increasingly popular method of extracting aggregates. The majority of dredging is from marine sources which have become more exploited in recent decades due to stricter environmental controls on land-based deposits (see Chapter 1). For instance, in the UK where the government has encouraged use of marine sources (subject to very strict controls) about 15 per cent of aggregates nationally derive from this source, although the proportion in southeast England, which produces the largest volume of concrete, is closer to 50 per cent. Marine aggregates are typically smooth-textured and rounded in shape, and due to their method of extraction are relatively free from dust, clay, and silt. They differ from land-based sources mainly in respect of the presence of seashells and salt, for which the precautions mentioned under beach deposits also apply. Marine dredged aggregates have had well over half a century of use in the UK during which time satisfactory performance has generally been proved.

Equipment used in marine dredging involves 'suction-hopper' dredging ships, barges, and the like, by which sand and gravel are pumped aboard. Occasionally, dredging techniques are used in land-based pits or quarries where material is being exploited below the water table, and where pumping to keep the excavation dry is either impractical or too expensive. Large excavators can often be used in these situations. Aggregates can be dredged also from rivers or lakes, although again environmental pressure on these operations is growing. Figure 2.6 shows an ocean dredger with the aggregate hoppers clearly visible.

Figure 2.6 Ocean dredger ship.

Hard rock quarries

As sources of granular deposits become depleted, particularly adjacent to areas of high demand such as urban centres, other sources of aggregates must be found. One source is recycled materials, dealt with in Chapter 7. Another is hard rock which is quarried and crushed. In some countries or regions, these materials are the only reasonable source of quality aggregates for concrete and have been used for many decades (Grieve, 2001). A picture of a large rock quarry is shown in Figure 2.7. It is possible to produce both coarse and fine aggregates from hard rock quarries. The properties and quality of the material will depend on several factors: nature of the parent rock (or rocks), for example whether it is massive or jointed, laminated or fissured; degree of weathering to which the rock has been subjected; methods of extraction from the quarry (blasting, mechanical ripping, rock-breaking techniques, etc.); and importantly the crushing and processing to which the rock is subjected.

Crushing

Rock crushing methods and techniques together with the nature of the rock itself govern the quality and properties of the product. In particular, particle shape can be beneficially or adversely affected by the crushing and this is more critical for crushed sands due to their strong influence on the plastic properties of concrete. The tendencies of many crushed sands to have elongated and flaky shapes can generally be reduced by use of appropriate crushing techniques.

Figure 2.7 Large rock quarry for concrete aggregates (photograph courtesy of ASPASA, via Gill Owens of C&CI).

On the other hand, certain rock types such as schists and slates with their laminated structures and marked cleavage planes have an inherent tendency to crush into flaky and elongated particles, and these materials should be avoided if possible.

Aggregate crushing methods that favour the production of good shapes include the removal of chips and fines during primary crushing, choke and closed-circuit feeding, use of corrugated crushing surfaces, and a low reduction ratio. ('Reduction ratio' is the ratio of the feed size to the product size.) However, in impact crushers, good aggregate shapes can be produced even with high reduction ratios. Types of crushing equipment are jaw crushers, roll crushers, disc or gyrosphere crushers, gyratory crushers, cone crushers, rod mills, and impact-type crushers such as horizontal impactors (e.g. hammer mill) or vertical impactors. A number of these crushers are illustrated in Figure 2.8 (Sims and Brown, 1998). Modern equipment such as gyratory and cone crushers and rod mills tend to produce better particle shapes. Washing and scrubbing may also be required for crushed aggregates.

Rock properties that influence the quality of the crushed product are hardness, fracture toughness, structure (fine- or coarse-grained or laminar), moisture content, and homogeneity. Regarding abrasiveness of rocks to the wearing surfaces of crushing equipment, the amount of quartz is most

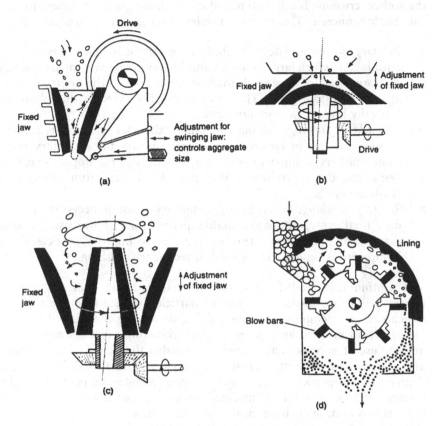

Figure 2.8 Schematic diagrams of some types of crusher; solid shading highlights hardened wear-resistant elements: (a) single-toggle jaw crusher; (b) disc or gyrosphere crusher; (c) gyratory crusher; and (d) impact crusher (reprinted from Sims and Brown, 1998 ©1997, with permission from Elsevier).

important since this mineral is harder than normal steel. Therefore wear rates increase with the quantity of free quartz in the rock and with its fracture toughness.

CRUSHED COARSE AGGREGATES

Where aggregates have to be crushed, coarse aggregates are often the most commonly produced. Even when natural gravels are the original source, some crushing may be necessary to reduce the particle size and render them suitable in the concrete mix. Crushing and re-blending with the natural material can improve concrete properties (e.g. strength) because of the presence of a proportion of more angular particles. If hard-quarried rock is the source, crushing is required to reduce the large quarry boulders to sizes suitable for concrete. This requires a series of stages in the crushing process:

a   *Primary crushing.* The first phase of crushing is known as primary crushing, in which large boulders and broken rock material are reduced to more manageable sizes. It may be carried out in the quarry itself or at the processing plant. The types of crushers used in this phase are generally jaw or gyratory crushers.

b   *Secondary crushing.* During this phase, the rock material is reduced to sizes for use in concrete, or in preparation for a tertiary stage. Cone crushers or impact breakers are used, with the impact machines being suited to produce better particle shapes from difficult-to-crush materials.

c   *Tertiary crushing.* Occasionally, a tertiary phase is necessary to produce final aggregates of acceptable quality, for instance when smaller than normal reduction ratios are required to improve particle shape. Equipment is similar to that used in secondary crushing.

If continuously graded aggregates are required, the final stage of the process will be to separate the crushed particles into suitable size fractions which can be re-combined during mixing to achieve the desired grading. However, for crushed aggregates, practice sometimes favours gap grading and the use of nominal single-sized stone rather than continuously graded aggregates. This is because crushed aggregates tend to exhibit a higher degree of inter-particle friction due to more angular shape and rougher texture, and elimination of intermediate sizes (from about 5 to 15 mm) reduces this friction creating more workable mixes.

CRUSHED FINE AGGREGATES

Crushed fine aggregates (sometimes called 'crusher sands') can be manufactured satisfactorily in modern crushing plants which frequently produce both coarse and fine aggregates. The tendency in many areas is still to prefer uncrushed fine aggregates, often because in the past the quality of crusher sands was poor. Nevertheless, there is increasing production and use of good quality crusher sands due to the depletion or non-availability of natural deposits.

There are certain advantages to crusher sands manufactured under modern controlled conditions that may not always be appreciated:

a   Modern crushing techniques can produce particle shapes equivalent or even superior to pit sources.

b   Controlled conditions can produce consistently uniform grading that may not be the case with natural deposits where gradings vary depending on the stratum of deposit being exploited at any given time.

c   Crusher sands are less likely to be contaminated with clay minerals and
    organic substances than are natural materials. For this reason, higher
    fines contents for crusher sands are often allowed in specifications,
    although care should be exercised when the source is certain shales or
    basic igneous rocks which may contain undesirable clays.

Properly produced crusher sands with their controlled properties may
have water requirements in concrete comparable to or even lower than natural sands. Also, the consistency of a crushed product is a major advantage
for modern concrete production plants.

Crushed fine aggregates may often be used on their own, but are frequently blended with natural sands for several reasons:

a   If the particle shape of the crushed material is less than ideal, blending
    with a better shaped natural sand can improve the overall sand quality
    and often impart a lower water requirement.
b   Blending may be necessary to achieve a desired overall grading that
    may not be possible with the crushed sand alone.
c   The overall economy of the mix may be improved by blending, assuming
    one or other sand source is significantly cheaper.

Waste rock dumps

Waste rock dumps may be suitable for aggregate production. These dumps
derive from mining or other heavy earth-moving activities such as excavation. They may contain a uniform type of rock, or they may comprise
mixed rock types. Their suitability for aggregate production will depend
on the nature of the rock, its state of weathering, and whether it is economical to process aggregates from the dump in competition with other
sources such as pit or gravel deposits. The rock dumps should be close to a
demand centre for aggregates, as remote dumps are unlikely to be attractive
as aggregate sources.

The advantage of waste rock dumps is that the primary breaking and
'winning' of the material has already occurred during mining or rock exca-
vations. The material requires only to be further processed or beneficiated
by crushing, screening, and possibly washing. A further benefit of the use of
these dumps is that environmental pressures are reduced by their removal.

Some caution needs to be sounded in respect of exploitation of waste
rock dumps. First, the rock types may be highly variable, and if one or
more unsuitable rock types are mixed in the dump, it might be very difficult
if not impossible to separate the materials (Ballim, 2000). Small quantities
of less desirable rock types may be acceptable provided their proportion
is reasonably uniform. Second, weathering of rock dumps may produce a
number of undesirable by-products: acid drainage waters where sulphides

are oxidized to sulphates and sulphuric acid; unstable clay minerals if basic igneous rocks are present; and other clays depending on the parent rocks. Pyrites are also frequently found in waste rock dumps, and these can cause unsightly staining in concrete. A waste rock dump from gold-mining operations containing mainly quartzite, an excellent concrete aggregate source, and also minor amounts of shale which tends to crush with poor particle shapes is shown in Figure 2.9.

Aggregate properties such as particle shape, surface texture, and grading will be covered in later chapters. However, it is instructive to examine how aggregate source and production affect these important properties. This is summarized in Table 2.6.

Aggregate beneficiation

Beneficiation as applied to aggregate production has a precise meaning: it refers to the selective removal of undesirable constituents in an aggregate. It is therefore the additional processing to upgrade or improve the quality of the raw material by a variety of means, most of which rely on gravity separation, or occasionally on centrifugal separation. Beneficiation is usually used to remove unsound, lightweight, or deleterious materials from aggregates, as well as removal of mica flakes by modified washing procedures. Beneficiation must be distinguished from the normal production

Figure 2.9 Waste rock dump from gold mine, with quartzite and minor amounts of shale (rock dump c.35-m high).

Table 2.6 How source and production of aggregates affect key physical properties

Source	Type of aggregate usually produced		Influence of source and production on aggregate properties		
	Sand	Stone	Particle shape	Particle surface texture	Grading
Pit	Yes	Some	Rounded to angular	Smooth to rough	Graded
Dune	Yes	None	Rounded	Smooth	Single-sized
River or beach	Yes	Some	Rounded	Smooth	Graded to single-sized
Rock quarry	Yes	Yes	Angular	Usually rough	Graded

Source: Table suggested by Dr Rod Rankine, C&CI.

processes of crushing, screening, and washing which are intended to provide proper gradation and cleanliness. In effect, the removal of unwanted clay and silt fractions can be regarded as 'beneficiation', although this is normally accomplished in conventional washing or scrubbing operations. Of necessity, beneficiation is expensive not least because some acceptable material is always lost, and producers would rather avoid this if possible.

Heavy media separation

This process permits the separation of lighter materials (e.g. coal or lignite) from the normal dense particles. Aggregates are passed through a medium containing a suspension of heavy minerals with a relative density that permits aggregate particle separation. This process can only be used when the undesirable particles have significantly lower relative densities than the aggregate.

Rising-current classification (reverse water flow)

This process has the same purpose as heavy media separation, but involves separating light materials such as wood and lignite in a high-speed upward moving flow of water that carries them away.

Hydraulic jigging

The jigging process is more suited to separate particles with only small differences in relative density. Water is pulsed upwards into a jigging box through its perforated base, to move the lighter material into a layer on top of the heavier material, from where it can then be removed.

Crushing

Crushing of rock boulders or cobbles is a normal production process which gives properly sized material. However, it can also be used as a beneficiation process since crushing removes soft and friable particles.

Washing and scrubbing

Finely divided material and surface coatings may be removed by this process.

Source variability

Natural aggregates, being of geological origin, exhibit variability. This variability is endemic to virtually all aggregate sources, and exists both spatially and in terms of time. Spatial variability refers to the fact that sources will vary from point to point in a deposit or quarry. In hard rock sources, this may involve a primary rock type with igneous intrusions, mineralized lenses, or variable joint patterns giving rise to varying degrees of in-situ weathering. A competent petrographer should be employed to examine the source and recommend on its acceptability. Particular care needs to be exercised with certain weathered basic igneous rocks (e.g. some basalts and dolerites) that can appear fresh and strong in situ, but rapidly weather to soft materials on exposure in stockpiles and the like. Such materials have been encountered over wide areas of South Africa, and can be detected in a test using ethylene glycol (Orr, 1979). In this test, swelling clays of the smectite group are readily and rapidly identified when the sample is soaked in ethylene glycol which reacts with the clays and causes disintegrative expansion. Selective quarrying can ensure that only favourable material is finally processed. Another frequent example is that of variable sedimentary layers in which certain strata may be more preferable for aggregate exploitation than others. In granular deposits, one can expect the particle sizes, gradings, and even type of material to vary, depending on the nature of original deposition or origin. For example, alluvial deposits tend to have layers of different fineness, depending on the velocity of the original depositing waters. Provided extraction methods and subsequent blending are linked to the nature of the material and its variability, a reasonably uniform final product can be achieved.

Variability also relates to the fact that the quality of material from an aggregate source will vary in time. This is a consequence of the spatial variability already discussed. As 'winning' of the source continues in time, so new and variable horizons or areas of the source will be exploited, giving in-time variability. In some cases, a uniform final product can be achieved by stockpiling and then re-combining the materials to eliminate variability to a large degree. This is particularly important for the production of high quality concrete, where source uniformity is essential.

A further source of variability relates to contaminants in the source, and these can vary spatially. Sims and Brown (1998) mention certain reactive pyrites in flint gravels in southeast England, which can cause unsightly staining of concrete. They also warn that even relatively small amounts of highly alkali-reactive material from veins or inclusions can seriously compromise the durability of an otherwise good aggregate source.

Sampling of aggregates

Aggregates must undergo various tests to characterize their properties and to determine appropriate use. Aggregates are used in vast quantities, and it is not possible to test entire lots destined for construction. Thus, it is essential that aggregates are sampled at regular intervals to determine their representative properties and characteristics.

Two different issues arise regarding sampling of aggregates. The first relates to the need for petrographic examination to determine the mineral composition of an aggregate source, while the second refers to the sampling of bulk aggregates from a stockpile or production stream in order to carry out laboratory testing. The latter is typical of tests on aggregates themselves, for example particle size analysis, as well as tests on concrete mixes incorporating the aggregates.

The two different situations require somewhat different approaches in order to arrive at a 'representative' sample. In the first case, the geologist or concrete petrographer will require small samples for microscopic and chemical analyses, while in the second case, the engineer or concrete technologist will require much larger samples for use in bulk tests. This section deals with the principles of sampling for both petrographic analysis and bulk aggregate testing.

It is stating the obvious that if an unrepresentative or carelessly obtained sample is tested, the results are not only worthless, they may in fact be detrimental if they are subsequently used to draw incorrect conclusions for implementation on a project. A well-devised sampling system will generally comprise four stages: (1) a Sampling Plan to obtain the necessary data as economically as possible, (2) actual selection and collection of samples according to standard procedures to ensure representivity, (3) testing of the samples, and (4) analysis and interpretation of test results. It is appropriate to review stages (1) and (2) here, as well as provide some guidance on sample sizes. (For further details on sampling, see ASTM D75; Pike, 1990; Barksdale, 1991; and Smith and Collis, 2001.)

Sampling plan

The purpose of a sampling plan is to establish the average material characteristics that are typical for the lot, and determine the nature and extent of variability. A sampling plan defines the number, size, and location or timing of samples necessary in order to represent the lot. It should also give procedures for sample size reduction in order to provide lab or test samples. Once initial data come to hand, it will be possible to detect trends and, in the case of bulk aggregate testing, make any necessary adjustments to production processes in order to ensure acceptable quality of materials.

As mentioned, a sampling plan aims at obtaining samples 'representative' of the source (e.g. quarry) or a larger batch of aggregates. Such 'representivity' can be established on two bases: using semi-objective measures of experience and judgment by a competent engineer or geologist who selects samples that are deemed representative; or by random sampling, the results of which can be subjected to statistical analysis. The outcome of a random sampling and statistical analysis scheme will be measures of the average property of a lot and of variability, generally in terms of a probability that can be calculated (ASTM D75; Pike, 1990; Smith and Collis, 2001).

Obtaining samples

The procedures for obtaining test samples have the aim of achieving as representative a batch of materials as possible. Since a completely consistent supply of aggregates from source, or an entirely homogeneous stockpile or consignment does not occur, perfect representivity is unobtainable. Aggregates, being natural materials, suffer more than most engineering materials from continuous variability, even when derived from a reputable supplier or quarry. Further, granular aggregates display segregation during handling, and it is important to avoid variations introduced by this phenomenon. Figure 2.10 illustrates the problem of coarse aggregate segregation in an aggregate stockpile from a conveyor. Consequently, a good sampling plan is needed, and every effort must be made to obtain as practically representative a sample as possible for testing. Procedures for sampling must avoid the bias of selecting either the best or the poorest materials.

Practical sampling entails taking a sample large enough to be representative within acceptable and known limits, but small enough to be conveniently handled. The principles of random sampling are set forth in ASTM D3665 which covers the determination of random locations or times at which samples can be taken. This is based on the time for the material lot to pass the sampling point and the number of samples required, and a random number table is provided. ASTM D75 (*Standard Practice for Sampling Aggregates*) and CSA A23.2-1A (*Sampling Aggregates for Use in Concrete*) cover the sampling of coarse and fine aggregates for preliminary investigation of a potential source of supply, as well as quality control of materials at source or on site. Guidance is given on how to obtain samples from a flowing aggregate stream (such as bins or belt discharge), from a

Figure 2.10 Coarse aggregate segregation during discharge from a conveyor (from Barksdale, 1991).

conveyor belt, or from stockpiles or transportation units such as trucks or barges. SANS 5827 gives similar information, and covers pre-sampling considerations such as an appropriate sampling plan, as well as making up laboratory samples. Good sampling techniques are summarized in Table 2.7.

Table 2.7 Sampling techniques for aggregates

Aggregate in stockpile	Take at least ten increments from different parts of the stockpile, working from the bottom of the stockpile upwards. Avoid sampling in segregated areas (extreme top or bottom) or from the surface.
Aggregate in bins or bays	Take at least eight increments evenly spaced over the area. Before taking the increments, remove and discard the top 100–150 mm of material.
Aggregate being loaded or unloaded from vehicles	Take at least two increments from each quarter of the load while loading or unloading.
Aggregate in vehicles	Dig a trench across each of three approximately equal sections, at least 300-mm deep and approximately 300-mm wide. Take four increments equally spaced from the bottom of each trench, pushing the shovel down vertically.
Moving belt conveyor	Stop the conveyor and remove approximately 1-m length of aggregates across the width of the belt.

Sampling for petrographic examination

When an aggregate source is subjected to petrographic or mineralogical analysis, a representative sample may be derived from the quarry or source operation directly, or it may come from processed material that has already gone through a certain degree of selection and homogenization. Sampling of quarry materials in situ is a specialized topic, outside the scope of this book. Sampling of already processed or partially processed materials will generally follow the principles of random sampling with progressive sample reduction if necessary, to arrive at an adequately representative sample of suitable size. This aspect is briefly reviewed below.

Petrographic work involves examining a representative sample for its different minerals, and describing the minerals in terms of their characteristics of interest to concrete making. The sampling problem is one of selecting a suitable sample size in regard to mineral distribution. The sample size required to obtain a representative sample within acceptable limits will depend on the number of different minerals, and whether they exist as free particles or embedded in other particles. Larger sample sizes are required as the proportion of a given constituent reduces. Statistics can define the relationship between accuracy (or representativeness), size of sample, and composition of sample. Clearly, as sample composition becomes more complex (e.g. samples with multiple constituents such as mixed gravels), so the sample size required also becomes larger.

Guidance on sample sizes is given in Tables 2.8–2.11, compiled from different sources. The problem in petrographic analyses is that the proportions of the different minerals are not known initially, so some degree of iteration in sampling and analysis may be required. For guidance, the minimum size of sample for dispatch to the laboratory, based on BS 812: Part 104, is given in Table 2.8. On arrival at the laboratory, aggregates are initially qualitatively examined for aggregate type and general characteristics. Thereafter, a representative test portion for quantitative analysis is obtained by sample reduction. The minimum mass of this test portion is

Table 2.8 Minimum size of sample for dispatch to the laboratory for petrographic examination

Maximum particle size (mm)	Minimum mass (kg)
50	200
40	100
20	13
≤10	2

Source: From BS 812: Part 104: 1994.

given in Table 2.9, and represents an accuracy of ±10 per cent (relative) for a constituent present at 20 per cent. Figure 2.11 can be used for the minimum test portion required for a ±10 per cent relative error for constituents at other concentrations.

REDUCING LARGE SAMPLES TO APPROPRIATE SIZE FOR PETROGRAPHIC ANALYSIS

For fine aggregates, it is usually no problem to obtain representative sample sizes for petrographic analyses. However, for coarse aggregates, and when

Table 2.9 Minimum size of test portion for quantitative petrographic examination

Nominal maximum particle size (mm)	Minimum mass of test portion* (kg)
50	100
40	51
20	6.4
10	0.8
5 or smaller	0.1

Source: From BS 812: Part 104: 1994.

Note
* Mass required to give a relative error of ±10 per cent for estimated proportion of a constituent of interest of 20 per cent, when using duplicate test portions.

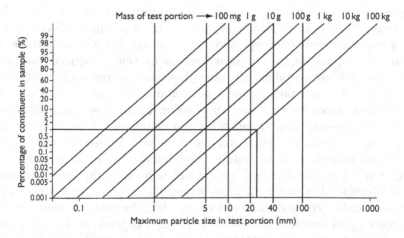

Figure 2.11 Mass of each test portion necessary to achieve a relative error of ±10 per cent for a given constituent when using duplicate test portions (from BS 812: Part 104: 1994).

the proportion of important minerals is small, large sample sizes will be
required if the sampling error is to be kept sufficiently low. For example,
Figure 2.11 shows that for 25-mm maximum particle size and a constituent
at the 1 per cent concentration level, 100 kg of sample will be required to
maintain an accuracy of ± 10 per cent. Clearly, this is impractical from the
analysis perspective, and consequently it is usually necessary in such cases
to reduce the sample size by crushing the original sample to a smaller size,
taking a new sample by the random method, and so on until a suitable size
is obtained for analysis. In these cases, the cumulative sampling error needs
to be considered. If e_1 is the relative error of the first sampling operation,
and so on for n sampling operations, then the cumulative relative error
will be

$$\Sigma e = \sqrt{(e_1^2 + e_2^2 + \cdots + e_n^2)} \tag{2.1}$$

The relative error increases with successive sampling operations. For example, if each sampling operation has a relative error of 10 per cent, this rises
to 20 per cent with four sampling operations.

ANALYSIS BY SEPARATION, OR PARTICLE OR POINT COUNTING

Once a representative sample (within the error limits achieved during the
sampling process) has been obtained, it can be analysed physically or chemically. For petrographic analysis, it is necessary to report the results on a
mass or volume proportion basis. This requires either separating the sample
by hand into its discrete particles (if different minerals are represented by
liberated particles), or counting constituent grains within a sub-sample by
eye or under a microscope, depending on grain size. For a quantitative estimation of composition, the sample is screened into closely spaced fractions.
Where minerals are mixed in individual grains, thin or polished sections are
used for microscopic examination. A point counting technique is then used,
and the error depends on the number of points counted and the proportion
of minerals present. Point counting on thin or polished sections is usually
done using an 'automatic' point counter. A separate key is assigned to each
constituent mineral and the slide is mechanically advanced one interval
along a grid counting line each time a key is pressed, the key depending
on which mineral is under the cross-wires. The total count for each key
is automatically recorded and used to calculate the mineral composition.
The pre-set grid interval (generally 0.2–1.0 mm) depends on the grain size
of the rock. A minimum of 1600 points is usually considered representative. This specialist technique must be undertaken by a competent geologist
or petrographer.

*Sampling and sample reduction for bulk aggregate
and concrete testing*

It is often necessary to obtain samples for testing aggregates or for use in a laboratory concrete mix from a larger stockpile, and then reduce their size to that required for the test or mix. When sampling from a stockpile the samples should not be taken from the surfaces, the base edges, or the top, due to problems of segregation; for example coarser fragments gather at the base. Samples should be drawn from the middle of the stockpile faces at several points around the pile and at least 200 mm below the surface. If sampling from a conveyor or from a hopper discharge, a number of increments are taken over a period of a day, these samples being remixed to produce a composite sample before further reducing the size of the sample. (If, however, in-production variability must be determined, then each increment should be tested.) Table 2.10 gives the number of increments required over a period of a day, the mass depending on the maximum size of the aggregate. Table 2.10 can be used for sampling from hopper discharges or from vehicle deliveries, in which case the increments should preferably be taken from different vehicles arriving during the day. If applied to stockpile sampling, Table 2.10 can also be used, although it is permitted to reduce the minimum number of increments to 10 for the larger aggregate sizes (50 mm and 64 mm), and proportionally 5 for the smaller aggregate sizes. Stockpiles should obviously be closely observed at all times to detect any unusual variation or characteristics of the material.

Table 2.10 Minimum masses for sampling of bulk aggregates

Maximum size present in substantial proportion (85% passing) (mm)	Minimum mass of each increment (kg)	Minimum number of increments	Minimum mass dispatched (kg)
>80	50	16	150
50–80	50	16	100
20–50	50	8	50
10–20	25	8	25
<10 (including fine aggregate)	10	8	10

Source: Compiled from information in Smith and Collis, 2001, CSA A23.2-1A, BS 812: Part 102: 1989.

Notes
1 ASTM D75 gives similar, occasionally slightly larger, values for mass dispatched than those given in the table above.
2 SANS 5827 provides guidance for sampling from stockpiles, bins, moving stream of aggregate, or in vehicles.
3 In general, the larger the maximum size of aggregate, the greater the sample mass required. Consequently, a somewhat greater sample can be taken at the upper end of the ranges given above.

TECHNIQUES FOR AGGREGATE SAMPLE REDUCTION

Composite aggregate samples obtained from larger stockpiles or con-
veyances need to be reduced to smaller sizes in preparation for testing.
A common and long-standing technique is that of 'cone and quartering',
whereby the larger sample is formed into a conical shape by piling all the
material onto the apex and allowing it to distribute itself with radial sym-
metry. In the case of fine aggregates, the material should first be dampened
to avoid segregation. The pile is flattened and halved vertically through
the original apex, the remaining half again being halved, to produce quarters of the original sample. Two diametrically opposite quarters are then
re-combined, the whole exercise being repeated as necessary until the
required sample size is obtained.

   Another technique for sample reduction is use of a riffler, shown in
Figure 2.12. This device has the advantage of providing greater consistency
and accuracy in sample separation than other methods. It consists of a
rectangular hopper that feeds the particles into a series of adjacent slot-
shaped apertures, each alternate aperture discharging in opposite directions.
Material deposited in the riffler is divided into equal parts collected in
separate receptacles on opposite sides of the riffler. The riffling procedure
can be repeated to produce sub-samples, with the rate of throughput
being quite high. ASTM C702 has details of the above sample reduction
techniques.

   A rotary sampler may also be used to recover small samples from large
lots, although such equipment would not be found routinely in most labs
due to their cost. The principle involves a series of identical hoppers
arranged around the periphery of a horizontal rotating disc. A constant
stream of aggregate particles is fed from a fixed feeder into each hopper in

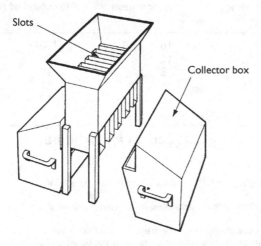

Figure 2.12 Schematic view of a riffler.

turn as they rotate. The speed of rotation and the rate of feed are governed by the sample size and the diameter of the disc. There are several variations based on these principles.

As mentioned earlier, sample reduction gives an accumulating standard error of measurement. There is evidence to suggest that quartering introduces higher variability in sub-samples than riffling (Goodsall and Matthews, 1970).

SAMPLE SIZE

Statistics show that the probability of selecting a sample that falls within acceptable limits (say ±10 per cent of the true value) increases very rapidly as the size of the sample is increased. However, practical limitations dictate that sample sizes be kept as small as possible. Table 2.11 gives the size of samples for various tests. An important part of the table is the sample sizes required for sieve analysis, where up to 50 kg of sample may be required for large size (63 mm) aggregate.

On a more fundamental level the following formula may be used to estimate the sample size required to carry out particle size analyses, assuming the particles are of similar density (Gy, 1982):

$$M = \frac{10^{-9}[((1/p) - 2)v + \Sigma p_i v_i]\rho}{CV^2} \tag{2.2}$$

Table 2.11 Sample sizes for various tests

Test	Minimum amount to be available at the laboratory
1 Durability	
a Nominal maximum size >12.5 mm	80 pieces
b Nominal maximum size <12.5 mm	3.6–5.4 kg
2 Sieve analysis	
Maximum size up to 63 mm	50 kg
50 mm	35 kg
40 mm	15 kg
28 mm	5 kg
20 mm	2 kg
14 mm	1 kg
10 mm	0.5 kg
6 or 5 or 3 mm	0.2 kg
Less than 3 mm	0.1 kg
3 Shape Indices of flakiness and elongation	800 pieces
4 Proportion of impurities	As for sieve analysis

Source: Based on information in Smith and Collis, 2001 and BS 812: Section 103.1: 1985.

where

M = sample mass in kg
ρ = particle density in kg/m^3
CV = coefficient of variation, that is, standard deviation/mean (usually taken as 5 per cent, or 0.05)

and

p = mass proportion of the fraction of interest
p_i = mass proportion of fraction i
v = average particle volume of fraction of interest (mm^3)
v_i = average particle volume of fraction i (mm^3)

and the summation is over all size fractions.

The average particle volume of a fraction bounded by two sieve sizes d_1 (mm) and d_2 (mm) is:

$$v = \frac{d_1^3 + d_2^3}{4} \, \text{mm}^3 \tag{2.3}$$

To use the formula, values of p are required beforehand! Thus, a preliminary assessment must be made, and the numbers refined for an accurate size estimate. Applying the formula to a hypothetical example of a coarse aggregate with sizes ranging between 5 and 20 mm, the sample sizes needed are given in Table 2.12, where the mass proportions of the different fractions (i.e. p_i) are as shown.

The 10–5-mm size fraction dominates the sample size required, which would be, say, 5 kg in this case. Equation (2.2) also indicates that the sample size is very dependent on the particular mass fraction of interest. For fine aggregates, the method gives very small sample sizes, so that in practice

Table 2.12 Calculation of sample sizes needed for sieve analysis of coarse aggregate between 5 and 20 mm (hypothetical example)

Fraction i	Size range (mm)	Mass proportion p_i	Average particle volume v_i (mm^3)	$p_i v_i$ (mm^3)	Required sample mass (kg)
1	20–15	0.65	2843.8	1848.5	0.9
2	15–10	0.25	1093.8	273.5	4.6
3	10–5	0.10	281.3	28.1	4.7
	Σ	1.00		2150.1	

the sample size will be determined by the standard sampling and testing devices used. In practice, standards specify minimum sample sizes required for representivity in testing, such as those given in Table 2.11.

Summary of national usage

National usage of various aggregate types in North America, Britain, and South Africa is summarized in Table 2.13. The table covers both crushed and non-crushed aggregates, the main sources, and usage in terms of the types of concrete construction in which the aggregates might be used and the approximate percentage of use, where such information is available. It is immediately obvious that aggregate type and usage vary enormously in different regions, reflecting source availability and local practices that have become established from decades of experience. The table helps to illustrate a point that cannot be over-emphasized: that good concrete can be made with a wide variety of aggregate sources and types. This is the main reason for concrete being ubiquitous in every country, and for its extensive use as the most important construction material globally.

Closure

To conclude this chapter, a few general comments are appropriate. The rocks that make up the vast bulk of natural aggregate production are, in the main, excellent source materials for concrete aggregates, provided they have been carefully selected and proved. However, they are all variable materials. This variability applies not only to the aggregate properties themselves, but also to the influences they have on the properties of concrete. Not all aggregate sources will be suitable for all types of concrete, and it is necessary for the engineer to exercise sound judgement and provide clear specifications for aggregates required to produce concrete for a given purpose or a specific quality. Many of these aspects are dealt with in the subsequent four chapters. It is inevitable that on occasions, sub-standard aggregates may be the only material available to the engineer for use in concrete. This may occur in remote areas or when economics precludes the use of a higher quality but more expensive material. The influence of such aggregates on the properties and quality of the resulting concrete will need to be assessed carefully. In the majority of cases, it will be found that perfectly satisfactory concrete can be made even with less than ideal aggregates. In a world of shrinking resources, where environmental pressures on resource exploitation are ever increasing, and where maximum economy of construction is constantly sought, the imaginative use of aggregate sources is very important.

Table 2.13 Summary of national usage of natural aggregate types

Origin of aggregates and approximate usage	
Crushed sources	*Non-crushed sources*

USA

Crushed stone aggregate is produced commercially in every state in the US except Delaware. In general, crushed stone operations tend to be very large, mainly due to the large investment required. Of the total production of crushed stone in the US, approximately 17 per cent (1990) was used as concrete aggregate.

Crushed stone aggregates are derived from a wide variety of parent bedrock materials. Limestone or other carbonate rocks account for approximately 75 per cent of rocks used for crushed stone; sandstone, granite and other igneous rocks make up most of the remainder.

Sedimentary rocks

Limestone and dolomite: Of the total production of crushed stone in the US in 1981, 74 per cent was limestone. Thus, carbonate rocks represent the bulk of crushed stone for concrete. Nearly every state has carbonate rock formations adequate to supply some concrete aggregate. Carbonate rock aggregates are most extensive in the central and eastern US, especially Kentucky, Tennessee, Missouri, and the Great Lakes Region, and are more scattered in the west.

Sandstone: This makes up only 3 per cent of the total US production of crushed stone, but is a major source of aggregate in some areas. Only non-friable and low porous sandstones are appropriate for concrete aggregates.

Sand and gravel aggregates are produced commercially in every state in the US. Approximately 50 per cent of the total coarse aggregate consumed by the concrete industry in the United States consists of gravels; most of the remainder is crushed rock. Sand and gravel commonly occur as river or stream deposits or in glaciated areas as glaciofluvial and other deposits. Natural silica sand is predominantly used as fine aggregate, even with most lightweight concretes.

Glacial materials: These materials are restricted to the northern US or high altitudes, and were deposited directly by the ice or as glaciofluvial deposits. Glacial deposits are generally called till. Due to the glacial processes, sand and gravel deposits are highly variable. Some material is too fine to be used as natural aggregate.

Alluvial materials: Large areas of the western United States contain alluvial deposits which are remnants of previous fresh water lakes. These are sources of large deposits of sand and gravel aggregates. The quality, physical properties, and grading of aggregates in these areas vary widely.

Marine or fluvial deposits – Atlantic coastal plain: The Atlantic coast from Long Island south and the Gulf Coast from Florida to Texas is covered with extensive deposits of sand, silt, clay, and gravel of varying thickness, of either marine or fluvial origin. Suitable sands and gravels are available for concrete aggregates, their quality depending on the rock from which they originated.

Igneous rocks: About 12 per cent of crushed stone produced in the USA is granite and about 8 per cent trap rock (i.e. gabbro, diabase, and basalt) (1981). Granites are widely distributed in the eastern and western US, with minimal exposures in the mid-west. Other igneous rocks are largely concentrated in the western US and in a few isolated localities in the east.

Metamorphic rocks: Metamorphic rocks (i.e. quartzites, schists, gneisses, and marbles) are found throughout the eastern Appalachian belt, Piedmont belt, Adirondacks, and in New England, the Lake Superior area, and scattered through the Cordilleran, Great Basin, and Pacific Coastal belts. Not all these rock types are suitable for concrete aggregates, however.

Natural aggregates, though widely distributed in the US, are not universally available for construction use, such as areas devoid of sand and gravel, and where overburden thicknesses are excessive, thus making quarrying impractical. In other areas, aggregates may not meet the physical, mechanical, or chemical requirements for use in concrete. In such cases, local aggregates may need beneficiation, or aggregates are imported or substituted by artificial aggregates.

Canada

As far as Canadian production of aggregates is concerned, the picture is essentially the same as in the US, since the industries are quite integrated. Canadian production of aggregates for concrete represents about 14 per cent of the total aggregate production, or about 33 million tons/year.

In the Maritime Provinces of Eastern Canada, some aggregates, particularly greywackes, are susceptible to a slow-to-appear form of alkali–aggregate reaction, and alkali–reactive aggregate sources occur in parts of British Columbia. Also, in Northern Ontario, there are deposits of dolomitic limestone which are susceptible to alkali–carbonate reactions.

Midwest region: Large portions of the mid-west are covered with soft, semi-consolidated sedimentary rocks. Surface exposures range from silt to sand and gravel, sand being most abundant. Gravels occur haphazardly throughout the region, usually in small deposits, e.g. as terraces on mountain flanks, and channel and terrace deposits of major rivers and streams.

Residual soils resulting from in-situ weathering of bedrock: Large portions of the US (including Arizona, the Carolinas, Kentucky, Oklahoma, Oregon, South Dakota, Tennessee, Texas, Virginia, Washington, Wisconsin, and Wyoming) comprise sands and gravels derived from in-situ weathering of bedrock. Properties depend on those of the parent bedrock, with composition ranging from nearly all clay minerals through mixtures of clay, silt, sand, and gravel, to nearly pure sand or sand and gravel.

Marine sand and gravel: Marine deposits on the continental shelves are large potential sources of sand and gravel, little exploited at present. Materials are generally sand-sized or finer, except for large deposits of sand and gravel off the coasts of New England and New Jersey, and scattered areas along both the Atlantic and Pacific coasts. They contain varying quantities of shell and require washing to eliminate salts.

Table 2.13 (Continued)

Origin of aggregates and approximate usage

Crushed sources	Non-crushed sources
United Kingdom	
Igneous rocks: Igneous sources include basalts, dolerites, and microdiorites, e.g. basalt in Midland Valley of Scotland, Antrim tertiary basalt in Northern Ireland, Whin Sill of northern England for dolerite. Granites and similar rocks are also used, e.g. western Scotland granite, where one source from a 'mammoth quarry' now supplies other parts of the UK, Europe, Middle East, and North America with crushed aggregates by sea.	*Glacial, fluviatile, and marine materials:* Most sand and gravel deposits derive from relatively recent (geologically) glacial, fluviatile, and marine processes: 'Outwash' sands and gravels (from retreating ice fronts) are found in East Anglia and northern England, and the so-called 'Plateau Gravels'.
	Floodplain or Terrace deposits: These comprise abundant sources of good sand and gravel aggregates, e.g. Thames Valley deposits and Trent Valley in East Midlands. There are some workable beach deposits.
Sedimentary rocks: These comprise limestones, e.g. Mendips, south Wales, Derbyshire, Pennines, and some sandstones, e.g. greywacke in Wales and Scotland, sandstone in south Wales and northern England, and some orthoquartzites. Metamorphic rock sources are used very little in the UK. Sixty per cent of all crushed rock aggregate in the UK comprises carboniferous limestone.	*Weak conglomerates:* A small proportion of aggregates derive from digging into earlier geological formations, e.g. weakly cemented Triassic quartzite conglomerate ('Bunter Pebble Beds') in the West Midlands, and east Devon. About 10 per cent of sand and gravel aggregate in England and Wales is from this source.
	Sea-dredged aggregates: These are composed mainly of flint of low porosity and absorption. Main areas of exploitation are the east coast between the Humber and Thames Estuaries, the English Channel, the Bristol Channel, and Liverpool Bay. Many of these deposits are derived from glacial deposition in the last ice age. They comprise between 10 and 15 per cent of concrete aggregates in the UK. In southeast England, they account for up to 50 per cent of aggregates in certain areas.
Sands and gravels: Many natural sand and gravel deposits in the UK contain cobbles and boulders larger than the maximum size normally used in concrete (nominally 20 mm, but up to 40 mm nominal maximum size). These are sometimes crushed to smaller sizes.	
South Africa	
The vast majority of coarse aggregate consists of crushed hard rock. In many cases, the same sources for crushed coarse aggregate are used to crush fine aggregate for concrete, and the demand for good quality crusher sand is steadily rising.	This section will deal exclusively with non-crushed fine aggregates for concrete.
	River sands: South Africa is a relatively arid country, and consequently there is a shortage of good quality river sands. One major source is

the Mgeni river in Durban, which has been used for concrete sand for decades, usually blended with pit sands to address its fines-deficiency. Mgeni sand is the preferred sand in Durban for concrete manufacture.

Beach sands: These are exploited for concrete near Durban, and in East London. They may require blending with other sands to improve grading, and may need washing to remove salts. Shell content up to 50 per cent has been noted, but it is generally less than about 20 per cent. In Durban, about 15 per cent of sand production is from beach sands.

Dune sands: These are mainly used in the Cape Town area, from sources on the Cape Flats, being beach-derived materials with a shell content of about 20 per cent. The majority of concrete in the Cape Town area uses dune sand. The sands are bunch-graded (i.e. relatively single-sized), and give rise to bleeding problems. Chlorides in these sands are not a problem. In Port Elizabeth, dune sands are used as blending sands with crushed materials.

Pit or Quarry sands: These sands derive from the in-situ weathered mantle, comprising either residual or transported materials. They usually require processing to remove coarse fragments and a proportion of the silt and clay. Weathered granite pit sands occur north of Johannesburg and are exploited in very large quantities. They tend to have rather high water requirement, and may contribute to increased concrete shrinkage.

Weathered granite sands are used for all types of construction, but it is difficult to make high quality concrete (e.g. HPC) with this material. Fine 'pit' sands are used in Durban as blending material for most

Quartzites: Vast quantities of quartzites have been crushed for many decades from the waste rock dumps of the gold mines. This has occurred in the Witwatersrand area, and the goldfields of the Free State and North West Province. Variable amounts of shale, siltstone, and basic igneous rocks (dolerites and diabase) are intermixed with the quartzites in the waste dumps. Pyrite may also be present. The quartzite can be alkali-reactive. Crushed quartzites have been the main source of coarse aggregate in Johannesburg and the Reef for decades, used in all types of concrete construction. These sources are now being depleted. Quartzites are also crushed for use as fine aggregates.

Other quartzites

a Rocks of the Table Mountain Group, exposed along the southern and eastern seaboards. This material has a moderate degree of alkali-reactivity. They provide roughly 75 per cent of coarse aggregate requirements in Durban; in Port Elizabeth, nearly all coarse aggregate, and a large proportion of fine aggregate, are from this source.

b Pretoria Quartzites. Pink or buff-coloured quartzites of the Pretoria Group. They have very low level of alkali-reactivity. They account for approximately 40 per cent of coarse aggregate in Pretoria area, and are also used for fine aggregate. These

c Quartzites and Sandstones of the Karoo Supergroup. These are widespread in the interior. Beaufort group quartzitic sandstone is a major source of coarse and fine aggregate in the East London area. Ecca Group sandstones are used in KwaZulu Natal midlands.

Table 2.13 (Continued)

Origin of aggregates and approximate usage

Crushed sources	Non-crushed sources
Basic igneous rocks: These exist in vast quantities throughout the region.	concretes. These are a source of high quality concrete sands. Siliceous pit sands are exploited to the north of Cape Town ('Klipheuwel' sands). They have low water requirement of about $170 l/m^3$.
Dolerites and Diabase: These refer to dyke and sill intrusive rocks, usually medium or fine-graded. Unweathered sources are high quality sources of aggregates. Weathered sources are suspect due to the presence of clay minerals. Crushed dolerite is the preferable source of coarse aggregate in vast areas of the central interior, e.g. Bloemfontein. Crushed dolerite sands are also used, but these tend to be harsh.	*Miscellaneous:* Many of the Karoo sands give rise to excessive shrinking and swelling of concrete and mortar. However, there are acceptable Karoo sands available, e.g. sands from Ecca Group, amongst others. Sands from Molteno sediments are used in the Free State (Bloemfontein) for concrete sand.
Gabbro and Norite: These occur as part of the Bushveld Complex, in an arc to the north of Pretoria. They are an acceptable source of concrete aggregate if unweathered, but like dolerites can cause segregation and bleeding in concrete due to high particle R.D. Used extensively as coarse aggregate in Pretoria North and Rustenburg areas.	
Basalt and Andesite: Enormous volumes of Karoo Basalts are available in the Drakensberg Escarpment area, and in Swaziland, Springbok Flats (Limpopo) and Namibia. Weathering of these materials can give rise to smectite clays which are dimensionally unstable, imparting high shrinkage to concrete. Thorough testing and petrographic examination of basalts is necessary before use in concrete. Basalt is used as source of concrete aggregate where other sources are not available. These materials were extensively used in construction of the giant Lesotho Highlands Water Scheme which included the large Katse Dam.	

Andesites in the form of Ventersdorp Eruptives cover large areas of southern Gauteng and North West Provinces, and northern areas of Northern Cape Province.

The rock imparts superior strength to concrete. Andesite is well known as an excellent source of crushed coarse aggregate in the southern Johannesburg area, where major exploitation occurs. Andesite is also used for crusher sand.

Granite: These are also widespread in South Africa, but less commonly exploited since sources are sometimes remote from areas of demand. Granites are used in the north of Johannesburg in large quantities for coarse aggregates, less so for fine aggregates. They are also used in the Cape Town area, particularly if AAR is of concern with local greywacke.

Greywacke: Developed by thermal metamorphism of argillaceous rocks of the Malmesbury Group, they are abundant in the Cape Town area. Particle shapes of crushed material may be rather angular and flaky. The rock is very susceptible to alkali-reactivity. Greywacke is extensively used in Cape Town as crushed coarse aggregate. 'Rock flour' from crushing is sometimes used as fine filler.

Miscellaneous

a *Tillite*: Cemented debris and pebbles of glacial origin. Highly variable. Extensively quarried and used as coarse aggregate in the Durban area, where about 25 per cent of concrete produced contains tillite.

Table 2.13 (Continued)

Origin of aggregates and approximate usage

Crushed sources	Non-crushed sources
b *Dolomite*: Occurs extensively over large areas of former Transvaal and Northern Cape. Used extensively in concrete sewer pipe construction. Crusher sands have remarkably low water requirements (from 175 to 190 l/m³). These dolomites impart high elastic modulus to concrete. They are quarried in Johannesburg and Pretoria areas where they are mainly used for precast concrete pipe manufacture, as both coarse and fine aggregate.	

Sources:
1 Main sources:
- UK: Eglinton, 1987; Sims and Brown, 1998
- South Africa: Grieve, 2001
- USA: Langer, 1988; Langer and Glanzman, 1993.
2 Sims and Brown (1998) give information on aggregate sources in areas such as the Middle East.

References

Ballim, Y. (2000) 'The effect of shale in quartzite aggregates on the creep and shrinkage of concrete – a comparison with RILEM Model B3', *Materials and Structures*, 33: 235–242.

Barksdale, R.D. (1991) *The Aggregate Handbook*, Washington, DC: National Stone Association.

Eglinton, M. (1987) *Concrete and Its Chemical Behaviour*, London: Thomas Telford.

Goodsall, G.D. and Matthews, D.H. (1970) 'Sampling of road surfacing materials', *Journal of Applied Chemistry of the USSR*, 20(12): 361–366.

Grieve, G.R.H. (2001) 'Aggregates for concrete', in B.J. Addis and G. Owens (eds) *Fulton's Concrete Technology*, 8th edn, Midrand: Cement and Concrete Institute.

Gy, P.M. (1982) *Sampling of Particulate Materials: Theory and Practice*, Oxford: Elsevier.

Kosmatka, S.H., Kerchoff, B., Panarese, W.C., Macleod, N.F. and McGrath, R.J. (2002) *Design and Control of Concrete Mixtures*, Engineering Bulletin 101, 7th Canadian edn, Ottawa, ON: Cement Association of Canada.

Langer, W.H. (1988) 'Natural aggregates of the conterminous United States', *US Geological Survey Bulletin 1594*, US Department of the Interior.

Langer, W.H. and Glanzman, V.M. (1993) 'Natural aggregate: Building America's future', *US Geological Survey Circular 1110*, US Department of the Interior.

Miller-Warden Associates (1967) 'Effects of different methods of stockpiling and handling aggregates', *Highway Research Board*, Washington: National Research Council.

Orr, C.M. (1979) 'Rapid weathering dolerites', *The Civil Engineer in South Africa*, 21(7): 161–167.

Pike, D.C. (ed.) (1990) *Standards for Aggregates*, Chichester: Ellis Horwood.

Sims, I. and Brown, B. (1998) 'Concrete aggregates', in P.C. Hewlett (ed.) *Lea's Chemistry of Cement and Concrete*, 4th edn, London: Arnold.

Smith, M.R. and Collis, L. (2001) *Aggregates: Sand, Gravel and Crushed Rock Aggregates for Construction Purposes*, Engineering Geology Special Publications 17, London: Geological Society.

Further reading

Best, M.G. (2002) *Igneous and Metamorphic Petrology*, 2nd edn, Oxford: Blackwell Publishing.

Blyth, F.G.H. and De Freitas, M.H. (1984) *A Geology for Engineers*, 7th edn, London: Edward Arnold Ltd.

Press, F., Siever, R., Grotzinger, J. and Jordan, T.H. (2004) *Understanding Earth*, 4th edn, New York: Freeman.

St John, D.A., Poole, A.B. and Sims, I. (1998) *Concrete Petrography: A Handbook of Investigative Techniques*, London/New York: Arnold/ Wiley.

Tucker, M. (2001) *Sedimentary Petrology*, 3rd edn, Oxford: Blackwell Publishing.

Chapter 3

Properties and characterization of aggregates

If concrete science and technology are to keep pace with developments in other fields of engineering, it is essential that we make the most effective use of concrete materials. It is necessary to move beyond simply knowing the properties of the aggregates to an understanding and application of the effects of their properties on concrete in its various forms. The link between aggregate properties and performance in concrete is not yet entirely understood in many respects. Aggregate properties have profound influences on concrete properties, and these influences need to be understood and appreciated.

Concrete aggregates are required to meet minimum standards of cleanliness, strength, and durability, and to be substantially free of deleterious substances. Materials that are soft, very flaky, too porous, or that can react detrimentally in concrete should be excluded. For this reason, thorough testing and examination, including petrographic examination, should be carried out before new or untried sources of aggregate are used. Aggregates for concrete are characterized by using standard tests. This ensures that the aggregates conform to minimum specification criteria. Characterization also permits equitable comparison amongst different aggregates, allowing the engineer to make proper selections. A further goal of characterization, not yet fully achieved, is to provide measurable properties which can be related to the aggregate performance in concrete. Aspects of this are dealt with in subsequent chapters.

This chapter discusses important aggregate properties – physical, mechanical, and chemical. It also deals with the granular nature of aggregates as well as undesirable properties and constituents. A section on aggregate petrography is included. The emphasis is on conventional normal-density aggregates, since these make up the vast bulk of aggregates used in concrete. Special aggregates, including low density (lightweight) and high density aggregates, are covered in Chapter 7. A list of standard aggregate tests (ASTM, CSA, BS, and SANS) is given in Appendix A, and reference to standards is therefore not always given in the text. (Chapter 8 also gives a summary of the provisions of standard specifications for concrete aggregates.)

Table 3.1 provides a summary of the properties of aggregates indicating their effects on, and significance to, relevant concrete properties, and gives the basis of measurement of the properties. Table 3.2 gives typical ranges of values of aggregate properties for generic aggregate types based on their petrological classification. These values will vary widely even within one aggregate group, and the properties must be obtained for any given aggregate. As would be expected, properties of hard, sound, fresh rocks, particularly igneous rocks, are superior to sedimentary or some of the metamorphic rocks. The table indicates the wide range of different rock types used as concrete aggregates.

Physical properties of aggregates

This section covers the important physical properties of aggregates. Not all these properties will necessarily be evaluated for all aggregates, with some of them (e.g. thermal properties) being rarely required. Others (e.g. surface texture) may only be assessed initially for a new or untried source of aggregate.

Porosity

Most conventional dense mineral aggregate particles have a measurable porosity and are able to absorb water. Porosity, p, is the internal pore volume as a proportion of the total volume of a solid,

$$p = \frac{V_p}{V_T}$$ (3.1)

where

V_p is the volume of internal pores
V_T is the total volume of the solid.

While the pore volume includes all pores, we can measure only the interconnected porosity in typical laboratory tests. Porosity is measured by drying a sample at $100–110\,°C$ to constant mass, and then saturating the sample in water. The standard saturation period is $24\,h$, though some aggregates will absorb water for longer due to the difficulty of removing air by diffusion from the pores of the aggregate. Vacuum-saturating the sample will saturate the pores more rapidly.

The significance of aggregate porosity lies in its effects on aggregate density and thus concrete density which is indirectly related to concrete strength and stiffness. Porous aggregates will have lower density, elastic modulus, and strength, extreme cases being lightweight and highly porous aggregates

Table 3.1 Aggregate properties in relation to their performance in concrete

	Significance	Effects on concrete	Basis of measurement	Typical standard tests
Physical properties				
Aggregate particles				
Porosity	Porosity influences absorption and density of aggregates	Can influence concrete strength, stiffness, permeability, and durability	Gravimetric, with oven-drying and saturation	ISRM and Absorption tests below
Absorption and surface moisture	Absorptive aggregates require possible precautions for water requirement, strength, and freeze–thaw resistance	Can influence concrete strength, movement properties, and durability	Gravimetric, with oven-drying and saturation	ASTM C127 (C.A.), C128, C70 (F.A.); BS 812: Part 109: 1990 SANS 5843
Water content (Moisture content)	Water content affects aggregate density and mix-water requirement	Can influence concrete strength and movement properties	Oven-drying followed by weighing	ASTM C70, C566 BS 812: Part 109: 1990 SANS 5855
Particle density	Density influences concrete density in plastic and hardened states	Influence on segregation, concrete strength, stiffness, and thermal properties. Used in mix proportioning	Mass and volume measurements, using volume displacement techniques	ASTM C127 (C.A.), C128 (F.A.) BS 812 Part 2: 1975 SANS 5844
Particle shape	Shape influences particle packing and internal aggregate interlock	Very important influence on concrete workability, and hence mix-water requirement. May also influence concrete strength – angular particles are preferred for improved strength. Flaky particles can reduce strength	Appearance, geometric ratios, and standard descriptors. Indirect measure by bulk density and void content – see below	Petrographic: ASTM C295 BS 812: Part 104: 1994 Flakiness: SANS 5847; BS 812: Part 105.1: 1989 Measures of shape: ASTM C29, C1252, ASTM D4791 (C.A.), D3398 CRD-C-119 (C.A.); CRD-C-120 (F.A.)

Particle surface texture	Surface texture can affect surface frictional properties in a mix, and therefore mix 'harshness'	Influences concrete workability and mix-water requirement (but less so than shape). Rougher texture improves aggregate bond	Usually assessed subjectively by examination or reference to other aggregates	Petrographic: ASTM C295 BS 812: Part 104: 1994
Moisture movements	Dimensional instability (e.g. excessive shrinkage) can render aggregates unsuitable for use in concrete	If excessive, can cause dimensional instability in concrete, and associated cracking	Strain measurements during wetting and drying cycles. Alternatively, assess effect of aggregates on shrinkage and expansion of mortar	ASTM C157 (mortar shrinkage) BS 812: Part 120: 1989 SANS 5836 (mortar shrinkage)
Temperature movements	Thermal contraction/ expansion of aggregates is significant for concrete properties	Can result in strain incompatibility between aggregate and paste, leading to cracking (e.g. low thermal coefficient materials)	Strain measurements taken during heating and cooling cycles	CRD-C-125 (coarse aggregate particles) Method of Verbeck and Haas (1950)
Thermal properties (specific heat, conductivity, diffusivity)	Thermal properties of aggregates strongly influence concrete properties, important for massive sections, and for large temperature differentials	Concrete can crack under large temperature differentials. Thermal properties also influence insulation properties of concrete	Measurement of heat flow	CRD-C-124: Specific heat only
Permeability	Rock permeability usually similar to that of paste; occasionally important to concrete permeability and durability	Influences permeability of concrete, therefore water-tightness. Can also influence freeze–thaw resistance	Permeability of aggregates not usually measured. Rock permeability measurements can be done in conventional manner	(Use standard geotechnical or rock mechanics techniques)

Table 3.1 (Continued)

	Significance	Effects on concrete	Basis of measurement	Typical standard tests
Bulk aggregates				
Bulk density and void content	These determine the paste or matrix volume required, thus the 'richness' and economy of a mix	Influence relative proportions of concrete ingredients (mix proportioning calculations). High bulk density gives improved technical properties and better economy	Mass of oven-dry sample in loose or dense packing in a container of known volume	ASTM C29; C1252 BS 812: Part 2: 1975 SANS 5845
Particle size distribution (grading), including fines content, fineness modulus	Grading represents the distribution of particle sizes in a mix, and is fundamental to the nature of concrete as a 'bound conglomerate'	Has important influence on workability and other plastic properties of concrete. Fines content (minus 75-μm fraction) crucial to cohesiveness and control of bleeding	Sample passed through standard sieves of different openings, to separate aggregate into size fractions. Quantity of fines and presence of clay can be determined	ASTM C136 (incl. max. size); C117 (Fines); D2419 (Clay) BS 812: Part 103.1: 1985 & Part 103.2: 1989 (Fines) SANS 201, 6241 (F.A.), 6244 (F.A. – Pipette Method)
Mechanical properties				
Strength (in rock sample form)	Aggregates contribute directly and indirectly to concrete strength and should not be the 'weak link'		Measured on specimens cut or cored from larger solid such as a boulder (but not frequently done)	UCS: ISRM UCS: ASTM D2938 (rock cores) Tensile: ASTM D2936 (rock cores)
Strength in granular form	Strength is easier to measure in granular form; however, this is not a fundamental strength value – depends for example on shape	Lower strength aggregates may limit concrete strength. Aggregate strength is crucial in HSC	Crushing or impact of sample in confining container; the 'fines' produced are measured	ACV: BS 812: Part 110: 1990 SANS 5841 10% FACT: BS 812: Part 111: 1990 SANS 5842 AIV: BS 812: Part 112: 1990 SANS 6239

			ISRM techniques
Elastic modulus	Stiffer aggregates generally produce concrete with superior mechanical properties	Aggregate modulus has a strong influence on concrete modulus, lesser influence on concrete strength	Measured on specimens cut from larger solid such as a boulder. (Poisson's ratio can also be measured)	
Abrasion resistance (resistance to degradation)	A surface property, important in cases of surface wear, e.g. trafficked or erosion areas, wear-resistant floors		Many different wear tests are available; test must be matched to abrasion condition	LA Abrasion: ASTM C131, C535 AAV: BS 812: Part 113: 1990 Attrition: ASTM C1137 Calif. Durab. Index: ASTM D3744
Surface hardness	Hard minerals in aggregates improve concrete resistance to abrasion and improve surface frictional properties	Aggregate abrasion resistance and surface hardness influence abrasion resistance of concrete as a composite. In addition, resistance to degradation is important in regard to possible aggregate breakdown during handling and mixing	Seldom used for aggregates. Adapted Vickers Test can be used; alternatively, petrographic techniques	Petrographic: ASTM C295 BS 812: Part 104: 1994
Fracture parameters	There is no direct application of these parameters in concrete design at present; of more use comparatively between aggregates	Indirectly influence fracture properties and strength of concrete	Tests done on notched prismatic specimens, e.g. notched beams or cylinders	ISRM (1988)

Table 3.1 (Continued)

	Significance	Effects on concrete	Basis of measurement	Typical standard tests
Chemical and durability properties (See also Chapter 6)				
Table 3.13 covers this aspect more fully. Chapter 6 likewise deals with aggregate durability issues in respect of hardened concrete. For completeness, three aspects are dealt with below.				
Sulphate soundness	Relates to physical effects of excessive volume change of aggregates under weathering actions	Unsoundness can cause disruption of concrete, ranging from superficial pop-outs to severe cracking	Wetting and drying in sulphate solutions, which imposes stress on aggregate pore structure	ASTM C88 BS 812: Part 121: 1989 SANS 5839
Freeze–thaw resistance	Poor freeze–thaw resistance of aggregates can cause major distress in concrete. Often associated with porous aggregates of sedimentary origin	Occurrence of disintegration or pop-outs, surface scaling, roughness, and/or D-cracking in slabs on grade, along joints and free edges	Measured on 'critically saturated' aggregates in a series of freeze–thaw cycles	Aggregates: AASHTO T103 CSA A23.2-24A Aggregate in air-entrained concrete: ASTM C666; C682
Alkali–aggregate reactivity (see also Table 6.3)	In many areas of the world (including Canada, UK, USA, SA) AAR represents a major threat to long-term performance of affected structures. Measures exist which can eliminate or control AAR in new concrete	Causes expansion and cracking when sufficient moisture is present. Effects vary from minor cracking to major structural breakdown, and also unsightly staining. Occurs mainly with silica-bearing aggregate, but can also occur with certain carbonate rocks	Petrographic detection; chemical methods; expansion or length change of mortars, concrete, or rock cylinders	Petrographic: ASTM C294; C295 BS 812: Part 104: 1994 Chemical: ASTM C289 Expansion (Mortar Bar): ASTM C227; C1260 BS DD 249 CSA A23.2-25A SANS 6245 Carb. Rock: ASTM C586, C1105

Undesirable, deleterious, and organic substances in aggregates: see Table 3.15.

C.A. = coarse aggregate; F.A. = fine aggregate.

Notes
1 ACI 221 R Committee Report has a useful table (Table 1.1) on aggregate properties, their significance, and test methods (ACI 221R-96, 2001).
2 CSA Standard A23.2 contains tests for aggregates that often differ only slightly from the corresponding ASTM tests. See Table 8.2 for a synoptic listing of relevant test methods.

Table 3.2 Typical ranges of properties of aggregates*

Aggregate (rock) type	Physical				Mechanical†						Chemical/Durability	
	Porosity (%)	Water absorption (%)	Apparent relative density	Coefficient of linear thermal expansion (10^{-6}/°C)	Elastic modulus (GPa)	UCS (MPa)	ACV (%)	10% FACT (kN)	AIV (%)	LA Abrasion (%)	Soundness (MgSO$_4$) (% loss)	Chemical stability
Siliceous rocks												
Quartzites												
UK	0.5–5.0 1.9–15.1	0.7–1.2	2.6–2.7	10.3–10.8	10–70 59, 73	280–390 124–423 (330)	14 16	293 140–250 (195)	17 13–22			
SA (crushed) US		0.3	2.69	8.5–12.5 11.0–12.5	61–92	127–373 340–470 252	7.6–15.8 (Tensile)	193–244		28		Good
Quartz												
US				11.5–12.0	67–77							
Sandstone												
UK	5–25	0.2–1.5 0.3–31.0 (1.0)	2.6–2.9 2.2–2.7	4.3–13.9	5–50 15	190–260 20–175 44–240	10–18 12–31 (15)	215–400 100–350 (200)	8–20 10–58 (18)	18	2 0–100‡ (4)	
SA US		1.8	2.54	10.5–12.0 10.5–12.0	3–53	131				38		Fair to Good
Gritstone												
UK	0–48		2.67		55	220	12		15	18.1		
Flint/Chert												
UK		0.4–7.6 (1.6) 0.1–0.7	2.4–2.7 2.4–2.6	7.4–13.1 7.3–13.1	55	200 205	14–31 (16) 17	160–320 (260)	16–24 (20) 17		0–6‡ (3)	
US		1.60	2.50		50–74					26		Poor

Table 3.2 (Continued)

Aggregate (rock) type	Physical				Mechanical†						Chemical/Durability	
	Porosity (%)	Water absorption (%)	Apparent relative density	Coefficient of linear thermal expansion ($10^{-6}/°C$)	Elastic modulus (GPa)	UCS (MPa)	ACV (%)	10% FACT (kN)	AIV (%)	LA Abrasion (%)	Soundness ($MgSO_4$) (% loss)	Chemical stability

Granite and Porphyry rocks

Aggregate (rock) type	Porosity (%)	Water absorption (%)	Apparent relative density	Coeff. lin. thermal exp. ($10^{-6}/°C$)	Elastic modulus (GPa)	UCS (MPa)	ACV (%)	10% FACT (kN)	AIV (%)	LA Abrasion (%)	Soundness ($MgSO_4$) (% loss)	Chemical stability
Granite												
UK	0.5–1.5; 0.4–3.8	0.2–1.9; 0.6–0.8 (0.6)	2.6–3.0	1.8–11.9	5–50; 47; 44–63	100–250; 114–257 (185)	20–25; 23–30; 13–32 (24)	150–190; 75–280 (155)	17–21; 23–26; 11–35 (21)	27	1–3; 1–31‡ (2)	
SA		0.3	2.65	6.5–9.7	70–79	235–303		155–280				
US				6.5–8.5	9–43	85–275 (180)				38		Good
Felsite												
UK					81	120–526; 377–453			12–15			
SA		0.8	2.66	7.0–9.2				305		18		Good
US						324						
Rhyolite												
UK			2.6–2.9	5.7–8.4								
Gneiss												
UK		0.4–0.7 (0.6)				94–235		230	24–28		1–2‡ (1)	
US		0.3	2.74	6.5–8.5		147				45		Good

Basic igneous rocks

Andesite

UK		0.9 / 1.4–7.9 / (3.0)		4.1–10.3 / 4.1–10.3		210–280	15–17 / 11–12	260 / 170–190 / (180)	11–16 / 12–31 / (18)	0–96‡ / (4)	
SA				5.5–8.0	101	516–538		458			
Basalt											
UK	0.5–1.5	0.4–1.1 / 0.4–5.4 / (1.8)	2.6–3.0	3.6–9.7	5–50 / 75 / 64–96	200 / 100–310	11–17 / 11–58	270–380 / 160–350 / (220)	16–18 / 10–22 / (19)	— / 0–16‡ / (2)	
SA		0.5		5.5–8.0							
US			2.86		64–69					14	Good
Dolerite											
UK		0.6–1.1 / 0.2–5.7 / (0.6)		3.6–9.7		180–190	11–19 / 11–36 / (19)	295–375 / 50–380 / (190)	9–17 / 9–40 / (18)	2 / 0–1‡ / (1)	
SA				6.1–7.2	78–83	288–331		291		12–16	Generally Good
Gabbro											
UK		0.5	2.95								
US		0.3	2.96	3.6–9.7		195	17	240	17–19	18	Good
Diorite											
UK		1.1–1.2									
US		0.3	2.92		94–103		15–20	190	13–20	22	Good
Diabase											
UK				3.6–9.7							
US		0.3	2.96			105–235				18	Good

Table 3.2 (Continued)

	Physical				Mechanical[†]						Chemical/Durability	
Aggregate (rock) type	Porosity (%)	Water absorption (%)	Apparent relative density	Coefficient of linear thermal expansion ($10^{-6}/°C$)	Elastic modulus (GPa)	UCS (MPa)	ACV (%)	10% FACT (kN)	AIV (%)	LA Abrasion (%)	Soundness (MgSO₄) (% loss)	Chemical stability

Carbonate rocks (Calcites and Dolomites)

Limestones

UK	5–20 0–37.6	0.2–7.5 0.5–0.7 (1.6)	2.5–2.8	0.9–12.2	2–70 76 67	93–241 130–190 35–250 165	18–27 19–31 (24)	90–290 110–250 (170)	14–28 14–43 (24) 9	22–30	1–10 0–71[‡] (6)	
SA				3.5–6.0								
US		0.9	2.66	3.5–6.0	3–84	159 90–270 105–200	2.1–6.2 (Tensile)			26		Good to Fair

Dolomite

UK				6.7–8.6	114	212–396						
SA				7.0–10.0				230		25		Good to Fair
US		1.1	2.70	7.0–10.0	21–62							

Marble

UK	0.5–2.0		2.6–2.8	1.1–16.0	51–244 40–100	100–250	25–28		16–21			Good
SA				4.0–7.0								
US		0.2	2.63	4.0–7.0	117				47			

Miscellaneous

	C1	C2	C3	C4	C5	C6	C7	C8
Greywacke (Hornfels)								
UK	0.5	2.7–3.0			340–370 / 340	10 / 11–15 / 11	380	9 / 16–17 / 17
SA			10.9	73	297–308		299	
Mixed gravels								
UK	2.07–3.44 / 1.1–3.15 / 0.9–4.50	2.7–2.8						
US					165–235			
Tillite								
SA			6.5					
Siltstone								
SA			6.8	25	125–226		240	
US				19–88				

Manufactured (Synthetic)

	C1	C2	C3	C4	C5	C6	C7	C8
Blastfurnace slag								
UK	2.0–4.8				90	33–42 / (34)	55–210 / (85)	15–33 / (28)
Haematite ore								
US				97–141				
SA		5.0				11.1–13.7	390–465	
Shale								
UK	10–30	2–2.7		5–25	10–100			
US			9.5–11.0	8–53				

Table 3.2 (Continued)

Aggregate (rock type)	Physical				Mechanical[†]						Chemical/Durability	
	Porosity (%)	Water absorption (%)	Apparent relative density	Coefficient of linear thermal expansion ($10^{-6}/°C$)	Elastic modulus (GPa)	UCS (MPa)	ACV (%)	10% FACT (kN)	AIV (%)	LA Abrasion (%)	Soundness ($MgSO_4$) (% loss)	Chemical stability
Typical values from Table 1.1, ACI 221R-96	Coarse agg.: 1–10%	Fine agg.: 0.2–2% Coarse agg.: 0.2–4%	1.6–3.2, but 2.5–3.0 more normal	Agg. particles: 2–16 · $10^{-6}/°C$	7–70 (Poisson's ratio: 0.1–0.3)	Rock cores: 70–276	14–30% (typical SA results)			15–50% loss	Fine agg.: 1–10% Coarse agg.: 1–12%	

Sources: Sources are quoted for the UK, SA, and US. The sources are generally representative of the country or region concerned, but not exclusively so. For example, some data quoted under UK may have derived from other worldwide sources. Nevertheless, the values give an indication of the ranges in the area considered.

UK sources: Sims and Brown (1998), Neville (1995), Kaplan (1959) and Teychenné (1978)
South African sources: Alexander (1990) and Fulton (2001)
US sources: Meininger (1994), Waddell (1974) and ASTM STP 169C

On occasions, the Sources used were quoting other sources.
ACI 221R-96 has a table 'Properties of concrete influenced by aggregate properties' that is useful. Values quoted from ACI 221R-96, without reference to particular aggregate types, are shown above.

Notes
1 Typical South African aggregates have absorption <0.5%.
2 Tests on several South African sands gave void contents ranging from 28 to 44%, mean value = 36% (Davis, 1975).
3 Typical UK aggregates have water absorptions from <1 to ~5%.
4 BS 8007: 1987 and BS 6349: 1984 recommend 3% maximum water absorption for water-retaining and marine structures.
5 Measurements of coefficients of water permeability of small (25 mm) aggregate pieces are not simple to perform. Available values are:
   • Dense trap rock (i.e. dolerite, gabbro, etc.): 3.5×10^{-13} m^3/(s · m^2 · MPa/m)
   • Granite: 2.2×10^{-9} m^3/(s · m^2 · MPa/m).
These values are similar to those for mature hcps of w/c = 0.38 and 0.71 respectively. Note that the porosities of these rocks (typically less than 1%) are considerably less than those of the corresponding hardened pastes (Source: Helmuth, 1994).

* Ranges are quoted as far as possible rather than averages. A single value usually represents a single determination. A single value in parentheses represents a mean value of the range immediately above it.
† Abbreviations are: UCS – Unconfined compressive strength; ACV – Aggregate crushing value; FACT – Fines aggregate crushing test; AIV – Aggregate impact value; LA Abrasion – Los Angeles Abrasion value.
‡ $MgSO_4$ after 5 cycles of ASTM C88 Test.

that require special mix proportioning and handling procedures. They may increase the permeability of the concrete to ions and fluids, particularly if their pore system is interconnected. Porous aggregates that are not fully saturated at the time of mixing can withdraw water from the mix resulting in a porous interfacial transition zone (ITZ) and weak interfacial bond. Conversely, in modern high strength concrete (HSC), aggregates with a higher than normal absorption can be used to provide a latent source of moisture to the hydrating matrix after initial hardening, so that strength development is enhanced and autogenous shrinkage is reduced. Porous aggregates can also help to reduce the disruptive expansion from alkali–aggregate reactivity (Collins and Bareham, 1987). Higher than normal aggregate porosity and absorption does not necessarily imply reduced concrete strength or durability.

Pore system of coarse aggregates

Winslow (1994) discusses this in great detail, making the point that the pore system of aggregates is probably their most important single feature. Pore space comprises a portion that is accessible from the surface and a portion that is completely isolated by the surrounding solid. The vast majority of aggregate pore space is likely to be accessible, in which case it will have a substantial influence on aggregate durability, in particular freeze–thaw durability which will be discussed later. The two main assessments of pore space are its volume and size. While volume of a pore space is usually reasonably unambiguous to measure, size is more complicated. Most aggregates contain pores that form a continuous yet tortuous system, making the definition of pore size difficult, since the pore space is more or less continuous. Pores of a given 'size' refer to that portion of the continuous system where the cross section is characterized by an appropriate size parameter. Size is therefore defined by the experimental technique used to determine it, and by how the analysis of the technique is modelled. Generally, when pores are being characterized for size, they are modelled as having a circular cross section with a characteristic diameter. In using such a characterization, the diameters of the hypothetical model pore space are assumed to respond to the measurement technique in the same way as does the actual pore space. Size–volume relationships are usually expressed as a cumulative distribution with the abscissa displaying the pore sizes and the ordinate showing the pore volume having either larger or smaller sizes. Each pore diameter has an associated pore volume representing that portion of the pore space that has either larger or smaller sizes. Other pore system parameters such as specific surface area and hydraulic radius, as well as experimental methods for characterizing the pore system, are discussed in Winslow (1994).

Absorption and moisture state

Aggregates that are porous can absorb water. Absorption is thus governed by porosity. For the pores of an aggregate particle to fill with water, the pores must be interconnected and open to the surface so that water from the exterior can penetrate the solid. This is not always the case, and what is measured is usually an 'apparent porosity' which does not account for the impermeable pores. This is not normally a problem, except that the density or strength of the aggregate may be affected. Absorption is measured in the same type of test as that for porosity, by allowing oven-dry aggregates to absorb water while being submerged and then measuring their mass in a saturated-surface-dry (SSD) condition. Absorption is expressed as the ratio of the increase in mass of an oven-dried sample after saturation to the mass of the saturated-surface-dry sample, in per cent.

Water absorption of less than 1 per cent will have little practical effect on concrete properties such as shrinkage and creep. Aggregates with absorptions much higher than 2–3 per cent should be treated as suspect and checked for their influence on concrete performance, for example whether they impart higher drying shrinkage to concrete. Absorption limits are rarely specified in standards, although project-specific specifications may if a suspect aggregate is likely to be used. Where a relationship exists between absorption and some other undesirable property such as poor frost resistance, water absorption limits may be specified for control and compliance purposes. Porous aggregates may adversely affect the resistance of concrete to freeze–thaw conditions, particularly if pores are smaller than 4 μm in size, and may increase concrete water permeability. The freeze–thaw resistance of aggregates will be covered later in this chapter.

Moisture state

Aggregate particles can assume different moisture states: *oven dry*, in which evaporable water is driven off at 100–110 °C; *air dry*, when aggregates dry to hygral equilibrium with the surrounding air and some moisture is retained by the aggregates; and *saturated-surface-dry* (SSD), which is the condition when the aggregate particles themselves are saturated, but there is no free or excess moisture on the surface. The condition of saturated particles with free surface moisture (*'wet'*) can also occur. Figure 3.1 illustrates these different moisture states.

The moisture state will influence aggregate density and therefore density is measured at some standard or specified moisture state. The most useful moisture state is SSD, since the aggregate will not take up water from the mix thereby adversely affecting workability, nor contribute excess water to the mix thus reducing the strength. The SSD state can be obtained for coarse aggregate particles by saturating them in water and then drying the surfaces with absorbent cloth, preferably in a stream of moving air (ASTM C127).

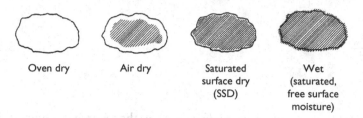

Figure 3.1 Moisture states of aggregates.

On completion of this process, indicated by the loss of sheen on the surface, the particles are immediately weighed in air and then in water to obtain their SSD density. For fine aggregates, the process is not quite so simple. A sample of saturated sand is progressively dried by being exposed to a gently moving current of warm air while being stirred. From time to time, the sample is lightly tamped into a truncated cone mould. The SSD state is reached when the fine aggregate slumps slightly on removal of the mould (ASTM C128). If excess surface moisture is still present, the sand retains its shape, and if it is too dry, it will flow and slump when the mould is removed. These states are illustrated in Figure 3.2 (Aïtcin, 1998). The test is best suited to natural well-shaped sands, and judgment is needed for angular aggregates such as crusher sands or sands with high fines contents which do not slump as readily.

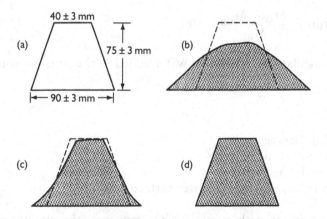

Figure 3.2 Determination of the SSD state for a sand: (a) The standard mini-cone used; (b) Sand having a water content below its SSD state; (c) Sand in an SSD state; and (d) Sand having a water content above its SSD state (from Aïtcin, 1998).

Aggregate porosity and absorption will govern its water content ω, defined by

$$\omega = \frac{M_W}{M_D} \qquad (3.2)$$

where

M_W is the total mass of evaporable water in the aggregate (including absorbed and free surface water)
M_D is the mass of (oven-dry) solids.

Water content is determined gravimetrically by heating a sample of aggregate to 100–110 °C and measuring the amount of water driven off.

Equation (3.2) can be expressed in other forms to define particular parameters for aggregates:

$$\text{Absorption (or absorption capacity)} = \frac{M_{SSD} - M_D}{M_D} \times 100\% \qquad (3.3)$$

which represents the maximum amount of water the aggregate can absorb

$$\text{Effective absorption} = \frac{M_{SSD} - M_{AD}}{M_{SSD}} \times 100\% \qquad (3.4)$$

which represents the amount of water required to bring an aggregate from the air-dry (AD) state to the SSD state

$$\text{Surface moisture} = \frac{M_{wet} - M_{SSD}}{M_{SSD}} \times 100\% \qquad (3.5)$$

which is used to calculate the additional water added to the concrete with the aggregate

where

M_{SSD} = Mass of SSD aggregate
M_D = Mass of oven-dry aggregate
M_{AD} = Mass of air-dry aggregate
M_{wet} = Mass of wet aggregate (having free surface moisture).

For a saturated aggregate with excess surface moisture, the amount of this excess moisture is sometimes called the 'moisture content' or 'surface moisture' of an aggregate (not to be confused with water content given by Equation 3.2). This moisture is important because, if present during batching and mixing, it contributes to the total water content of a mix and will therefore increase the water/cement (w/c) ratio. It is difficult to measure this quantity and so the total water content of an aggregate is usually assessed for mix proportioning and batching purposes. Provided the absorption of the aggregates is low (less than 1 per cent), reasonable consistency in mixing water requirements can be achieved.

In an aggregate stockpile, the water content of stone typically varies from 1 to 3 per cent, and of fine aggregate from 3 to 6 per cent, but this depends on weather conditions and the position in the stockpile from which the sample is taken. For mix proportioning calculations, the SSD mass of aggregates is used. The water content of an aggregate stockpile is measured or estimated and the mass of aggregate batched into a mix is adjusted for the excess water content. Likewise, the water added to the mix is adjusted (reduced) to account for the water associated with the aggregates.

A practical consequence of the water content of fine aggregates is that *bulking* occurs in these materials, that is moist sand will occupy a larger volume than completely dry sand since the packing of the particles is influenced by surface tension forces, causing voids between the sand particles in a manner similar to flocculation in clay particles (Dewar and Anderson, 1992). The bulking of sands is important in mix proportioning and in batching calculations and mixing operations, since bulked sand has a lower unit weight than in the dry state. Typical bulking curves are shown in Figure 3.3a. Bulking depends on moisture content and grading of the sand with fine sands bulking more than coarse sands for the same moisture content. This has an important influence on bulk density of the sand, shown in Figure 3.3b. This reinforces the need to use mass batching rather than volume batching if consistent concrete quality is to be achieved.

Bulking is not usually an issue for stone particles, since in the larger sizes the surface tension forces between particles are insufficient to hold the particles in a loose packing configuration.

Density

Density ρ of a solid is defined as the ratio of its mass to the volume it occupies, that is

$$\rho = \frac{M_S}{V_S} \tag{3.6}$$

where
M_S and V_S are the mass and volume of the solid, respectively.

The density of aggregates is used as the basis for classifying them into 'normal weight', 'lightweight', or 'heavyweight' aggregates. There are several measures of density that must be considered.

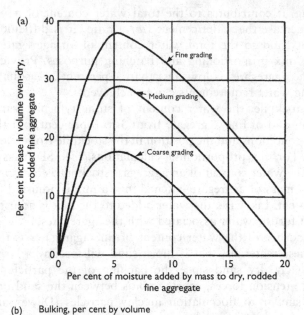

(a)

Per cent increase in volume oven-dry, rodded fine aggregate

Fine grading

Medium grading

Coarse grading

Per cent of moisture added by mass to dry, rodded fine aggregate

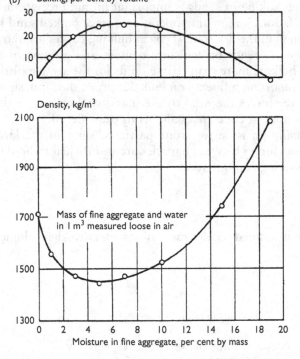

(b) Bulking, per cent by volume

Density, kg/m³

Mass of fine aggregate and water in 1 m³ measured loose in air

Moisture in fine aggregate, per cent by mass

Figure 3.3 (a) Typical bulking curves for concrete sands; (b) Bulk density of sand affected by bulking (from Kosmatka *et al.*, 1995).

Absolute density

This is the density of a solid excluding the internal enclosed pores, that is it is the density of the solid material only. This is not normally used in concrete technology, since it involves pulverizing the material to eliminate the enclosed impermeable pores.

Apparent density

This refers to the density of a solid, including the impermeable pores, but excluding the permeable (or capillary) pores. The difference between these two types of pores is that the capillary pores lose their trapped moisture at 100–110 °C whereas the impermeable pores do not. The apparent density is calculated from the oven-dry mass and the SSD volume of the aggregate, measured using Archimedes Principle or volume displacement techniques. A variation of the apparent density can be measured using the SSD mass of the aggregates, rather than the oven-dry mass, and yields a slightly higher value. This is the value most frequently and easily determined and is an important value in concrete technology, since it is included in calculations of the *yield* of a mix (i.e. the mass of constituents required to make up a given mix volume).

Relative density (RD)

The relative density (RD) of a solid is its density divided by the standard density of water (1000 kg/m³). It is a relative measure of density, useful since it is independent of the units used for absolute density. In concrete technology, the relative density frequently refers to the ratio of apparent density to density of water. The term 'relative density' is preferred to 'specific gravity' which is sometimes used to denote the same property. Conventional aggregates have relative densities that vary from about 2.2 to 3.0.

Bulk density

This is the density of the material in bulk granular form, that is the mass of aggregate particles occupying a certain volume:

$$\rho_{bulk} = \frac{M_T}{V_T} \tag{3.7}$$

where

M_T is the total mass of the granular sample (the mass of solids)
V_T is the total volume occupied by the sample.

The bulk density depends on the packing of the particles, which in turn depends on characteristics such as particle shape, surface texture, and grading. The measurement of bulk density includes the void content of the aggregates (see later) and is carried out using aggregates in a dry condition.

Two practical measures of bulk density, and therefore of particle packing and void content, are Loose Bulk Density (LBD) and Consolidated Bulk Density (CBD). LBD tends to occur in stockpiles where the material is loosely tipped in place. Of more importance to concrete is the CBD which represents a practical upper bound for bulk density of the aggregate. It is measured in the laboratory by consolidating a mass of aggregate in a steel container by vibration or by rodding, and then measuring the overall volume including the void volume occupied by the sample.

For a sample of graded aggregates, an estimate of the bulk density in terms of the 'granular packing density' can be obtained from a formula proposed by Caquot (1937):

$$\rho_{bulk} = 1 - 0.47 \left(\frac{D_{min}}{D_{max}} \right)^{0.2} \tag{3.8}$$

where
D_{min} and D_{max} are the minimum and maximum aggregate sizes, corresponding to the sizes representing 10 and 90 per cent passing, respectively.

Void content

Void content applies to an assemblage of particles, usually a collection of particle sizes of irregular shape. These particles do not fit together perfectly, leaving voids between them. The volume of these voids is extremely important in a concrete mix because it has to be filled with cement paste or matrix. The void volume is affected by the particle shape, size distribution (grading), and packing efficiency. The bulk density of aggregate varies inversely as the void content.

The concept of void content is illustrated in Figure 3.4 (Kosmatka *et al.*, 1995). The left-hand container is filled with large aggregate particles of reasonably uniform size and shape (i.e. 'single-sized'). The centre container has an equal volume of small aggregate particles, also of uniform size and shape. The graduated cylinder under the containers indicates that the volume of water required to fill the voids in each aggregate sample is the same for both cases, that is each aggregate sample has the same void content which is independent of particle size. When portions of the two samples are mixed and placed in the right-hand container, the void content decreases shown by the smaller volume of water in the right-hand graduated container. Thus, a combination of different particle sizes reduces the overall void content, which is an important objective of mix proportioning.

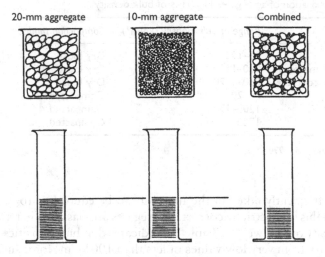

20-mm aggregate   10-mm aggregate   Combined

Figure 3.4 Illustration of void content of aggregate particles (from Kosmatka et al., 1995).

The level of liquid in the graduates, representing voids, is constant for equal absolute volumes of aggregates of uniform but different size. When different sizes are combined, the void content decreases. The illustration is not to scale.

The void content, e, is defined as

$$e = \frac{V_V}{V_S} \tag{3.9}$$

where

V_V is the volume of the (external granular) voids
V_S is the solid volume of the bulk sample.

Tightly packed particles will have a lower void content.
From equations (3.6), (3.7), and (3.9), it can be shown that

$$\rho_{bulk} = \frac{\rho}{1+e} \tag{3.10}$$

Unit weight

The concept of unit weight also applies to an assemblage of aggregate particles, that is to the bulk material. The unit weight γ of a bulk material is the bulk density expressed in weight (i.e. force) units, thus

$$\gamma = \rho_{bulk} \cdot g \tag{3.11}$$

Table 3.3 Classification of aggregates in terms of bulk density

Classification	Range of bulk density (kg/m³)	Compaction mode
Insulating	96–196	Dry loose
Lightweight for masonry	880–1120	Dry loose
Lightweight for concrete	880–1120	Dry loose
Air-cooled slag	>1120	Compacted
Normal weight	1200–1760	Compacted
Heavyweight	1760–4640	Compacted

Source: After Landgren, 1994.

'Bulk density' is frequently taken in the literature to be equivalent to 'unit weight' although this is not strictly correct. Aggregates are classified in terms of their bulk density or unit weight. Table 3.3 indicates that bulk densities of aggregates can vary from very low values of less than 100 kg/m³ for insulating aggregates to values in excess of 4500 kg/m³ for heavyweight aggregates.

NOTE ON TERMINOLOGY

In North American usage, terminology differs from that used above which reflects UK and SA usage. ASTM C127 and C128 use the term 'specific gravity' to refer to aggregate particles, with three values being defined: (1) 'Apparent specific gravity', which is the ratio of the oven-dry mass of an aggregate to the mass of an equal volume of water, where this volume is restricted to the volume of aggregate solids plus any enclosed impermeable voids (excluding the volume of water absorbed into the permeable pores). This is a fundamental measure, but of no great practical value in concrete technology; (2) 'Aggregate bulk specific gravity' which is equivalent to the 'apparent relative density' measured on oven-dry aggregates described previously; (3) 'Aggregate bulk specific gravity (SSD basis)', which is equivalent to apparent relative density measured using the SSD mass of the aggregates, as described previously. ASTM's use of the term 'bulk' can be confusing; in the convention of this book, 'bulk' is taken to refer to the aggregate in 'bulk' form, that is granular form.

Significance of density

Aggregate density in relation to concrete performance is important in several ways. First, most mix designs are based on a measure of aggregate bulk density, typically CBD of the coarse aggregate fraction. The higher the CBD of an aggregate the lower its void content, and consequently the less volume of paste required to fill the voids. This usually translates not only into better economy for the mix, but also results in improved technical properties if adequate workability is maintained. The apparent density (SSD) of the aggregate particles is required in mix proportioning calculations and to determine mix yield. Aggregate density also has an important influence on concrete density, since aggregates occupy up to 75 or 80 per cent of the volume of concrete. This has implications for formwork pressures during casting of the plastic mix and the amount of self-weight that the structure must carry. The strength and stiffness of the hardened concrete are indirectly related to its density, and aggregate density may affect these concrete properties. Occasionally higher density concrete is required, for example in dams and structures for radiation shielding, and higher density aggregates help to achieve this requirement. However, higher density aggregates *per se* do not necessarily increase the concrete strength. Table 3.2 shows typical values for the relative densities of aggregates.

Particle shape

Particle shape refers not only to the basic shape of aggregate particles, but also to other measures such as angularity, flakiness, and so on. Particle shape can be quantified and classified by measuring the dimensions of particles, that is length, width, and thickness. This is easier to do for coarse than fine particles.

Shape can be described in terms of three geometric properties: 'sphericity', 'roundness', and 'form' (Powers, 1953; Galloway, 1994). The first two are illustrated in Figure 3.5a, and were derived for descriptions of sedimentary rock boulders. 'Sphericity' is a measure of how closely the particle approaches a spherical shape, while 'roundness' describes the sharpness of the edges and corners. 'Form' (expressed as a form or shape factor) describes the relative proportions of the three axes of a particle. These properties depend on the source and nature of the aggregate. For example crushed aggregates may vary from well shaped in the sense of cubical and sub-angular particles to highly angular particles, or flat and elongated, flaky particles, depending on the rock and the type and efficiency of the crushing equipment. Rocks with natural bedding planes such as certain metamorphosed shales or sandstones tend to produce flaky particles of low sphericity during crushing. Synthetic aggregates such as crushed metallurgical slags can also have very poor particle shapes. On the other hand, natural gravels will tend to be more spherical and have rounder edges due to wear. Roundness primarily depends on the strength and abrasion resistance of the rock, and the degree of wear to which the particle has been subjected. Figure 3.5b provides standard descriptions that may be of more practical use for concrete technologists.

(a)

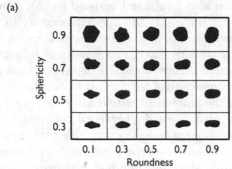

Sphericity = nominal diameter/maximum intercept
Roundness = average radius of corners and
edges/radius of maximum inserted circle

(b)

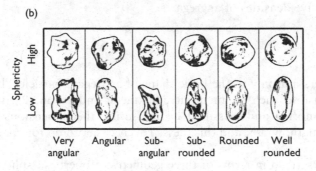

Figure 3.5 Sketches for the visual assessment of particle shape: (a) Derived from
measurements of sphericity and roundness; (b) Based on morphological
observations (from Sims and Brown, 1998).

A classification of roundness is given in BS 812 (BS 812: Part 102: 1989)
reproduced with modifications in Table 3.4. The closer to spherical and
rounded the aggregate particles are, the more desirable it is for concrete
from the perspective of workability and low water requirement. Flaky and
elongated aggregates can trap bleed water underneath particles as well as
increase the water requirement. Figure 3.6 provides a series of photographs
illustrating particle shapes.

Angularity

This term denotes the roundness (or lack thereof) of particles. It is assessed
by measuring the void content of the aggregate, and then calculating the
angularity number as

$$\text{Angularity number} = \text{percentage voids} - 33 \tag{3.12}$$

Table 3.4 Particle shape classification

Classification	Description	Examples
Rounded	Fully water-worn or completely shaped by attrition	Gravels and sands derived from marine, alluvial, or windblown sources
Irregular	Naturally irregular, or partly shaped by attrition and having rounded edges	Other gravels, typically dug from pits
Angular	Possessing well-defined edges formed at the intersection of roughly planar faces	Crushed rocks of natural or artificial origin; talus rocks
Flaky	Material in which the thickness is small relative to the other two dimensions	Poorly crushed rocks, particularly if derived from laminated or bedded rocks; other laminated rock
Elongated	Material, usually angular, in which the length is considerably larger than the other two dimensions	
Flaky and elongated	Material having the length considerably larger than the width, and the width considerably larger than the thickness	Poorly crushed rocks, as above. Poor processing techniques can exacerbate the undesirable shape, and vice versa

Source: After BS 812: Part 102: 1989.

where

'33' represents the typical void content of single-sized, well-rounded beach gravel, an effective 'baseline' for void content.

The test is covered in BS 812: Part 1: 1975, but is rarely used.

Flakiness

Flakiness Index (FI) refers to the proportion of flaky particles in a sample, by mass. It is measured as the percentage of the mass of stones that will pass slots of specified width for the appropriate size fraction of the aggregate (SANS 5847; BS 812: Part 105.1: 1989). Flakiness Index indicates the aggregate interlocking properties. Limits according to BS 882: 1992 are 50 per cent for uncrushed gravel and 40 per cent for crushed rock or gravel, while in South Africa a limit of 35 per cent is generally used. The test is not of great significance to the concrete-making properties of the aggregate. While a flakier aggregate may increase the harshness of the mix, this can be largely overcome by appropriate mix proportioning (Perrie, 1994). In practice, concrete can be made with aggregates of virtually any

Figure 3.6 Illustrations of particle shapes. Shapes (according to BS 812: Part 102: 1989) are, clockwise from top left: Rounded, Irregular, Angular, Flaky, Elongated, Flaky and Elongated.

shape, but requirements of workability and compactibility will always be very important.

Flatness and elongation

'Flatness' and 'elongation' are defined as follows (ASTM C 125):

1   A flat particle has width/thickness ≥3
2   An elongated particle has length/width ≥3

Such particles tend to occur more commonly with crushed aggregates.

Significance of particle shape

Shape, particularly for fine aggregates, is an important property since it has a strong influence on the plastic properties of concrete. Rounded, less angular particles are able to roll or slide over each other in the plastic mix with less resistance compared with flaky and angular particles. Poorly shaped particles also tend to induce aggregate interlock in the mix, resisting compactive effort. Concrete technologists have various terms for these effects; for example, a mix is regarded as 'harsh' when it resists efforts to compact it, usually due to poorly shaped aggregates.

Shape also influences the void content and packing density of aggregates, with lower compacted densities and higher void contents resulting from poorly shaped particles. As our ability to measure and characterize shape improves, for example using image analysis techniques, it will be possible to obtain more useful parameters that can be used in, say, mix proportioning (Dilek and Leming, 2004). The higher the CBD of coarse aggregate in particular, the more aggregate can be accommodated in a mix, which is usually an advantage for economics and technical quality. The combined effects of aggregate interlock and packing density cause aggregate shape to have an important influence on the *water requirement* of a concrete mix. This is dealt with in more detail in Chapter 4.

Aggregate shape may also influence hardened concrete properties. For example, higher concrete strengths usually result from the use of more angular although not flaky aggregates, provided full compaction can be achieved. This is because cracks are forced to follow more tortuous and complex paths around angular aggregates, and because angular aggregates induce a higher degree of internal friction. For concrete where flexural strength is important, angular and rough particles are preferred. On the other hand, there is little sense in using angular aggregates if this will result in a concrete that is difficult to compact, or one in which the water requirement is unacceptably high leading to an excessive paste content and an uneconomical concrete.

Certain standards for aggregates prescribe limits for particle shape properties, the limits to Flakiness Index for SANS 1083: 1994 and BS 882: 1992 mentioned previously being an example. ASTM C33 has no limitations on particle shape for coarse or fine aggregates.

Chapter 5 discusses newer developments in modelling concrete microstructure. Modelling of particle shape is particularly complex due to 3-D randomness of the particles. It is possible to mathematically model particle shape in 3-D and represent these shapes graphically (Garboczi, 2002). The techniques used are X-ray tomography and spherical harmonics, and existing particles are analysed to produce shape models. These modelled particles can then be randomly incorporated into higher order multi-particle computer models for concrete properties. An example of modelled particles is given in Figure 3.7 for a standard concrete sand. At present, the computer

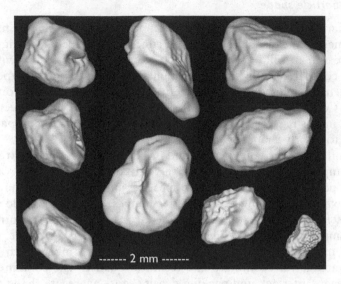

Figure 3.7 Fine aggregate particles modelled mathematically and presented in 3-D
        (diagram courtesy of Dr E. Garboczi).

power needed to undertake this kind of representation would generally render it impractical, but in time, such mathematical and numerical approaches
will become more feasible.

Particle surface texture

Surface texture is another important property affecting the performance
of aggregates in plastic and hardened concrete. Surface texture depends
on hardness, grain size, pore structure, and texture of the parent rock, as
well as the amount of wear on the particle that may have either smoothed
or roughened it. Surface texture is difficult to define objectively, usually
being described in terms such as 'smooth' or 'rough' with varying degrees
in between (Galloway, 1994). There are two independent geometric properties to describe surface texture: (1) the degree of surface relief, that is the
roughness, and (2) the amount of actual surface area per unit of plane projected area. These properties are of more value in fundamental research on
texture than in practical concrete technology (Ozol, 1978). Different visual
and quantitative assessments of surface texture are given in the literature
(Wright, 1955; Porter, 1962; Orchard, 1973; Li *et al.*, 1993) but are not
of great practical value. Experience is the best guide, with new aggregates
being evaluated in comparison with known sources. Standard descriptions
of surface texture which are of more practical value are provided in BS 812
(BS 812: Part 102: 1989), given in Table 3.5 with typical examples.

Table 3.5 Surface texture of aggregates

Surface texture	Characteristics	Examples
Glassy	Conchoidal (i.e. curved) fracture	Glassy or vitreous materials such as slag or certain volcanics
Smooth	Water-worn or smooth due to fracture of laminated or fine-grained rock	Alluvial, glacial or windblown gravels and sands; fine-grained crushed rocks such as quartzite, dolomite, etc.
Granular	Fracture showing more or less uniform size rounded grains	Sandstone, coarse grained rocks such as certain granites etc.
Rough	Rough fracture of fine- or medium-grained rock containing no easily visible crystalline constituents	Andesite, basalt, dolerite, felsite, greywacke
Crystalline	Containing easily visible crystalline constituents	Granite, gabbro, gneiss
Honeycombed	With visible pores and cavities	Brick, pumice, foamed slag, clinker, expanded clay

Source: After BS 812: Part 102: 1989.

Figure 3.8 illustrates different surface textures. These conditions apply to 'macrotexture' with the scale of roughness measured at the millimetre (mm) level. It is possible to assess texture at different levels of scale, and microtexture (at the level of micrometers or even nanometers) can occasionally be important to concrete performance. For example, aggregates with macroscopically similar macrotextures, but different micro-roughness, can develop different bonding between paste and aggregate, thus affecting concrete strength (see Chapter 5).

Like particle shape, surface texture is profoundly influenced by the source and production of the aggregate. Natural gravels subjected to attrition tend to be relatively smooth, while crushed materials will fracture with surface textures characteristic of their composition and mineralogy. These can vary from highly glassy and smooth surfaces for volcanic glasses (e.g. opal) to coarse and rough textures for coarse-grained granites or sandstones. Surface coatings in the form of adhering clay or dust particles can mask a natural surface texture and if present in excessive amounts can markedly increase water requirement of a mix since they tend to be absorptive.

Rough surface texture increases the total surface area of an aggregate and increases the internal friction between aggregate particles during compaction. These effects tend to increase mix-water requirements, since the former requires more water to wet the aggregate surface while the latter promotes 'harshness' in a mix. On the other hand, rougher textures can

(a)

(b)

Figure 3.8 Surface texture of different aggregates. Textures (according to BS 812: Part 102: 1989) are: (a) various, e.g. smooth (centre), rough (upper right); (b) rough and honeycombed (air-cooled slag) (Photo Courtesy of Slagment).

lead to better bonding between aggregates and paste and can enhance the mechanical properties of concrete (Kaplan, 1959).

ASTM D 3398 contains a practical test, the 'Particle Index Test', for assessing the combined effects of particle shape and texture of aggregates. Primarily used for road asphalt mixtures, the test determines the effects of shape and texture on the compaction and strength of the aggregates. The test involves measuring the voids in a contained aggregate sample under different degrees of compaction and using these values to determine a Particle Index which can be used for design of mixtures.

Grading

Grading refers to the particle size distribution and is a characteristic of aggregates in their granular form. Aggregate grading is very important in relation to the plastic properties of concrete. Well-graded and well-shaped aggregates give workable mixes that are readily transported, placed, and compacted. Cohesive mixes are ensured provided sufficient fine material is present, and particularly in low cement content mixes this requires an aggregate grading with an adequate amount of 'fines'. Having stated the obvious, it is also necessary to stress that good concrete can be made even with poorly graded or poorly shaped aggregates, provided the mix properties can be matched with the concreting operation and the structural application. Grading and its effects on plastic concrete is important not least since the hardened properties of concrete cannot be fully realized if the concrete is unworkable and difficult to compact.

Grading of aggregates governs the amount of voids that must be filled by paste as well as the surface area of aggregates that needs to be coated with paste. Several types of grading are common for concrete aggregates, illustrated in Figure 3.9. Uniform or single-sized aggregates (Figure 3.9a) contain large volumes of voids between the particles, whereas continuous grading (Figure 3.9b) in which a range of sizes is present decreases the void space and reduces the paste requirements. Using a larger maximum aggregate (Figure 3.9c) can also reduce the void space. Occasionally, gap-graded (Figure 3.9d) or no-fines gradings (Figure 3.9e) are used as well.

Grading of aggregates is not a 'property' as such. However the combined effects of particle shape, surface texture, and grading largely govern the plastic properties of concrete and none of these properties can be considered in isolation of the others. Grading is a much-researched and well-documented topic (Powers, 1968; Galloway, 1994; Neville, 1995; Sims and Brown, 1998; Grieve, 2001). The treatment of grading here will cover definitions of grading and its measurement and theory, and deal with those aspects that are important and relevant in respect of concrete performance. Certain detailed aspects will be addressed later in this chapter and in other chapters. The treatment will focus on practical aspects of grading for concrete mixtures. However, it is possible to treat grading from a theoretical or an empirical and numerical basis, to arrive at measures of overall grading. For example, fineness modulus can be calculated from knowledge of certain particle sizes, as can specific surface of the aggregates. Average particle size of the complete grading can be computed. These numerical computations of gradings and particle size characterizations can be used for optimizing gradings, for designing suitable blends of aggregates and for specification purposes. Examples of these and other approaches are given in Popovics (1979) and Powers (1968).

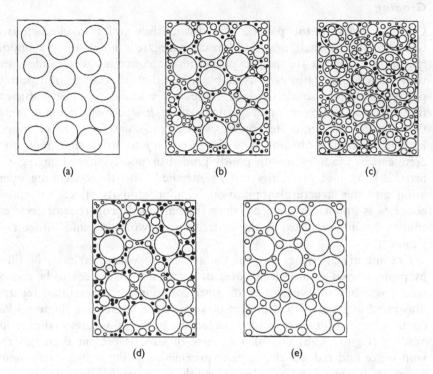

Figure 3.9 Schematic representations of aggregate gradations in an assembly of aggregate particles: (a) uniform size; (b) continuous grading; (c) replacement
of small sizes by large sizes; (d) gap-graded aggregate; (e) no-fines grading
(from Mindess *et al.*, 2003).

Definition and measurement

The grading of an aggregate is the quantitative distribution of its various
particle sizes in terms of the proportions passing through sieves with square
openings of different standard apertures (or sizes). Grading is determined
by a sieve analysis on a sample of aggregate, in which a series of standard
sieves are nested or stacked one on top of another with increasing aperture
size from bottom to top, and through which a sample of aggregate is passed
from the top, usually aided by shaking or vibrating the sieves. Procedures are
given in various standards (see Table 3.1). Table 3.6 gives typical standard
sieve sizes used in North America, UK, and South Africa. Sieve sizes range
from as large as 125 mm (or larger) down to 63 μm, and cover coarse and
fine aggregate sizes. Not all the sizes are necessarily used during routine sieve
analysis. For example, the intermediate sizes 53, 26.6, 13.2, and 6.7 mm
are frequently omitted.

Table 3.6 Sieve sizes according to different standards

ASTM standard sieve numbers	ASTM E11*	BS 812: Part 103.1: 1985	CSA A23.2-2A	SANS 201
		Coarse aggregate sieves		
	125 mm			
			112.0 mm	
4 inch	100			
			80.0	
3†	75.0	75.0 mm		75.0† mm
2½		63.0		
			56.0	
				53.0
2	50.0	50.0		
			40.0	
1½†	37.5	37.5		37.5†
		28.0	28.0	
				26.5
1	25.0			
		20.0	20.0	
3/4†	19.0			19.0†
		14.0	14.0	
				13.2
1/2	12.5			
		10.0	10.0	
3/8†	9.5			9.5†
				6.7
	6.30	6.30		
		Fine aggregate sieves		
		5.00	5.00	
No. 4†	4.75			4.75†
		3.35		
			2.50	
8†	2.36	2.36		2.36†
		1.70		2.00
			1.25	
16†	1.18	1.18		1.18†
		850 μm		
			630 μm	
30†	600 μm	600		600† μm
		425		425
			315	
50†	300	300		300†
		212		
			160	
100†	150	150		150†
	75	75	80	75

Notes
* Selected common values.
† Standard sieves.

The definition of coarse or fine aggregates varies in different countries. In general coarse aggregate (also called 'stone') has sizes larger than about 5 mm and vice versa for fine aggregate (also called 'sand'), with the requirement for fine aggregate also that it be retained on the 75-μm sieve. A typical sand grading analysis is given in Table 3.7 and is shown graphically in Figure 3.10 (Sample A), together with the ASTM C33 grading limits for

Table 3.7 Example of sieve analysis of a fine aggregate (sand) (Sample A in Figure 3.10)

BS sieve size	ASTM sieve no.	Mass retained (g)	Percentage retained	Cumulative percentage retained	Cumulative percentage passing
10.0 mm	3/8 in.	0	0.0	0.0	100.0
5.00 mm	4	0	0.0	0.0	100.0
2.36 mm	8	38	9.3	9.3	90.7
1.18 mm	16	113	27.6	36.9	63.1
600 μm	30	79	19.3	56.2	43.8
300 μm	50	111	27.1	83.3	16.7
150 μm	100	47	11.5	94.8	5.2
<150 μm		22	5.4	–	–

Total = 410 g   Total = 281

Fineness modulus = 281/100 = 2.81

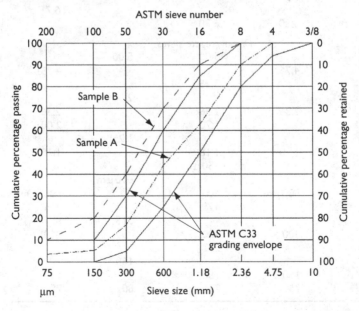

Figure 3.10 ASTM C33 grading envelopes for natural sand, and grading of two sand samples (Sample A from Table 3.7).

fine aggregates. Such a curve is termed a 'grading curve'. 'Fines' are generally considered as material passing a small sieve size, typically 80, 75, or 63 microns. The 'fines content' has a profound influence on plastic properties of concrete (see Chapter 4). Certain fines can be beneficial to concrete properties particularly if they are of the same nature as the parent rock and do not contain excessive clay material. However, excessive amounts of clay can be detrimental depending on the nature of the clay. For example, smectites absorb large amounts of water leading to reductions in strength and increases in drying shrinkage. The content of fine material can be determined by a sedimentation method which involves dispersion of the fines and determination of the suspended material (see Grading section in Table 3.1). Fines can also be present in coarse aggregates, often as coatings adhering to the particles; wet sieving can release this material.

The definition for *maximum size of aggregate* is important to avoid confusion. ASTM C125 defines it as the smallest sieve opening through which the entire amount of aggregate is *required* to pass. The ASTM definition for *nominal maximum size of aggregate* is similar: the smallest sieve opening through which the entire amount of aggregate is *permitted* to pass. This definition is perhaps an unnecessarily restrictive condition for coarse aggregate size, and a more practical definition may be the sieve size on which 15 per cent of the particles are retained. Canadian Standard CSA A 23.1 defines it as the standard sieve opening immediately smaller than the smallest sieve opening through which all the aggregate passes, a definition that is consistent with the South African definition. These definitions indicate that some confusion can arise when this term is being referred to, particularly if the standard used for definition is not given. This may affect the interpretation of specifications when these are applied in different countries. These definitions also have implications for structural design, since there are limits to the maximum size of aggregate allowable in concrete construction. For example, the maximum aggregate size should not exceed about one-fourth of the minimum formwork dimension, or three-quarters of the clear distance between reinforcing bars. It should also not exceed the specified concrete cover to reinforcement.

In general, it is an advantage to have as large a maximum aggregate size in a concrete mix as possible, since this reduces the total surface area per unit volume of aggregate which the paste has to cover. This will also reduce the water requirement of the mix. High strength concretes usually have a restriction on maximum aggregate size, typically 19 mm.

Grading curves and grading envelopes

The concept of a 'grading curve' was introduced previously. It is a graphical representation of a sieve analysis and depicts the particle size distribution of an aggregate sample (e.g. Figure 3.10). The ordinate represents the cumulative percentage passing while the abscissa shows the sieve size plotted to a logarithmic scale. Provided the sieve sizes are in a constant ratio (normally 1/2 in a standard series), the log plot shows these sieve openings at a constant spacing.

When aggregate grading curves are plotted consistently on identical graph paper, it is possible to check the compliance of an aggregate with the specification requirements very rapidly and easily. Requirements in specifications are normally given as grading limits (upper and lower) which when plotted on a grading chart are called a 'grading envelope'. An aggregate would normally be required to have a grading falling within the envelope. An example is shown in Figure 3.10, where the ASTM C33-specified grading envelope for natural sand is given, together with the gradings of two sand samples. Sample A falls within the grading envelope and would therefore be acceptable (assuming it meets other criteria of course). Sample B, however, falls above the upper envelope curve, indicating a finer sand. In terms of the specification, this sand may not be acceptable, but does not need to be rejected out of hand before trials are conducted on its behaviour in a mix. It may not be economically feasible to reject such a sand, and it may be possible to use the sand on its own or blended with a coarser sand to make it acceptable. ASTM C33 permits sands outside the recommended grading limits to be used provided similar aggregate from the same source has a demonstrated performance record. If a record does not exist, then the sand may still be used if the concrete made with the aggregate has shown equivalent properties to an 'acceptable' material. It must be stressed that 'insistence on narrow limits of grading may result in the rejection of locally available materials and consequent increases in costs. With thorough testing, it may be shown that many aggregates outside such limits do not need to be rejected' (Grieve, 2001).

Standard 'acceptable' gradings are often used in the industry to facilitate routine mix designs, and more importantly to check for consistency on delivery. This ensures that the performance of the plastic mix will not vary too widely in practice. Standardized aggregate gradings can be found in national standards. For example, BS 882: 1992 gives standardized coarse and fine aggregate gradings as well as alternative grading schemes for 'graded' or 'single-sized' coarse aggregate. ASTM C33 likewise provides grading limits for coarse and fine aggregates, and allows for continuously graded or nominally single-sized coarse aggregate. Gradings for concrete production are very much a function of local practice. National standards can be consulted where necessary and further detail is provided in Chapter 4.

Stone grading

A typical stone grading for a nominal single-sized 19-mm stone is shown in Figure 3.11. It can be seen that the sample contains sizes other than

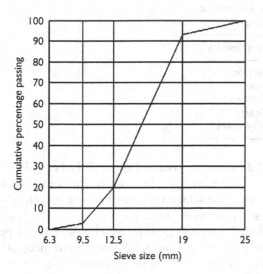

Figure 3.11 Typical grading curve for nominally single-sized 19-mm stone.

19 mm. It is thus not truly single-sized, but usually contains a small amount of over-sized material and a somewhat larger amount of under-sized material. Allowable grading limits recognize this fact. Based on the CSA definition, a 19mm nominal maximum stone size must have all the aggregate pass through the 25-mm sieve, which is the case for the stone grading in Figure 3.11. Usually, neither the sand nor the stone grading has a major effect on concrete water requirement, but the sand grading has a very important influence on the workability, cohesiveness, and bleeding properties of plastic concrete (see Chapter 4).

Fineness Modulus (FM)

This dimensionless parameter is used as a single number to characterize and evaluate a grading and is effectively a measure of the average particle size (or more strictly, the logarithmic average particle size). It is a parameter of the particle size distribution, obtained by adding together the cumulative percentages of material retained on each of the *standard* sieves, excluding the 75-μm sieve (i.e. 150, 300, 600 μm, 1.18, 2.36, 4.75 mm and onwards) and dividing the sum by 100. As a rule, gradings having the same fineness modulus will require a similar quantity of water to produce a mix of the same consistency. However, the water requirement depends not only on the average particle size as given by the fineness modulus, but more particularly on the percentage of finer material (<300μm) in the grading. An example of

Table 3.8 Categorization of fineness of concrete sands, using FM

FM	Sand fineness
<1.0	Very fine
1.0–2.0	Fine
2.0–2.9	Medium
2.9–3.5	Coarse
>3.5	Very coarse

the calculation of FM of a sand sample is given in Table 3.7. FM calculations apply equally to stone as to sand samples, although fineness moduli are usually only computed for sands, for which they can be used as a measure of sand fineness or coarseness as suggested in Table 3.8.

In North American practice, the typical grading envelopes result in permissible limits of FM between 2.3 (Fine) and 3.1 (Coarse). An alternative categorization into coarse (C), medium (M), or fine (F) sands is given in BS 882: 1992 for fine aggregates, based on the percentage passing the 600-μm sieve, as follows:

$$\left.\begin{array}{ll} C & 5-54\% \\ M & 25-80\% \\ F & 55-100\% \end{array}\right\} \text{passing 600-}\mu\text{m sieve}$$

This arrangement has significant overlap between the categories and wide permissible ranges. It has the drawback that the aggregate is characterized by only one size range.

Fineness Modulus can also be represented, to a suitable scale, by the area lying above a grading curve. Sand with a finer grading will have a curve lying towards the upper region of a grading chart, with a smaller area above the curve. The numerical value of the FM will then also be smaller and vice versa for coarser sands. It should immediately be obvious that there will be an infinite number of gradings that could all have the same FM, so that this parameter is not a unique or single measure of grading. Nevertheless, it is a useful measure and finds application in the proportioning of concrete mixes.

Gap-graded aggregates

In many parts of the world, use of graded aggregates, that is those graded continuously from the coarse to the fine sizes, is the norm (e.g. UK, most parts of North America, Europe). However, in other countries, 'gap-graded' aggregates are more commonly in use. Gap grading is when one or more intermediate size fractions are omitted (Shacklock, 1959). In South Africa, a predominance of coarse and fine crushed aggregates of less than ideal shape has led to the practice of using gap-graded aggregates, in which certain particle sizes are eliminated, particularly in the 5–10-mm range. In other cases, gap grading may be used if available sands are particularly fine and lack coarser fractions, or to obtain uniform textures in exposed aggregate concrete. It is usual for a nominally single-sized coarse aggregate to be used with a graded fine aggregate, the 'gap' therefore existing between the stone and sand sizes, although there are other possibilities. The smallest stone size is larger than the biggest sand size, and there is no overlap between the grading of the sand and stone. The 'gap' in a gap-graded aggregate is represented on a grading curve by a horizontal line extending over the size range that is absent. A typical gap-grading curve, representing the sand and the stone in Figures 3.10 and 3.11 respectively in the ratio (by mass) of 0.7:1.0 (sand:stone) is shown in Figure 3.12, where the 'gap' in the grading is clearly seen. Figure 3.13 shows photographs of sawn cross sections of a gap-graded concrete mix and a continuously graded concrete mix, in which the difference between the aggregate sizes present in the two mixes is illustrated.

Gap-grading has certain advantages. It helps to avoid the phenomenon of 'particle interference' which occurs when the distance between larger particles is too small to allow the passage of smaller particles. Thus, the distribution of sizes affects the flow characteristics of fresh concrete. It also has the practical advantage of minimizing the number of aggregate stockpiles required. However, for mixes of higher workability (slump exceeding about

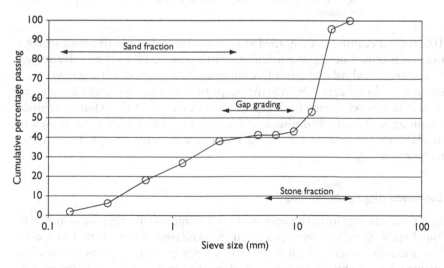

Figure 3.12 Gap-grading curve for combined sand and stone of Figures 3.10 and 3.11 (0.7:1.0 ratio by mass, respectively).

Figure 3.13 Photos showing aggregates in 68-mm diameter concrete discs with 19-mm nominal coarse aggregates. *Top LH*: Gap-graded, angular stone; *Top RH*: Gap-graded, rounded stone; *Bottom*: Continuously graded, angular stone.

100 mm), a continuously graded stone may be preferable to avoid segregation. In pumped mixes in particular, continuous grading is an advantage.

The technical advantages of continuous and gap-graded aggregates are summarized in Table 3.9. Whether gap- or continuously-graded mixes are used depends very much on accepted local practice, on the relative technical advantages of each, and ultimately on economic considerations (based on availability, storage and batching requirements, etc.) that may outweigh technical issues.

Combined aggregate grading

In a concrete mixture, the aggregates act compositely, that is, in the combined state. Grading envelopes given in standards, when plotted as a combined grading curve, often show deficiencies of certain particle sizes. For example, a combined grading based on the medians of the ASTM C33 grading envelopes for coarse and fine aggregates would show a lack of

Table 3.9 Technical advantages of gap-graded and continuously graded mixes

Gap-graded aggregates	*Continuously graded aggregates*
• Less chance of particle interference • Greater sensitivity of slump to changes in water content, which aids in more accurate control of mixing water • Greater responsiveness to vibration of stiff mixes	• Less segregation of higher slump mixes • Less sensitivity to small changes in water content, which is an advantage where uniform workability is required • Improved pumpability especially at higher pressures • Improved flexural strength due to the increased surface area of graded stone

sizes around the 10-mm range. Such 'gaps' may represent a less than ideal combined grading, and can on occasions give rise to high water demand, poor workability and pumpability, and so on. Notwithstanding this, it is common practice in some countries to utilize gap-graded aggregates for reasons discussed earlier.

Figure 3.14 illustrates an 'ideal' combined grading which results in the spaces between the coarse aggregate particles being filled with smaller aggregate particles rather than with mortar, generally giving a more economical mix. However, the workability of such mixes can be compromised

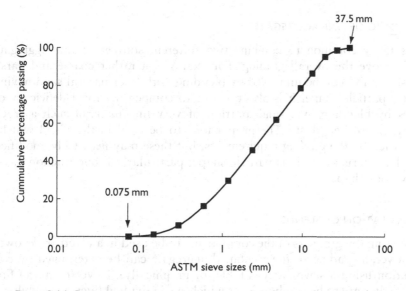

Figure 3.14 Optimum combined aggregate grading for concrete (from Kosmatka *et al.*, 2002).

if aggregate interference becomes a problem. Such gradings would usually be constructed by special screening of aggregates and then blending of the appropriate sizes. For this reason such ideal gradings are seldom achieved in practice. Nevertheless, gradings such as in Figure 3.14 can have the benefit of reducing the amount of cement required in a mix, and can give workable mixes particularly if particle shape and texture are good (Shilstone, 1990).

Blended or combined aggregates

Aggregates are normally delivered to site graded and ready for use in the mix. However, on occasions it will be necessary to blend aggregates with different gradings from two or more stockpiles to produce an acceptable overall aggregate grading. It is rare that blending would be done separately; usually the various fractions are charged into the mixer where blending would occur automatically during mixing.

BLENDING OF COARSE AGGREGATES

Assuming there is a variety of aggregate stockpiles, each pile having a certain size fraction, it is a relatively simple matter to combine the fractions to achieve a desired grading. This is done for continuously graded coarse aggregates, although economics dictates a practical limit to the number of stockpiles on site or at a batching plant. The proportions of the various size fractions would be governed by the requirements of the grading curve.

BLENDING OF FINE AGGREGATES

It is fairly common to combine two different sources of fine aggregate to improve the overall grading. For example, a rather coarse and harsh crushed sand can be improved by blending with finer natural sand having better particle shape. It is also possible to compensate for a deficiency of fines by blending in a small portion of very fine material such as rock flour or crusher 'dust'. The proportions to be used in the blend will be governed by the grading requirements, but these may need to be modified by characteristics such as particle shape, particularly if one fraction has a very poor shape.

CALCULATION OF BLENDS

Assuming the gradings of the components to be used in a blend are known, the overall grading of the combined aggregate can be determined by calculation using a sieve analysis table, or graphically. If two (or more) fine aggregates are to be combined, for which the individual fineness moduli are known, the resulting overall FM can be calculated by arithmetic proportion of the individual FMs in relation to the sand proportions. For example, if two sands with fineness moduli of FM_1 and FM_2 are to be blended in the proportion of 70 per cent of sand 1–30 per cent of sand 2, the overall FM of the blend will be

$$FM_{blend} = (0.7 \times FM_1) + (0.3 \times FM_2) \tag{3.13}$$

where sands have different particle densities, the calculations of the mix proportions and of the yield should be based on the absolute volumes of the sands.

A further problem arises when a specified overall grading is to be achieved, either for a combination of fine or coarse aggregates or for a continuously graded aggregate from separate fine and coarse components. In this case, it will only be possible to match exactly the desired over-all grading at a limited number of points on the grading curve. If three separate fractions are to be combined, the overall desired grading can be matched at three points, but it is possible to solve the problem by matching at two points. The matching point or points chosen may be intermediate, for example in the 1–5-mm range, or may be one of the end points such as 150 μm. The decision of which point or points to use will depend on aspects such as aggregate shape and texture that will affect workability, whether more or less fines are required to ensure cohesiveness, and so on. As an example, consider the fine aggregate of Table 3.7 and the 19-mm nominal maximum size coarse aggregate of Figure 3.11. These gradings are repro-duced in Table 3.10. It is desired to match the median grading for 20-mm nominal maximum size 'All-in aggregate' from BS 882: 1992, which is also given in Table 3.10, column 6. To achieve this, an intermediate aggregate size will be needed, in this case a coarse aggregate in the 5–15-mm size range. A hypothetical aggregate grading for this is also given in Table 3.10. (For simplicity, it is assumed that all the aggregates have the same (or very similar) relative density; if this is not the case, then adjustments to the volumetric proportions must be made.)

To achieve the desired grading in column 6 of Table 3.10, assume we match the grading at the 5- and 20-mm (nominal) points, for which the total percentages passing are 45 and 97 respectively. Let p, q, and r be the proportions of fine, 5–15-mm size, and 19-mm nominal maximum size aggregates respectively. For the requirement of 45 per cent passing the 5-mm sieve, we have:

$$1.00p + 0.05q + 0.00r = 0.45(p + q + r)$$

while for the requirement that 97 per cent passes the 20-mm sieve, we have:

$$1.00p + 1.00q + 0.93r = 0.97(p + q + r)$$

Table 3.10 Technique for combining aggregates to obtain a desired grading

BS sieve size (mm)	ASTM sieve no.	Cumulative percentage passing								
		Fine agg. from Table 3.7	Coarse agg. 5–15 mm	Coarse agg. from Figure 3.11	Median grading, 20-mm max. size 'All-in agg.' BS 882: 1992	[3] × 1.0	[4] × 0.29	[5] × 0.97	[7] + [8] + [9]	[10]/2.26
37.5	$1\frac{1}{2}$ in.	100.0	100	100	100	100.0	29.0	97.0	226.0	100.0
20.0	$\frac{3}{4}$ in.	100.0	100	93	97	100.0	29.0	90.2	219.2	97.0
14.0	–	100.0	95	20	–	100.0	27.6	19.4	147.0	65.0
10.0	$\frac{3}{8}$ in.	100.0	67	3	–	100.0	19.4	2.9	122.3	54.1
5.00	4	100.0	5	0	45	100.0	1.5	0.0	101.5	44.9
2.36	8	90.7	–	–	–	90.7	–	–	90.7	40.1
1.18	16	63.1	–	–	23	63.1	–	–	63.1	27.9
0.600	30	43.8	–	–	–	43.8	–	–	43.8	19.4
0.300	50	16.7	–	–	–	16.7	–	–	16.7	7.4
0.150	100	5.2	–	–	4	5.2	–	–	5.2	2.3
<0.150	>100	–	–	–	–	–	–	–	–	–
[1]	[2]	[3]	[4]	[5]	[6]	[7]	[8]	[9]	[10]	[11]

Solving these two equations in terms of ratios gives:

$$p:q:r = 1:0.29:0.97$$

This gives the proportions for combining the three aggregates. It remains to multiply the values in columns 3, 4, and 5 of Table 3.10 by the relevant proportions given, as shown in columns 7, 8, and 9 of the table. These three columns are added together to give column 10, the values of which should be divided by the sum of $1 + 0.29 + 0.97 (= 2.26)$. The final grading is given in column 11. For this case, the grading is well matched at the selected points, but is slightly lacking in material of the finer sizes.

The procedure can also be carried out graphically, using a square on which the percentage passing is marked along three sides (Neville, 1995). In the case of three aggregates as above, the two coarse aggregates are combined graphically first, and this combination is then combined graphically with the fine aggregate, using another square as before. The advantage of the graphical method is that it is possible by inspection to see whether a grading within a required envelope can be achieved with the given aggregates, and to determine the allowable range of proportions.

Grading and packing theories

Granular materials can pack together in various arrangements, from loose packing to dense packing. These arrangements depend on the nature of the particles, particularly their shape and size distribution, described by the grading. Packing of granular materials and their grading are interrelated, since the grading will affect the packing efficiency.

It is possible to achieve dense packing of granular particles by two types of grading:

1   By achieving the densest possible packing of a given maximum size of particles, and then ensuring the voids between the particles are filled by smaller sizes. This will result in the 'gaps' between the particle sizes being large, since the particle size required to fill a void in a dense packing arrangement of a certain size is considerably smaller than the larger particles. The particle size distribution would be one of large 'gaps' between successive particle sizes, as shown schematically in Figure 3.15.
2   By overfilling the void spaces of any particular particle size by the next smallest size. This will give dense packing and will also have a full range of sizes present in the grading, and is more typical of what happens in concrete mixtures. This arrangement also reduces internal friction and permits greater mobility of the particles.

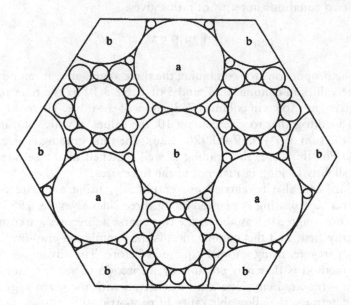

Figure 3.15 Schematic of the concept of dense grading (from Young *et al.*, 1998).

Grading theories for obtaining the 'best' grading are based on the packing characteristics of granular materials. Fuller and Thompson in 1907 determined a grading curve giving minimum voids according to the expression:

$$P_t = \left[\frac{d}{D}\right]^{0.5} \tag{3.14}$$

which was later generalized to the expression

$$P_t = \left[\frac{d}{D}\right]^{q} \tag{3.15}$$

where

P_t = fraction of total solids finer than size d
D = maximum particle size
and q may have values between 0 and 1.

Equation (3.15) gives rise to the so-called Fuller parabolic grading curve.

The Fuller and Thompson Grading Theory was discussed by Powers (1968) who gives relationships between the grading curves and parameters such as fineness modulus. However, as Powers points out, parabolic gradings do not permit a prediction of the optimum grading for a given combination of cement content and maximum aggregate size, and thus are not necessarily useful to the practising engineer.

A void content can be determined from the above expressions and, provided all particle sizes below D are present, depends only on q, approaching zero as q approaches zero. It might be considered desirable to have as dense a grading as possible in order to minimize the void content and thus the paste content to fill the voids. However, in practice, such an approach does not work. As Mindess *et al.* (2003) point out, from workability considerations a certain proportion of fine material is required. Between 2 and 10 per cent of the fine aggregate must pass the No. 100 sieve ($150\,\mu$m) and between 10 and 30 per cent must pass the No. 50 sieve ($300\,\mu$m). The recommended grading curves for fine and coarse aggregates given in, for example, ASTM C33 or CSA 23.1 are not parabolic (see Figure 3.10). In addition, the packing density decreases as the average particle size decreases, so that the lowest practical value of q is approximately $1/2$.

The fact is that concretes of suitable workability can be made with a wide variety of aggregate gradings, and there is no 'ideal' grading that will suit all circumstances. The ultimate test of a grading will be whether the water requirement for the concrete mix is acceptable leading to an economical mix, and whether other important mix requirements such as a lack of segregation and ease of finishing are met.

Notwithstanding these comments, modern high performance and ultra high performance concretes such as Self-Compacting Concrete, Very High Strength Concrete (>100 MPa), and Reactive Powder Concrete require very careful control of overall grading including aggregate and binder components to achieve optimum particle packing. Packing theories and models are used in designing these mixtures, and indeed proportioning such mixtures to achieve their high performance characteristics cannot be done without regard to overall grading. Most idealized packing models consider only geometrical parameters of the particle systems, such as particle size and possibly particle shape, and may only consider spherical, mono-sized particles. In reality, many factors influence packing density, including size, shape, surface texture, particle size distribution (PSD), the method of compaction, and the wall effect. The wall effect is of major importance when considering the packing of small particles on the surface of larger particles: the porosity at the surface of the larger particles is higher than in the bulk section, extending out from the surface to a range of up to five diameters of the smaller particles. (This is one of the important causes of the ITZ in concrete.)

PACKING MODELS

Packing models are characterized at the simplest level by the following parameters: the packing density (the volume fraction occupied by the solids), the porosity (the volume fraction of the voids), and the inverse of the packing density which is called the specific volume of the system. From these definitions, the following relationship applies:

$$\phi = 1 - p \qquad (3.16)$$

where

ϕ is the packing density
p is the porosity of the mixture.

Johansen and Andersen (1991) reviewed packing models of several authors: Furnas (binary systems), Aïm and Goff (including the wall effect), Toufar and co-workers (large particles discretely dispersed in a matrix of smaller particles), and de Larrard and co-workers with their Linear Packing Model and Compressible Packing Model (multi-component mixtures). Figure 3.16 summarizes various packing models, categorizing them as either discrete models (systems containing two or more discrete particle sizes) or continuous models (all particle sizes present) (Kumar V and Santhanam, 2003).

According to Johansen and Andersen (1991), comparisons between experimental results and model predictions show that most models reasonably predict actual packing densities, but differ in their accuracies depending on the geometries of the systems tested. For 'real' systems involving a range of particle sizes characterized by particle size distributions, the models generally consider a series of relevant size fractions. The most convenient way of presenting the results is in a ternary diagram showing the packing densities in terms of iso-density (or iso-porosity) contours. A typical outcome of a comparison between experimental results by Andersen and model estimations of the packing densities of ternary mixtures of cement, quartzite sand (<2 mm), and crushed coarse granite aggregate (8–16 mm) is shown in Figure 3.17a and b. Relatively good agreement exists between experimental and predicted results, with packing densities being somewhat overestimated by the model.

This type of approach can be used to predict and control the flow properties of mixtures and hence the placement of fresh concrete. Andersen found a good relationship between concrete rheological properties and packing densities. Ternary diagrams showing packing densities can be used to estimate the highest packing density for a combination of materials, and these can be optimized for other mix requirements.

APPLICATION TO MIX PROPORTIONING

Particle packing theories or models that show merit in terms of concrete mix proportioning are those of Toufar *et al.* (1976), Dewar (1986, 1999),

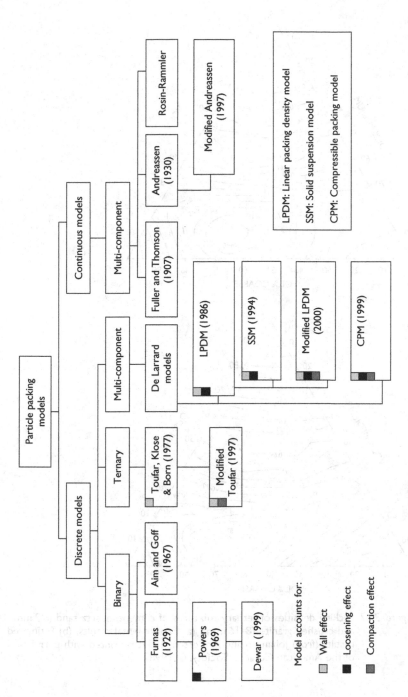

Figure 3.16 Particle packing models (adapted from Kumar V and Santhanam, 2003).

(a)

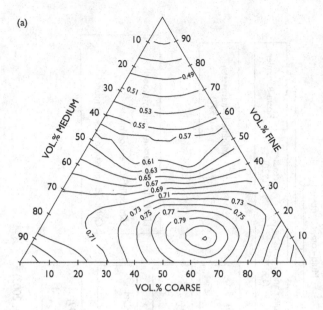

(b)

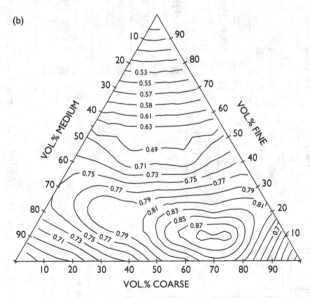

Figure 3.17 Packing densities of ternary mixtures of cement, quartz sand (<2 mm), and crushed granite (8–16 mm). (a) Experimental results; (b) Estimated results (from Johansen and Andersen, 1991; reprinted with permission of the American Ceramic Society, © 1998).

and de Larrard (1999). De Larrard's Linear Packing Model (LPM) and Compressible Packing Model (CPM) both require the particle size distribution (i.e. grading) for all constituent materials, including the cementitious phases. The model permits the packing density to be calculated as well as the selection of a compaction index appropriate for the particular compaction method. These all have the aim of determining the void ratio (or packing density) resulting from different combinations of materials, given their physical properties, and hence their optimum combination to minimize void content while retaining an adequately workable mix.

Table 3.11 gives several computer packages based on the different models that are available to assist in mixture proportioning. These can be used to simulate a large range of variables, before resorting to actual trials in the lab or on site.

Various models have been compared by Jones *et al.* (2002) using laboratory tests and published data. They conducted a series of tests using maximum aggregate sizes of 10 and 20 mm, two different sands, and a number of cement extenders. Void ratios of different aggregate combinations were measured, and the results were applied to mix proportioning calculations using the models. All the models gave broadly the same output, although it was noted that proportioning concrete mixtures for minimum void ratio produced harsher mixes than normal. Figure 3.18 shows results for an aggregate combination of a 10-mm stone and sand with FM of 1.81 (bulk densities of 1530 and 1650 kg/m³ respectively.) In this case, the Dewar, Toufar, and modified CPM models agree well with the test data, while the CPM over-estimated and the LPM under-estimated the actual voids ratio. Overall mean differences of the various models between measured and calculated void ratios over all series varied from 2.4 to 5.5 per cent. However, for any given combination of particles, the various models differed in their ability to predict the packing density.

Table 3.11 Computer programmes for mixture proportioning, based on particle packing models

Programme (software)	Model utilized	Source
EUROPACK	Modified Toufar	www.gmic.dk
RENE-LCPC	De Larrard models	ciks.cbt.nist.gov
4 C Packing	Modified LPDM	www.danishtechnology.dk
LISA	Andreassen models	www.silicafume.net
MixSim	Powers/Dewar	www.mixsim.net

Source: Adapted from Kumar V and Santhanam, 2003.

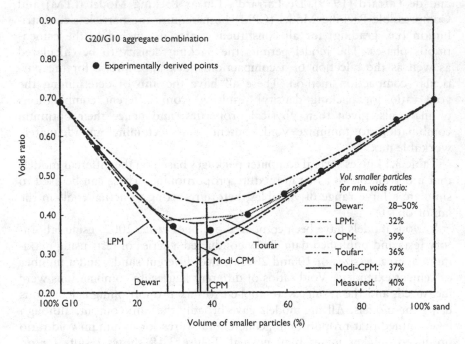

Figure 3.18 Comparison of different particle packing models to obtain minimum void ratio with 10-mm aggregate and sand (from Jones *et al.*, 2002; courtesy RILEM Publications s.a.r.l., France).

Quantifying the combined effects of particle shape, texture, and grading

It is useful to quantify the *combined* effects of particle shape, texture, and grading. Such quantifications provide practical engineering measures of these combined effects, particularly as they relate to the plastic properties of concrete. They provide a first assessment of the likely performance of the aggregate combination in fresh concrete. One assessment has already been discussed previously in this chapter: the consolidated bulk density (CBD) of an aggregate, which is a practical measure of the packing density of the aggregate. Two further methods are: (1) measuring the flow rate of the aggregate through an orifice and (2) measuring the percentage voids once the material has fallen into a container under controlled conditions. For the latter method, three procedures are standardized in ASTM C1252. Two procedures use graded fine aggregate (standard grading or as-received grading), and the other uses three individual size fractions (ranging from 2.36 to 300 μm) each tested separately. The procedures all involve passing dry fine aggregate through a funnel from a fixed height into a 100-ml cylindrical measure, striking off the cylinder, and determining the mass of fine aggregate in the cylinder. Uncompacted void content can be calculated knowing the relative density of the dry material. For a standard grading, or where as-received gradings are constant, the uncompacted void content gives an indication of the aggregate's angularity, sphericity, and surface texture compared with other fine aggregate of the same grading. The higher the uncompacted void content, the less workable the concrete would likely be. Since these methods measure the combined effects mentioned, the results should correlate with parameters such as standard water requirement (SWR) of a set of aggregates, and this is in fact so (Wills, 1967). Reference was made earlier to a test in ASTM D3398 which also assesses the combined effects of particle shape and texture on the compaction characteristics of an aggregate.

Dimensional stability (moisture and thermal movements)

The dimensional stability of aggregates refers to moisture-related (shrinkage and swelling) and temperature-related (thermal contraction and expansion) movement properties. These properties may be measured on representative rock prisms (cores or cut samples) that are subjected to wetting and drying tests or heating and cooling tests, with measurements of linear strain. Occasionally volumetric strain must be measured if anisotropy is present (Lane, 1994). Such measurements are not suitable for fine aggregate, unless it has been derived by crushing from rock of known properties. Fine and coarse aggregate properties can be indirectly inferred by testing mortar bar specimens, provided the corresponding properties of the paste are known (Verbeck and Haas, 1950).

   Moisture and thermal movement properties of aggregates have a profound influence on the corresponding properties of concrete. Consequently they are covered in Chapter 5 which also deals with a particular problem of aggregate moisture instability in the form of shrinking aggregates. Chapter 5 covers other thermal properties of aggregates (thermal conductivity, thermal diffusivity, and specific heat), and fire resistance of aggregates and concrete. Table 3.2 contains data on the coefficient of thermal expansion of common rock types.

Permeability

Permeability refers to the ability of a material to transport liquids and gases under a pressure gradient. Aggregate permeability is important since it may influence the overall permeability of the concrete. Measurements of the permeability of intact rock samples depend on their size which governs the frequency of internal flaws. Typical values for rocks range over several orders of magnitude, from around 10^{-10} to 10^{-14} m/s. Table 3.12 shows data for various rock types, in which the equivalent w/c ratio of

Table 3.12 Permeabilities of rocks and cement paste

Type of rock	Permeability coefficient (m/s)	w/c ratio of hydrated cement paste of same perm.
Trap rock	2.47×10^{-14}	0.38
Quartz diorite	8.24×10^{-14}	0.42
Marble 1	2.39×10^{-13}	0.48
Marble 2	5.77×10^{-12}	0.66
Granite 1	5.35×10^{-11}	0.70
Sandstone	1.23×10^{-10}	0.71
Granite 2	1.56×10^{-10}	0.71

Source: From Powers, 1958.

mature pastes of the same permeability varies from 0.38 to 0.71. Pastes and rocks therefore have similar permeability ranges. Typical concrete water permeabilities vary from 10^{-11} to 10^{-13} m/s, which is the same order of magnitude as rocks, although concretes are generally somewhat more permeable overall. Chapter 6 deals with the influence of aggregates on concrete permeability in more detail.

In high performance concrete (HPC), aggregate permeability may be more crucial than in normal concrete. A newer trend is to use more absorptive aggregates with a reserve of moisture, which is beneficial for curing the material. Aggregate permeability is also important in respect of freeze–thaw soundness of aggregates, and this is covered later in this chapter.

Mechanical properties of aggregates

The mechanical properties of aggregates comprise different strengths as well as parameters related to fracture, stiffness, and resistance to abrasion or attrition amongst others. Such properties are important to engineers, since the performance of concrete will depend to some degree on these aggregate properties, and more detail will be provided in subsequent chapters. Table 3.1 summarizes the significance of the various mechanical properties and the basis of their measurement. Table 3.2 provides typical values.

Strength of rocks

The strength of solids is normally measured on regular prismatic specimens (e.g. cylinders, cubes, beams) specially prepared for testing. Aggregates exist in granular form and for most aggregates the particle size is not appropriate for conventional strength testing. Information on the strength of the aggregate source material can be useful, particularly when analysing concrete as a multi-phase material or when evaluating new sources of aggregates.

Data on various types of strength are given in Table 3.2, including the following:

a   Unconfined Compressive Strength (UCS), which typically is measured on rock cores.
b   Aggregate Crushing Value, 10 Per Cent Fines Value (10% FACT), and Aggregate Impact Value, which are measures of granular strength.
c   Los Angeles Abrasion, which reflects the impact and abrasive strength of aggregate particles.
d   Tensile strength for a few cases, where data were available. Tensile strength is usually measured in an indirect tension test, typically a cylinder-splitting test.

The data in Table 3.2 indicate that aggregate source materials are generally strong in compression and relatively weak in tension, identifying them as brittle.

These tests represent simple approaches to obtaining characteristic material parameters. More complex testing yielding richer information is carried out in the rock mechanics field, but is not necessarily of great value to concrete engineers for characterizing aggregates. As with all natural materials, rocks display wide variability of properties, often complicated by features such as joints and bedding planes and defects such as cracks or inhomogeneities. Figure 3.19 corroborates this using UCS data obtained from tests

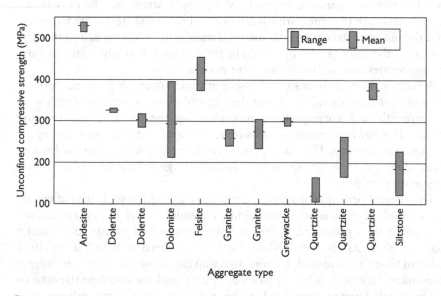

Figure 3.19 Unconfined compressive strength (UCS) of rock cores (for SA aggregates) (from Alexander and Davis, 1991).

on rock types used in South Africa for crushed aggregates. Fine-grained massive rocks such as andesite, felsite, dolerite, and dolomite generally gave superior UCS values. Bedded and jointed rocks such as quartzite and silt-stone, and coarse-grained rocks such as granite had values near the lower end of the range. Once crushed into aggregate particles, rock flaws or weak-nesses may be less important, although it has been found that inherent cracks due to blasting and stress-relief can influence the strength of concrete made with such aggregate (Ballim and Alexander, 1988).

Figure 3.19 shows that most rock strengths exceed the compressive strengths of conventional and even high strength concretes. For this reason, concretes can be made with a wide range of different aggregate types, and it is unusual that conventional concrete strength will be limited by aggregate strength unless this is less than about 50 MPa. Obviously, once concrete strengths exceed 80–100 MPa, selection of aggregates with higher strength becomes far more crucial.

Strength of aggregates in granular form

Aggregates exist in granular form. Consequently, measures of aggregate granular strength are more useful for assessing aggregate performance in concrete. In many cases, it is not possible to obtain aggregate specimens in prismatic form of a size sufficient to determine strength, so that only measures of granular strength are available. Strength in granular form has thus become the 'standard' measure of aggregate strength. The drawbacks of such tests are that they are usually limited to coarse aggregate particles, that the inherent strength of the material itself is not measured, and that the nature of the tests (generally crushing tests) does not really relate to how the aggregates interact in the concrete matrix.

A further indirect means of assessing granular strength is to incorporate the aggregates into a concrete, and then compare the concrete strength with the strength of a corresponding mix using aggregates of known perfor-mance. If a lower strength is measured, this may indicate poor aggregate strength and quality. This can often be confirmed by inspecting the broken faces of the concrete specimens for unusually large percentages of fractured aggregate particles.

The general principle of aggregate granular strength tests is that of crush-ing a sample of the coarse aggregate in a confining vessel and measuring the amount of fines produced. Either samples are crushed to a predetermined load, or alternatively the crushing load is measured at which a specified value of 'fines' is produced. Factors affecting the measured values are aggre-gate shape, flakiness, inherent flaws or cracks, and the inherent strength of the material. Strength measured in this way also gives some indication of the ease of crushing of the material and its performance during transporting and stockpiling and is thus useful to aggregate producers. Common tests are discussed below, and typical values are given in Table 3.2.

Aggregate Crushing Value (ACV) test

An approximately 2-kg sample of 10–14-mm sized coarse aggregate is crushed at a standard load (400 kN achieved in 10 min). The ACV is the amount of fine material (minus 2.36 mm) produced by crushing, expressed as a percentage of the sample mass. Lower ACV values represent aggregates more resistant to crushing. In the UK and South Africa, this test is no longer widely used, having been replaced by the 10% FACT test. This is because the ACV is insensitive to the strength of weaker aggregates (with ACV values greater than about 25) which tend to crush before the maximum load is achieved, whereafter they simply compact.

10 Per Cent Fines Aggregate Crushing Test (10% FACT)

This test uses the same aggregate size fraction as the ACV test (10–14 mm), but a number of nominally identical samples are subjected to different loads applied at a controlled rate, and the 'fines' value measured in each case. The 'ten per cent fines value' is the load in kN required to produce 10 per cent of fines by mass of the test sample, and higher values represent more resistant aggregates. Typically, a 10% FACT value less than 50 kN would represent an aggregate too soft and unsuitable for concrete work. A photograph of the apparatus for the 10% FACT and ACV tests is shown in Figure 3.20.

Figure 3.20 Apparatus for 10% FACT and ACV tests. Note confining steel cylinder and plunger, compacting rod, and different sieve sizes.

Aggregate Impact Value (AIV)

An identical sample to the ACV test is subjected to 15 blows by a small hammer in a standard set-up. The AIV is the amount of fine material produced expressed as a percentage by mass of the original sample. AIV values less than about 20 per cent represent impact resistant materials, while a limit on the maximum value of 45 per cent for other concretes is suggested (BS 882: 1992). This test is suited to situations where concrete surfaces with exposed aggregates are subjected to wear by impact.

Los Angeles abrasion test

Although this test is titled an 'abrasion' test, it is really a measure of aggregate impact and crushing strength since the mode of action involves impact of particles against steel balls, against each other, and against the walls of the steel container. The inherent strength of the particles as well as flaws and cracks, and to some degree shape, will influence test results. The test is discussed in more detail later, but it is appropriate to include it here as a test for granular strength.

Correlations amongst tests

ACV AND 10% FACT TESTS

For ACV values in the range 14–30 per cent and 10% FACT values in the range 100–300 kN, a correlation has been determined as

$$ACV = 38 - 0.08 \times (10\% \text{ FACT}) \tag{3.17}$$

where
ACV is expressed as a percentage, and 10% FACT in kN (Weinert, 1980).

UCS AND 10% FACT TESTS

Since these two tests represent quite different measures of strength, correlations might be weak. Typical results for tests on 12 different aggregate types are shown in Figure 3.21, with the linear regression based on mean values. There is a reasonable correlation (coefficient of correlation = 0.77) despite the spread of results and the limited data set. The correlation is influenced by the small number of high values associated with andesites and felsites.

There are no reported correlations between ACV or 10% FACT results and the strength of concretes incorporating the aggregates, and in any event such a relationship would be extremely tenuous. However, for HSC, aggregates with 10% FACT values exceeding about 170 kN (i.e. ACV less than about 24 per cent) are recommended (Alexander and Davis, 1991).

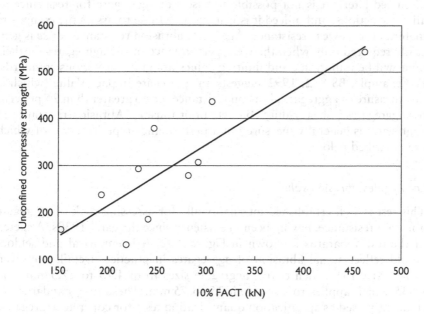

Figure 3.21 Correlation between UCS of rock cores and 10% FACT values for SA aggregates (from Alexander and Davis, 1991).

Abrasion, wear resistance, and hardness of aggregates

The abrasion resistance and surface hardness of aggregates are of little or no practical importance in the majority of concrete applications. However, in special cases these aggregate properties may govern the performance of concrete. The first case involves concrete surfaces exposed to severe abrasive forces, such as concrete pavements and trafficked slabs-on-grade, dam spillways and stilling basins, concrete canals carrying gravel- and silt-laden water, heavily trafficked pedestrian areas, and abrasion-resistant floor toppings. Second, aggregate abrasion resistance may be an important factor in assessing the performance of an aggregate during production and transportation, since aggregates will probably be exposed to their roughest treatment in these processes. Weak and friable aggregates may break down during batching, mixing, and handling, leading to loss of workability and decreased air content due to increased fines.

Abrasion of aggregate particles in hardened concrete will only occur once the aggregates are partially exposed at the surface either due to wear of the initial concrete surface or purposely in the case of exposed aggregate finishes. Wear of the aggregates can take many forms, such as rubbing, grinding, scratching, impact, and hydraulic erosion (Alexander, 1985). Impact resistance is a special case, and this property is sometimes called the 'toughness' of an aggregate (not to be confused with the fracture toughness discussed later). It is not possible to assess an aggregate for resistance to all these actions, and indeed it is usually preferable to assess the composite material for its wear resistance. Aggregate abrasion resistance *per se* is generally required only when the wear performance of the aggregates on their own will be important, and limiting values are given in various standards. For example, BS 882: 1992 suggests an Aggregate Impact Value (which is one measure of aggregate abrasion resistance) of no greater than 25 per cent for aggregates in heavy-duty concrete floor finishes. Abrasion resistance of aggregates is generally measured in a performance-type test, two of which are described below.

Los Angeles abrasion value

This test is well established internationally for assessing rock or aggregate abrasion resistance, having been in existence since the early 1930s. A sketch of the test apparatus is shown in Figure 3.22. It is empirical and seldom would reflect actual abrasion of aggregates in practice. Details are given in ASTM C131 which covers aggregate sizes from 4.75 to 37.5 mm, and C535 which applies to sizes from 25 to 75 mm. These two standards are frequently used as specification quantification tests for concrete aggregates. In reality, aggregate particles are subjected to impact during the test by being rotated in a closed steel drum together with a charge of steel balls. The test has the advantage over most impact tests that a larger and hence more representative sample may be tested (5 kg for C131, 10 kg for C535). The LA abrasion value is computed as the difference between the original mass of the sample and the final mass with the fines excluded, as a ratio of the original sample mass, expressed as a percentage. The test can be carried out on both fine and coarse aggregate samples. More resistant aggregates will therefore produce smaller values, and a maximum of 50 per cent is required

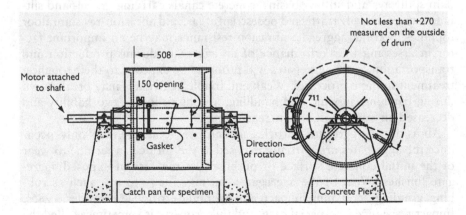

Figure 3.22 Los Angeles testing machine.

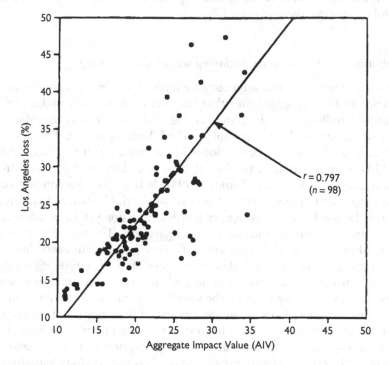

Figure 3.23 Correlation between LA abrasion test and AIV test for coarse aggregates in Ontario (from Senior and Rogers, 1991).

for coarse aggregates in all types of construction according to ASTM C33. Figure 3.23 indicates that the correlation between LA abrasion test results and aggregate impact value (AIV) is reasonably good, so that the AIV test could be regarded as a practical substitute for the LA abrasion test (Senior and Rogers, 1991).

Rogers and Senior (1994) criticize the LA test on the grounds that it is only a good predictor of mechanical breakdown; certain aggregates failing the test make satisfactory concrete without breaking down in the mixer, and vice versa. They recommend use of the Micro-Deval test (see p. 128). Experience indicates that it is inappropriate to use the LA abrasion test or the sulphate soundness test to evaluate degradation of aggregate in concrete during mixing, handling, and placement.

Aggregate Abrasion Value (AAV)

This rarely used test, sometimes specified for aggregates intended for concrete wearing surfaces, involves subjecting selected coarse aggregate particles to abrasion against a rotating steel lap charged with standard quartz sand, and measuring average mass loss.

Other abrasion or attrition tests, including wet abrasion testing

Sometimes it is useful to test an aggregate under wet abrasion conditions, which helps to identify aggregates that may break down into undesirable fines during handling, batching, and mixing. One such test is the Micro-Deval attrition test similar in principle to the LA abrasion test. This test is of European origin and has been adopted as a standard in Canada (CSA A23.2-23A) as an alternative to the sulphate soundness test (see p. 136) for fine aggregate (Rogers and Senior, 1994). It is a wet abrasion test on fine aggregates, and measures the amount of weak friable material present. It can also be used for coarse aggregates but has not yet been adopted as a standard for testing concrete coarse aggregate. The test determines aggregate abrasion loss in the presence of water and an abrasive charge, and provides information for judging the suitability of a fine aggregate subject to weathering and abrasive action when service information is not available. It is executed by rotating the sample in a stainless steel jar with a charge of steel balls and determining the amount of fines generated (material less than 80 μm). Experience with this test in Ontario indicates that results correlate with sulphate soundness and water absorption of fine aggregates (Rogers et al., 1991). It has a major advantage over the sulphate soundness test (ASTM C88) of much lower test variability. Rogers and Senior (1994) have advocated abandoning the sulphate soundness test on fine aggregates because of its poor precision. Other correlations, including those with AIV, AAV, and PSV tests are discussed in Senior and Rogers (1991).

Further tests to assess fine aggregates that break down easily by attrition are discussed in the report of ACI Committee 221 (ACI 221R-96, Cl. 3.3). These include the Corps of Engineers test method (CRD-C-141), the NAA-NRMCA attrition test (ASTM C1137), and the California Durability Index Test (ASTM D3744). The first two induce aggregate attrition by measuring the amount of material passing the 75-μm sieve produced after agitation of a fine aggregate–water slurry. The California durability index measures the tendency of coarse or fine aggregates to produce undesirable clay fines when degraded. The test involves agitating the sample in a mechanical washing vessel and then measuring the minus 75-μm fraction. For fine aggregates, an additional period of mechanical sieving is required. The durability index for coarse or fine aggregate is calculated from equations presented in the method, and can be used for specification purposes. A test for assessing the abrasive effects of waterborne gravel, rock, and so on over concrete surfaces in hydraulic structures is covered in ASTM C1138. In this test, the surface of a cylindrical concrete specimen is subjected to the abrasive effect of steel balls in vigorously agitated water for a total test period of 72 h, the volume loss being recorded at intervals during the test.

Hardness of aggregates

Surface hardness of aggregates is important only where aggregates are likely to wear in service. In large measure, it controls the wear properties of aggregates. The classic definition of hardness is its resistance to scratching, indentation, or rubbing. It depends on the hardness of the individual mineral grains in the aggregate and may therefore vary across a surface, particularly if it is coarse-grained. For this reason, some grainy aggregates tend to pit as well as scratch under abrasive conditions. Hardness can be measured in various ways: Vickers and Rockwell Hardness (for indentation); Shore Hardness (for rebound); Moh's Hardness (for scratch); and Dorry Hardness (for wear or abrasion). However, the more practical approach for concrete aggregates is via performance-type tests such as those described above.

Fracture properties

Comparatively little work has been done on characterizing the fracture properties of aggregates, often because it is not always clear how these parameters would be used in practice. Nevertheless, with progress being made in developing concrete strength and fracture computational models and to aid in understanding the composite nature of concrete, it is helpful to know something of aggregate fracture properties. As in the case of aggregate strength, it is difficult to measure fracture properties on small discrete particles of aggregate. Consequently fracture tests are generally carried out on rock specimens prepared from boulders or blast rock. This process tends to eliminate natural defects or planes of weakness in the rock, and the results may not be entirely representative of the aggregate. Furthermore, there is often a size effect associated with larger specimens (Bazant and Planas, 1997).

Brittle materials such as rocks can be characterized by fracture parameters derived from linear elastic fracture mechanics (LEFM). These parameters describe the strength of the material in the presence of stress-raising flaws and cracks. Details of fracture parameters based on LEFM are given in Chapter 5, where the necessary background to the fracture approach is also discussed. Fracture parameters for rock materials are given in that chapter.

Two fracture parameters that can readily be measured for aggregate-type materials are fracture toughness K_c and fracture energy (or specific work of fracture) R_c. Fracture toughness is a parameter relating to the critical stress field surrounding a crack while fracture energy refers to the

Figure 3.24 ISRM fracture test on notched rock core.

energy required to cause a crack to grow in a material, this energy usually being provided by the external load. It is also possible in a fracture test to measure elastic modulus E which will usually be an average of the compressive and tensile moduli. A practical way of measuring these parameters is by means of the ISRM Test Method I, using drilled rock cores (ISRM, 1988). In this method, a chevron-notched cylindrical beam is tested in centre-point loading. A photo of a test in progress is provided in Figure 3.24. The load–deflection relationship is measured from which the fracture parameters can be obtained (Mindess and Alexander, 1995). Typical values for rocks, where available, are provided in Table 5.3 from which it can be seen that dolomite is more brittle than andesite or granite. For the particular materials in Table 5.3, the rock fracture values are about 4–6 times higher for K_c, and 4–10 times higher for R_c compared with an OPC paste ($w/c = 0.3$), indicating the substantially higher fracture toughness of most rocks or aggregates. However, interface values for tests on composite cement/aggregate specimens tend to be much closer to the plain paste values, and this too is discussed in Chapter 5.

Elastic properties

Knowledge of the elastic modulus of aggregates, E_a, is useful for understanding the role of aggregates in influencing concrete elastic modulus and strength, and for application in computational models. This material parameter is also required for optimum design of rock blasting. Most aggregates derived from hard competent rock display linear stress–strain behaviour over the range of stresses to which they are subjected in concrete. For highly porous or non-isotropic materials which do not exhibit this behaviour, an appropriate E_a based on a chord or secant value can be defined.

It is difficult and impractical to measure E_a on aggregate particles. Elastic modulus is easily measured on rock cores drilled from boulders or other representative larger pieces of aggregate material (Alexander, 1993). Alternatively, E_a can be found from flexural tests on notched or unnotched beams. ISRM test methods can assist.

Selected information on elastic properties of aggregates is given in Table 3.2. Large variability characterizes the values for any given rock type, and it will always be necessary to test unknown or new sources of materials. Figure 3.25 gives typical values for South African aggregates. Dolomites tended to have very high E_a of about 116 GPa, approaching 60 per cent that of steel. In contrast, siltstone had a mean E_a value of only 25.2 GPa, not much greater than that of hardened cement pastes. Not much work has been done on the tensile modulus of elasticity of aggregates (Johnston, 1970a,b).

Aggregate elastic modulus has a strong influence on concrete stiffness, and in companion concrete tests using the aggregates shown in Figure 3.25, dolomite aggregate concrete had substantially higher elastic moduli than siltstone aggregate concrete. There is no direct relationship between aggregate (rock) strength and elastic modulus (Kaplan, 1959). Table 3.2 confirms this fact. Occasionally it may be preferable to have a low modulus aggregate, provided its strength is adequate, in order to produce a more resilient or 'yielding' concrete. An example would be situations where large movements

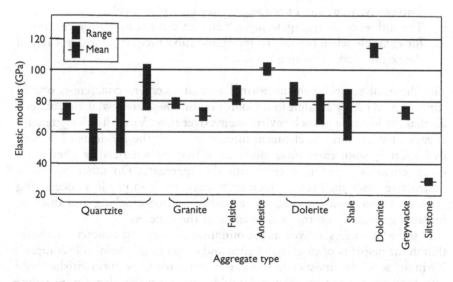

Figure 3.25 Aggregate elastic modulus results for South African rock types (from Alexander, 1990).

are expected and cracking is to be minimized, which occur in concrete slabs on grade or concrete pavements in hot dry conditions.

Very few aggregates exhibit creep at typical stress levels in concrete, and aggregate creep is therefore seldom considered. Aggregates can, however, have a substantial influence on concrete creep, and this is dealt with in Chapter 5.

Not many of the mechanical properties discussed above are actually used in concrete or aggregate specifications. Practice in the US and Canada tends to evaluate aggregates more by way of their actual performance in concrete, and to rely on past experience (Meininger, 1994). On the other hand, European and British practice utilizes properties such as ACV or 10% FACT values in specifications. One exception in North America is the LA abrasion value which is extensively used.

Chemical and durability properties of aggregates

The chemical properties and resistance of aggregates are important in respect of:

a   Chemical compatibility and stability between the constituents of aggregates and the high pH environment of hcp.
b   Possible long-term chemical reactions between aggregates and cement paste, adverse alkali–aggregate reactions being an important example.
c   Resistance of aggregates to external aggressive chemical attack on the concrete, for example acid attack where the chemical resistance or, conversely, solubility of the aggregates may play a role.
d   The influence of aggregate ingredients on cement hydration reactions, for example when organic or inorganic substances in aggregates retard the setting and hardening process.

The chemical environment in normal portland cement concrete is one of concentrated highly alkaline ionic solutions and there are few, if any, materials that are immune to such environments over time. Virtually all aggregates engage in some form of chemical interaction with the products of cement hydration. In some cases these interactions may be beneficial as when additional chemical bonding occurs with the aggregate. On other occasions destructive reactions occur, which invariably result in products occupying volumes larger than the original reactants causing cracking and distress, impacting seriously on the performance of the concrete.

Long-term studies as well as examination of very old concretes indicate that in the majority of cases aggregates and cements are chemically compatible in the sense that most concretes can be expected to perform satisfactorily over long periods of time. Blended binders incorporating slag, fly ash, silica fume, or other pozzolans create chemical environments less aggressive to sensitive aggregates thus reducing chemical incompatibility problems, and use of these binders should be encouraged. Nevertheless, there are occasions when incompatibilities arise between aggregates or their constituents and the hardened cement phase, leading to premature breakdown. It is therefore important to understand the chemical properties of aggregates and their interaction with concrete. Chapter 6 provides further details on aggregate interaction in concrete; this section will outline important aspects relating to the aggregates themselves.

One way in which aggregates may compromise the durability of concrete is by contributing undesirable constituents to the mix. These may comprise soluble salts, reactive minerals, soft and friable particles, organics, and so on. This subject is covered later.

Chemistry and mineralogy of aggregates

Aggregates reflect the chemistry of their constituent materials. They comprise minerals varying from relatively light-coloured felsic (acidic) types such as quartz and feldspar to mafic (basic), relatively dark-coloured minerals such as pyroxenes and amphiboles. Table 2.2 (Chapter 2) shows rock-forming minerals found in natural aggregates.

Certain aggregates when exposed to the atmosphere can undergo chemical reactions such as oxidation, hydration, and carbonation, and these reactions may be associated with volume increase. Damage to concrete is, however, usually limited since aggregates embedded in the concrete are unlikely to be affected. Surface pop-outs and staining can arise from these effects, a well-known instance resulting from pyrite oxidation – see Table 3.15 and Chapter 6.

It is not merely the chemistry of an aggregate that influences its interaction with the cement phase in concrete. The nature of the constituent minerals and crystals is also important. For example, silica in aggregates can exist in a stable lattice form (quartz), or occasionally display strained, unstable lattices that have higher thermodynamic energy, rendering such materials far more likely to engage in reactions with the cement paste. The most common example of this is the well-known alkali–silica reaction problem, dealt with in Chapter 6.

Aggregate reactions

The resistance of an aggregate to an aggressive external chemical substance will depend on the nature of the aggregate itself and the type of aggressive agent (e.g. acid, alkali, sulphate, or other aggressive salts). Table 3.13 summarizes interactions with typical aggregate types, based on their parent rock or source material. It is important to stress that attack of aggressive

Table 3.13 Resistance of aggregates to aggressive chemicals

Aggregate group and common types (see also Tables 2.1–2.4)				
	Acid igneous	Basic igneous	Sedimentary: Siliceous	Sedimentary: Calcareous (carbonates)	Miscellaneous
	Granite, gneiss, granodiorite, pegmatite, quartz-diorite, felsite, granophyre, rhyolite, trachyte, etc.	Andesite, basalt, diabase, dolerite, diorite, gabbro, norite, peridotite, etc.	Quartzite, sandstone, greywacke, arkose, chert, flint, etc.	Limestone, dolomite, variations of these such as dolostone, marble.	Phyllite, schist, hornfels, tillite.
Acids	Acid igneous aggregates are generally acid resistant.	Basic igneous rocks are generally acid resistant.	Siliceous aggregates are generally acid resistant.	Calcareous aggregates react readily with acids, and this may be used to good effect as sacrificial aggregates under acidic conditions. E.g. use of such aggregates in concrete sewers may prolong life by spreading the acid attack between aggregate and cement paste.	Resistance to acids depends largely on any carbonate content.
Alkalis	Certain acid igneous aggregates may contain alkali-susceptible silica minerals. See Chapter 6.	These aggregates do not contain free silica, and are therefore generally immune to alkali attack.	Siliceous aggregates may be susceptible to attack by alkalis, depending on the nature of the silica minerals. See section on AAR in Chapter 6.	Alkalis can attack certain forms of calcareous aggregates, typically those containing clay, such as certain argillaceous dolomitic limestones or argillaceous calcitic dolostones. See Chapter 6.	Resistance to alkalis depends on amount and nature of silica minerals present; certain of the above rock types have been known to be susceptible to alkali attack, particularly fine-grained varieties.

Others   Other chemical reactions involving aggregates, and which may be deleterious to concrete, are: (i) Hydration of anhydrous minerals, e.g. MgO (magnesia) (not common in natural aggregates); (ii) Soluble constituents, e.g. sulphates; (iii) Oxidation and hydration of iron compounds; (iv) Reactions involving sulphides and sulphates; (v) Reactions of synthetic slag aggregates with reactive components or those containing metallic fragments. Aggregates themselves can contribute deleterious substances to concrete, e.g. sulphates. See Table 3.15 for further detail on this and other aspects.

chemicals occurs on the composite material, that is the aggregates embedded in a cement matrix, and therefore the resistance to attack is a function of the resistance of the composite and not purely the aggregates. Furthermore, attack by aggressive agents will usually be far more rapid and severe on the cement component than on the aggregate. Consequently, any aggregate breakdown will often be of little consequence. There are of course obvious and important exceptions to this: alkalis in concrete can react deleteriously with some aggregates; on the other hand, the use of calcareous aggregates in acidic conditions where the aggregates may dissolve sympathetically with the cement paste is usually beneficial to concrete performance (SA-CSIR, 1959).

Other chemical reactions that involve aggregates include hydration of anhydrous minerals (e.g. hydration of calcium and magnesium oxides, typically in steel slag aggregates), cation exchange and volume change in clays and other minerals, soluble constituents, oxidation and hydration of iron compounds, and reactions involving sulphides and sulphates. Most of these problems are covered in Table 3.15 or elsewhere in this section. Materials that may cause such reactions can usually be detected by petrographic examination or other standard tests.

Another feature of aggressive chemical attack is that it might be possible for the aggregate particles to become coated with products from reactions between the cement paste and an aggressive agent, thus partially protecting the aggregates. In the other extreme, highly soluble and internally fractured calcareous aggregates may be dissolved by aggressive acids long before the cement paste phase is removed. Chapter 6 provides details on some of these aspects.

Soundness of aggregates

Although 'Soundness' strictly refers to a physical property of aggregates, it is discussed here because it has relevance for aggregate durability. 'Soundness' is a term broadly defined as the ability of aggregate to resist excessive changes in volume as a result of changes in physical conditions, such as freezing and thawing, thermal changes at temperatures above 0°C, and alternate wetting and drying. It relates to the physical competence or 'physical durability' of the material which is an important property since lack of such durability can severely compromise performance of concrete containing the aggregate. 'Soundness' should therefore be distinguished from volume changes due to chemical reactions between aggregates and their environment. Aggregates are regarded as unsound when volume changes induced by physical effects result in deterioration of the concrete in the form of surface scaling, pop-outs, and cracking. A typical pop-out due to a coarse slag aggregate particle that exhibited unsoundness due to unhydrated CaO (lime) and MgO (periclase) is shown in Figure 3.26.

(a)   (b)

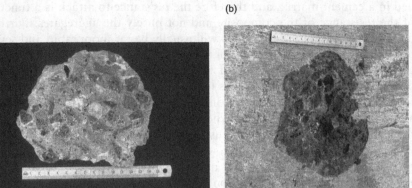

Figure 3.26 Pop-out due to unsound coarse slag aggregate particle: (a) typical pop-out; (b) concrete substrate. Both photographs shown with 15-cm scale rule (photographs courtesy of Dr R. Oberholster).

Being ill defined, soundness is difficult to measure, a fact complicated by the long-term nature of durability processes. Therefore, it is indirectly assessed in various ways. For example, the aggregate granular strength tests discussed previously (ACV, 10% FACT, and AIV) are usually carried out on dry materials; comparative testing of identical companion samples in a saturated state can provide a measure of physical soundness. This use of 'Wet Abrasion Testing' was discussed earlier in the chapter, where a number of different tests were mentioned such as the Micro-Deval test and the California Durability Index test. Other indirect measures of soundness are discussed below.

Sulphate soundness

This test is widely used internationally, having first been proposed in 1931 and approved as an ASTM standard test in 1963 (ASTM C88: BS 812: Part 121). The test purports to measure the aggregate's soundness, that is its resistance to weathering. It consists of subjecting an aggregate sample to alternate cycles of immersion in a saturated sulphate solution and oven drying. This causes salt crystals to precipitate in permeable pore spaces and the crystals then exert an expansive force on the aggregate. Either sodium or magnesium sulphate can be used, the latter generally being more aggressive. The test result is in terms of mass loss of the sample after washing and sieving. A visual examination of the degraded samples, best done on the plus 19-mm fraction, is also usually carried out to describe the nature of breakdown, which may be flaking, splitting, crumbling, and granular disintegration. The type of distress is categorized and the number of affected

Table 3.14 ASTM C33 soundness limits for coarse and fine aggregates

	Maximum weighted overall loss after 5 cycles (%)	
	Coarse aggregate	Fine aggregate
Magnesium sulphate soundness	18	15
Sodium sulphate soundness	12	10

particles compared with a count taken of the same particles in the original test sample.

The sulphate soundness limits in ASTM C33 are given in Table 3.14 in simplified form. The limits vary according to fine or coarse aggregate and the type of sulphate being used in the test. The coarse aggregate limits apply only to certain categories of construction or exposure conditions. Compliance with the limits in ASTM C33 does not necessarily guarantee satisfactory performance of the aggregates in concretes under freeze–thaw conditions, nor does failure to achieve the limits inevitably lead to performance failures. These limits are simply to be regarded as indicators where possible additional investigation such as petrographic examination is necessary.

Careful interpretation of the results of the sulphate soundness test is required (Forster, 1994). The test represents a severe environment and not all aggregates that fail the test are necessarily unsuitable in concrete. The ASTM standard specification for concrete aggregates (ASTM C33) allows use of both fine and coarse aggregates that exceed the specified limits of the test if it can be shown that the aggregate has performed satisfactorily in concrete of comparable properties under similar service conditions. It is claimed that the internal expansive force derived from the rehydration of the salt upon re-immersion simulates the expansion of water on freezing. The physical actions involved during the test include not only expansive salt crystallization pressure, but also forces of disintegration due to wetting and drying, and heating and cooling, making the outcome of the test a complex interaction of different effects. There is no direct theoretical basis to expect the test results to correlate strongly with actual freeze–thaw disintegration in situ. While some limited correlations have been found, it is unwise to rely too heavily on the test results in deciding on the suitability of an aggregate. The amount of scatter in correlation data, as well as relatively poor precision of the test method, also raises caution (Rogers and Senior, 1994). Many have questioned the relevance of the test in regard to its ability to adequately simulate frost resistance of aggregates (Rogers et al., 1989). Nevertheless, extensive data from its long history of use over many decades permit assessments of soundness to be made.

Experience with the sulphate soundness test on South African aggregates is that it is not highly selective but provides a rough indication of the physical durability of an aggregate subjected to aggressive environments. Direct correlations between the test results and aggregate performance in service are tenuous at best. Therefore, when an aggregate is to be evaluated based on this test, it should be done by comparison with an aggregate of similar mineralogical composition and geological history and which has a proven service record (Grieve, 2001).

Referring to a particular case study, soundness of basalt aggregates for the giant Lesotho Highlands Water Scheme was assessed by several methods. These aggregates were known occasionally to exhibit dimensional instability due to 'rapid weathering' (Orr, 1979). A test referred to in Chapter 2 using ethylene glycol immersion was used. This test will rapidly identify unstable aggregates, but represents a very severe condition. Somewhat inferior aggregates embedded in concrete tend to be protected and often give satisfactory performance.

Freeze–thaw soundness

Freeze–thaw mechanisms and damage

Aggregates may be susceptible to freeze–thaw damage depending on their pore structure and their absorption and permeability properties. The presence of certain deleterious clay minerals or related minerals such as chlorite can also induce freeze–thaw damage (Higgs, 1987). However, the primary mechanism of freeze–thaw damage in aggregates relates to the freezing of water in the pores of aggregates under conditions of severe frost action. Any solid with a porous system that can imbibe water and then freeze will be subject to expansive pressure. The governing factor will be the degree of saturation, since partially saturated pore space can permit the relief of freezing pressures. Thus, the concept of a 'critical level of saturation' for aggregates is very important. The physics of expansion of water on freezing dictates that for simple systems this critical level of saturation is about 91.7 per cent, and even for more complex structures of most aggregates this critical level is still near 90 per cent (Winslow, 1994).

In general, aggregate pores are much larger than those in cement paste and are readily filled with water. Aggregates that are susceptible to freeze–thaw damage can absorb sufficient water to the point of critical saturation mentioned above when they cannot resist the freezing pressures of water. Damage due to freezing in an aggregate's pore system arises either because of the simple expansion of water upon freezing or because of the hydraulic pressure of liquid flow induced by ice formation, or both. In addition, water expelled from an aggregate during freezing can disrupt the surrounding paste by its hydraulic pressure, without necessarily damaging the aggregate itself. Winslow (1994) makes the point that 'the propensity of an aggregate for causing difficulties by any of these mechanisms is associated with its pore volume and with the permeability of its pore system and with nothing else'.

Based on these mechanisms, aggregates can be classified into three groups in respect of their freeze–thaw behaviour: (1) aggregates having such a small pore volume (generally less than about 0.5 per cent) that freezing expansion results in strains too small to cause cracking; (2) aggregates in which the hydraulic pressure generated by the flow of water ahead of a freezing front in a critically saturated pore system is able to crack the aggregate itself. Such aggregates must have a sufficiently low permeability and a long enough flow path to generate high pressures; (3) aggregates with a sufficiently high permeability to prevent hydraulic pressures being large enough to crack the aggregate (although distress might be caused to the surrounding matrix due to this hydraulic pressure). Thus for critically saturated aggregates, the potential for freeze–thaw damage is governed by the permeability of the aggregate and the length of the flow path within it. From this it should be clear that smaller aggregates are far less likely to suffer distress. Both the ability of an aggregate to become critically saturated and its permeability are functions of the aggregate's pore size distribution. Unfortunately, quantitative relationships between these parameters are generally unavailable.

The important conclusion is that aggregate freeze–thaw problems will develop only if the aggregate becomes critically saturated. In addition very small and very large pore sizes are unlikely to cause a problem. Studies have concluded that durability is related to the volume of a segment of the pore space lying within certain size limits, these limits being between 0.004 and 0.04 μm at the lower end, to around 0.1 μm or so at the upper end. A relationship has been derived in which the volume and median diameter of the pore space having a diameter greater than 0.0045 μm are shown to be important (Kaneuji et al., 1980). This relationship gives rise to a so-called expected durability factor (EDF) as follows:

$$\text{EDF} = \frac{A}{\text{PV}} + B \cdot \text{MD} + C \tag{3.18}$$

where

        EDF = expected durability factor
         PV = pore volume with diameter $> 0.0045\,\mu\text{m}$
      MD = median diameter of pore volume with diameter $> 0.0045\,\mu\text{m}$
A, B, and C = constants.

An EDF in excess of about 40 has been found to be associated with durable aggregates. Figure 3.27 shows results for a given set of aggregates in which non-durable aggregates fall in the upper right-hand portion of the diagram.

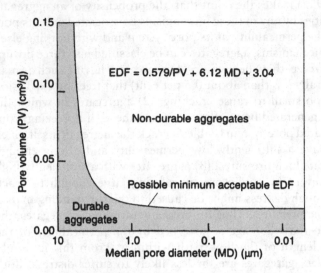

Figure 3.27 Classification of freeze–thaw durability of aggregates using the expected durability factor EDF (from Kaneuji *et al.*, 1980; © 1980, reprinted with permission from Elsevier).

Equation (3.18) attempts to quantify the pore system and relate it to the freeze–thaw durability of an aggregate.

Further research is needed to assist in a better understanding of this phenomenon and to arrive at better quantifications of potential freeze–thaw durability.

Suspect aggregates concerning freeze–thaw are porous cherts, shales, some limestones (particularly laminated ones), and some sandstones (Neville, 1995). A common characteristic of rocks with a poor record is their high water absorption although many other durable rocks may also have high absorption. Therefore, it is not simply total porosity that is important, but also pore structure and pore size, as previously discussed. Smaller pores are more damaging than an equal volume of larger pores in which ice pressures can more easily be relieved.

Freeze–thaw resistance also depends on particle size, as discussed, and for any given freezing rate and aggregate type there is a critical size above which the particles will fail under repeated freeze–thaw cycles if critically saturated. This critical size is related to the maximum distance water must flow to reach the exterior of the particle to relieve freezing pressures. For this reason, fine aggregates do not usually contribute directly to freeze–thaw deterioration of concrete. It is the larger coarse aggregate particles having higher absorption or porosity, with pore sizes in the range of 0.1–4 μm, that are most easily saturated and result in freeze–thaw damage. For fine-grained aggregates with low permeability (e.g. cherts), the critical particle size may be within the range of normal coarse aggregate sizes. For coarse-grained materials, the critical size may be sufficiently large to be of no consequence (Kosmatka *et al.*, 1995).

Aggregates that are susceptible to freeze–thaw damage disintegrate and cause surface pop-outs as well as cracking of concrete, in particular the so-called 'D-cracking' in slabs-on-grade along joints and free edges (see discussion in Chapter 6). This effect is caused by delamination of the aggregate within the concrete in slabs subjected to freezing and thawing with moisture constantly available on one side. Figure 3.28 illustrates a fractured carbonate aggregate that was judged the cause of D-cracking in concrete.

FREEZE–THAW TESTS

The sulphate soundness test (ASTM C88) discussed above is still used by the majority of specifiers of concrete aggregates in North America as the index test for frost resistance. It is however unsuitable for this on the grounds of simulating a different mechanism (soluble salt crystallization) and of poor precision (Rogers and Senior, 1994). A freeze–thaw test for unconfined aggregates is outlined in an AASHTO procedure (AASHTO T-103 (1982)) and in the Canadian Standard A23.2-24A. The Canadian test is used for coarse aggregates for which information is needed in the absence of service

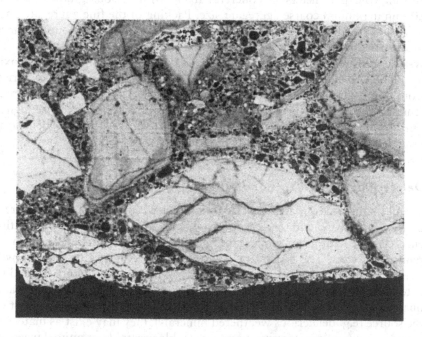

Figure 3.28 Fractured carbonate aggregate particles in concrete giving rise to 'D-cracking' (Stark, 1976).

records of the material exposed to actual weathering conditions, in order
to judge their soundness under freezing and thawing action (five cycles
of −18 to 20°C). A weighted average percentage loss of material on a
series of sieves (40–5 mm) is determined. Experience with the test in Canada
indicates that it has significantly lower multi-laboratory variation than the
sulphate soundness test (Rogers *et al.*, 1989).

Alternatively, ASTM C666 evaluates the freeze–thaw performance of
aggregate in air-entrained concrete by rapid cycles of freezing and thawing
between 4.4 and −17.8 °C. All these tests are empirical, and rely for their
validity on the competence of operators in reputable laboratories in order
to produce reliable results, and on a comprehensive database covering a
wide range of materials in order that comparative assessments can be made.
The test results cannot easily be used as acceptance/rejection criteria for
new or untried aggregate sources. Further testing of such sources incorporated into a 'real' concrete will be necessary. Ultimately, only a satisfactory
service record in concrete can prove the durability of an aggregate. Further
discussion is given in Chapter 6.

Evidence for the importance of the grading of fine aggregates for concrete
durability in terms of freeze–thaw resistance was presented by Okkenhaug
and Gjorv (2003). They studied the influence of grading of natural fine
aggregates, consisting mainly of quartz and feldspar, on total air content
and air-void parameters of concretes made with various gradings and a
20-mm maximum coarse aggregate size. Packing of the fine aggregate particles, governed by the shape of the grading curve and the amount of fines
(<125 μm), had a critical influence on the development of the air-void
system, with total air contents in the fresh concrete varying from approximately 4.5 to 9 per cent, while the specific surface of the air-voids varied
from approximately 20 to 40 mm^2/mm^3. Convex grading curves (representing finer grading) produced improved air-void systems; likewise higher
fines contents and greater total amount of fine aggregate were beneficial
provided convex grading curves were used.

Deleterious constituents and salts in aggregates

Various substances in aggregates can interact with the cement in concrete
or have otherwise undesirable effects. For example, sulphates may influence cement hydration, while excessive chlorides can incur corrosion of
reinforcing steel. The section on aggregate petrography that follows will
stress that new or untried aggregate sources must be thoroughly studied to
determine their suitability for use. This applies particularly to detecting and
quantifying undesirable constituents. These may occur naturally as part of
the source (e.g. deleterious weathered minerals), they may exist as materials
alien to a source (e.g. presence of unacceptable vegetable or animal waste in
surface deposits), they may be subsequent contaminants after the aggregate

has been processed and stockpiled (e.g. contamination from the base material or deposit of unwanted substances on a stockpile), or they may result from contamination during transport as sometimes occurs when trucks carrying aggregates have been used to transport other materials like fertilizers. In general, these substances are detrimental to aggregates in concrete.

Sims and Brown (1998) summarize the adverse effects that can occur in concrete by undesirable constituents as:

(1) chemical interference with the setting of cement, (2) physical prevention of good bond between the aggregate and cement paste, (3) modification of the properties of the fresh concrete to the detriment of the durability or strength of the hardened material, (4) interaction between the cement paste and the aggregate which continues after hardening, sometimes causing expansion and cracking of the concrete, and (5) weakness and poor durability of the aggregate particles themselves.

(Sims and Brown, 1998)

Deleterious materials generally relate to certain minerals and rocks, organic substances, and undesirable soluble salts and other chemicals in aggregates. Information on these groupings is given in Table 3.15 which has relevant explanatory notes. Table 3.16 gives allowable limits for chlorides in aggregates. Tests for deleterious substances are discussed in Forster (1994), but substantial information is contained in Table 3.15. Figure 3.29 shows petrographic micrographs of concrete containing pyrite deriving from the aggregates, which may have detrimental influences.

A word on sulphate-containing aggregates is appropriate here. The most common form of naturally occurring sulphate is gypsum ($CaSO_4 \cdot nH_2O$) which exists in various forms depending on the amount of water associated with the mineral (n can vary from 0 to 2). Gypsum is however only sparingly soluble. Other sulphates that can occur in aggregates are the readily soluble forms of sodium and magnesium sulphate. Sulphates chemically attack the cement hydrates, particularly the aluminates to give tricalcium sulphoaluminate or ettringite. This occupies more than twice the molecular volume of the aluminate and when formed in hardened concrete results in expansive forces which exceed the tensile strength of the concrete, causing cracking. However, certain sulphates such as barium sulphate (the mineral barytes) are of such low solubility that they do not contribute to sulphate attack. Thus, barytes is often used as a heavy aggregate in concrete for nuclear shielding. Sulphate-contaminated aggregates are a problem in the Middle East region and special precautions are needed to avoid or minimize their use (Eglinton, 1987). Draft European standards take into account the nature of the sulphate in different aggregate types, with appropriate limits. For example, a limit of 0.2 per cent SO_3 by mass for natural aggregates and 0.1 per cent SO_3 by mass for slags and other artificial aggregates is proposed.

Table 3.15 Deleterious constituents in aggregates

	Likely occurrence in aggregates, and typical examples	Undesirable effects in concrete	Minimization or avoidance of undesirable effects
I Deleterious rocks and minerals in concrete aggregates			
Metallic ores: *Sulphide-containing* Sulphides of Iron Iron pyrite: (FeS_2)[1]	Found in rocks of all types; occurs widely as accessory mineral (usually <1%, sometimes up to 5%) in basic igneous rocks. Brass yellow colour. In UK, it occurs in gabbros, metamorphic limestones, and sand and gravel deposits (e.g. flint gravels) in southern England. South African examples: Karoo dolerite, BIC gabbro, norite, felsite; in sedimentary rocks: gold-bearing reefs and other conglomerate deposits (e.g. Reef quartzites – normally not more than 0.5% by mass, but occasionally up to 3%); Tygerberg greywacke. Widespread in greenstone belt rocks in Barberton region and Zimbabwe.	Oxidation to sulphates can occur, then further decomposition to hydroxide, particularly in fine crushed aggregates. Sulphates cause expansion by reacting with cement compounds. Often result in surface pop-outs. Warm, humid conditions are worst. Coarse crushed aggregate less or non-problematic. In UK, pyrites were found to oxidize to brown iron hydroxide causing unsightly staining, even in small quantities.[2]	Set limit at 0.25–0.4% maximum. Pyrite deriving from mining waste to be treated with caution. In Sweden, sulphide minerals limited to <1%.
Marcasite and pyrrhotite	Marcasite much less common than pyrite. Found mainly in sedimentary rocks. Pyrrhotite found in many types of igneous and metamorphic rocks. Pyrite and pyrrhotite have caused sulphate attack on concrete in Sweden.	As with other sulphides, can oxidize to sulphate and sulphuric acid, also forming iron oxides and hydroxides. Marcasite is particularly problematic in warm humid conditions, producing brown staining and volume increase.	

Mineral	Occurrence / description	Effect	Recommendation
Copper pyrite	Known as chalcopyrite. Presence usually indicated by traces of green mineral malachite. In South Africa, it occurs in some rocks of BIC (gabbros, diabase), and in various gneissic and shaly rocks.	Combines readily with moisture to form sulphates which attack cement hydrates.	Avoid completely.
Others: zinc blende (sphalerite); lead glance (galena); sparry iron ore, etc.	These may be deleterious to concrete, due to presence of sulphides which can be oxidized.		
Sulphate minerals	Frequently occur in arid areas where evaporation exceeds precipitation. In this case, surface materials have undesirable salts concentrated in them. Gypsum ($CaSO_4 \cdot 2H_2O$) is the most common naturally occurring mineral. Others are Alunite ($K_2Al_6(OH)_{12}(SO_4)_4$), Epsomite ($MgSO_4 \cdot 7H_2O$). Water-soluble forms of sodium and magnesium sulphate are more harmful, e.g. saline Middle East aggregate deposits. Also the so-called alkali deposits of North America.	Cause decomposition of concrete through reaction with aluminates (ettringite formation) and calcium hydroxide in cement. Magnesium sulphates also attack cement silicates. (see also Section 3 of this table).	Limit gypsum to 0.25% by mass of coarse aggregate, according to USBR.
Other undesirable minerals Minerals with large amounts of Ferrous Iron (for Ferric Iron minerals, see note)[3]	Oxidation of ferrous iron occurs in some igneous rocks: a Certain granites with biotite mica b Ferromagnesian minerals in some dolerites, where highly ferriferous fayalite olivine may degrade to black chlorophaeite.	Effects vary: a Oxidation causes unsightly staining b Ferromagnesian mineral oxidation can result in volume increase and expansion.	Avoid deleterious varieties.

Table 3.15 (Continued)

	Likely occurrence in aggregates, and typical examples	Undesirable effects in concrete	Minimization or avoidance of undesirable effects
Hornblende	May be associated with some metamorphosed dolerites, gabbros, and gneisses.	Certain highly ferriferous hornblendes may degrade especially in saline moist conditions to produce reddish limonitic material which forms undesirable coatings on aggregates.	Avoid deleterious varieties.
Mica	Muscovite and biotite are common in granites, gneisses, and sandstones. Micas form distinctive platy crystals. Problem is most severe if mica occurs as free flaky grains in sands. Less or non-problematic if well bound in coarse aggregate. Mica in form of muscovite more harmful than in biotite form.	Two effects may occur: 1 Flaky grains in sand increase water requirement and can reduce strength and durability of concrete. Strength reduction up to 5% per 1% mass of muscovite. 2 Biotite can rapidly alter by weathering to sericite (fine-grained variety of hydro-muscovite) and illite clay. Problems then as above.	Limit to 3–5% or less, possibly more in non-critical situations.[4]
Clays and altered minerals	These can occur when certain weathered argillaceous rocks and ferromagnesian minerals in basic igneous rocks and greywacke produce highly active clay minerals: active smectite clays, montmorillonites, and illites, which expand or shrink with changing moisture content. These may be present as lumps, or as aggregate coatings. (Non-active clays are not harmful provided they are not excessive.) 'Culprits' may be argillaceous carbonate rocks, and certain weathered basalts (South Africa, Australia, and South America), dolerites, gabbros, greywackes.[5]	Clays and weathered minerals increase water requirement in fresh concrete, and give dimensional instability in hardened concrete particularly excessive shrinkage. These minerals also lead to weaker and more permeable concrete. Other non-detrimental fines such as 'rock flour' can be beneficial to a certain degree, and if not excessive. Relevant tests are given in Note 6.	Petrographic examination can identify undesirable clays, and these should be limited. ASTM C33 limits amount of clay lumps and friable particles to 3% of fine aggregate, and 2–10% in coarse aggregate, depending on use of concrete.

	Occurrence	Effect on concrete	Recommended limits
Miscellaneous	This includes synthetic glass, clinker, chert, magnesium and/or calcium oxide, soil, hydrocarbons, wood chips, etc.	Effects depend on particle constituents, but vary from retarding cement setting and hydration reactions, to expansion and disruption of the concrete.	Small quantities may be acceptable depending on purpose of concrete. Experience and judgment necessary.
Alkali-susceptible minerals	These minerals comprise reactive quartz and silica minerals, as well as certain reactive carbonates, which react with alkalis in the hardened concrete to cause damaging expansion. This subject is dealt with in Chapter 6		
Aggregates with undesirable physical properties			
Coal and lignite	Bituminous coals and lignites associated with coal-bearing deposits can occur in some natural sands and gravels. This can often be associated with shales. Staining of roof tiles in UK has occurred due to coal particles in sands.	Can reduce concrete strength due to softness and swelling. Lignites can cause unsightly staining on exposed surfaces in the presence of water.	Limit to 0.5% in sand or 1% in stone (ASTM C33). Limit to 0.5% for aesthetic concrete. Determined by material that floats in a liquid of RD = 2.0.[7]
Chalk and other soft or friable particles	Weak chalk aggregates sometimes occur, e.g. in UK, clay lumps are usually soft and weak (for other clay problems, see previously in this table).	Friable particles can break down into smaller grains during transporting or mixing. Can reduce concrete strengths and lead to cracking.[6]	Limit to 3% by mass of fine aggregate; 2–10% by mass of coarse aggregate, depending on use (ASTM C33). Do not use in concrete subject to abrasion.
Shells	Sea-dredged or beach or dune deposits or marine aggregates. In UK, sea-dredged aggregates. In South Africa, dune sands (Cape Flats) and Port Elizabeth; beach sands, Durban and East London.	Generally, not problematic in sands unless present in excessive quantities. Can be flaky or have hollow shape, affecting workability.[8]	Limit broken shell content to 30% by mass maximum (SANS 1083-1976), less if unbroken shell. BS 882: 1992 limits on coarse (>10 mm) aggregate are 8%, on 5–10 mm < 20%, which may be over-restrictive. No limit in BS on shell content of fine aggregate (<5 mm).

Table 3.15 (Continued)

	Likely occurrence in aggregates, and typical examples	Undesirable effects in concrete	Minimization or avoidance of undesirable effects
Absorptive and micro-porous materials	Some micro-porous flints in England are liable to freeze–thaw damage. Provided these are hard and competent, they may not directly reduce concrete strength and durability.	Some porous aggregates may be susceptible to freeze–thaw damage, depending on pore size and structure of pore system.	Limit amount of highly absorptive material.
Particle coatings	Coatings can occur on otherwise dense, competent aggregates. They comprise silt and clay encrustations, calcium carbonate and silica (which are chemically precipitated), gypsum, and hydrated metallic and iron oxides. Some coatings of non-deleterious materials may be removed during mixing and, provided they do not adversely affect the workability, are acceptable. Some aggregate coatings are, however, chemically reactive, e.g. sulphates and phosphates.	Soft, friable coatings cause high water requirement of mixes, poor paste–aggregate bond, and volumetric instability. Others may react with the cement hydration products (e.g. alkalis) or cause staining or efflorescence of the concrete. Certain coatings may be entirely innocuous, e.g. dust coatings that are removed during mixing.	Limit amount of coated particles. A petrographic examination can provide information on extent of coatings and likely effects.

2 Undesirable organic matter in aggregates

	Likely occurrence in aggregates, and typical examples	Undesirable effects in concrete	Minimization or avoidance of undesirable effects
Organic materials	These may occur in natural sands (less likely in gravels or crushed aggregates); vary from plant and vegetable matter (roots, decaying leaves, seeds, etc.), to animal wastes, food processing wastes, etc.[9] Decaying organics often produce humic acid, tannic acid, and lignins. Even very small amounts (typically less than 1%) may have significant effect.	Organics interfere with chemical reactions, leading usually to a retardation of setting and hardening and an adverse effect on strength. Discolouration of the concrete or mortar may also occur.	Check for possible presence using standard tests.[8,9] Carry out mortar or concrete tests to check effect on strength. Avoid contaminated sources, or blend with acceptable sand.

3 Deleterious soluble salts and other chemicals in aggregates

Sugars	Widely distributed in plant matter, e.g. roots, gums, fibres, seeds, sap. May occur in sands, usually as an external contaminant, or a contaminant during transport. In industrial situations, sugars occur as sucrose, glucose, maltose, or lactose.[10]	Can have a dramatic effect on retardation, extending the setting times even for days, or completely inhibiting reactions as sugar contents approach 1% by mass of cement. Fermented sugars in the form of alcohols can have similar effects.	Limit amount of soluble material or avoid altogether.
Salts affecting cement hydration	1 Certain salts cause acceleration of hydration, e.g. chlorides. These can subsequently cause efflorescence, or steel corrosion. Other chemicals are sometimes purposefully used in admixtures to accelerate setting and hardening. 2 Some chemicals can cause retardation of set and slow strength gain, e.g. humic acid, sugars. These chemicals are sometimes purposefully used in admixtures. 3 Sodium chloride in concrete increases the reactive alkali content which may exacerbate AAR or potential AAR.	Cause set acceleration and rapid strength gain, which may be undesirable. Set retardation and slow strength gain can be very detrimental to construction, and may reduce later strength.	
Salts causing efflorescence	Many salts can cause efflorescence, e.g. soluble chlorides and carbonates.	Cause unsightly whitish deposits on concrete surface.	Limit amount of soluble material, or avoid altogether.
Salts destructive to hydration products	Sulphates in aggregates can react with hydration products and destroy their properties, e.g. sodium and magnesium sulphates. Gypsum-bearing materials can be troublesome, particularly in arid areas, e.g. Middle East; also, some	Expansive reactions (ettringite formation) or ion-exchange reactions destroy the binding ability of the hydrates, softening, cracking, and weakening the concrete. Note: Portland cement has a controlled amount of sulphate present (gypsum),	Limit total sulphate from cement, aggregates, and mixing water to no more than twice the amount of sulphate from cement alone. Suggested safe limit: 0.4% for

Table 3.15 (Continued)

	Likely occurrence in aggregates, and typical examples	Undesirable effects in concrete	Minimization or avoidance of undesirable effects
	crushed carbonate rock aggregates of evaporative origin. Deleterious sulphates can sometimes be present as coatings to the aggregates.	added during manufacture, as a set-controlling agent. Typically 2–2.5% as SO_3.	fine and coarse aggregate, and 4% limit in concrete, by mass of cement.
Salts causing steel corrosion	These are mainly chlorides, which may occur in unwashed sea-dredged aggregates, particularly sands, or unwashed beach and dune deposits. May also occur in arid areas with high salt residuals in the sand, e.g. Mid East. Some crushed carbonate rocks of evaporative origin may be high in chlorides, difficult to remove by washing.	Chlorides in aggregates and concrete are highly aggressive to embedded reinforcing steel, causing corrosion, cracking and spalling, and occasionally collapse (see Table 3.16 for allowable limits).	Ensure marine-associated aggregates are thoroughly washed before use. Test for soluble chlorides to check if they are within limits (see Table 3.16).
Aggregates containing releasable alkalis	Alkalis are naturally present in concrete due to the cement component. Additional alkalis which may derive from the aggregates can be problematic. These must be 'releasable' alkalis to be potentially deleterious. Many rock-forming minerals contain alkalis (e.g. feldspars), but not all are 'releasable'. Some granites and volcanic rocks (e.g. some basalts in New Zealand) are problematic.	Additional alkalis raise the overall alkali content of concrete, and may result in expansion due to AAR (see Stark and Bhatty, 1986; Goguel, 1994).	Petrographic examination can assist in identifying releasable alkalis. AAR tests can also be carried out.

Sources: Eglinton (1987), Grieve (2001), Neville (1995), and Sims and Brown (1998) were used as primary sources for this table. Interested readers should consult these and other sources for more details.

Notes
1 Ability of iron pyrite to oxidize in normal environments depends on structure and purity of the mineral. Sulphides normally oxidized by weathering near surface in South Africa, but occur at surface in Canada where ice age glaciers have removed pre-existing weathered rocks. Oxidation of pyrite also promoted by presence of adjacent grains of iron-bearing sulphides (e.g. pyrrhotite) that cause galvanic reactions. Since oxidizing sulphides produce sulphuric acid, once started the process becomes rapid.

2 Presence of sulphate depends on degree of oxidation of sulphides to sulphate. Not all forms of pyrite are reactive. Since decomposition occurs in lime water, suspect aggregate can be checked in this way: reactive aggregates rapidly produce blue-green precipitate of ferrous sulphate, which on exposure to air (further oxidation) changes to brown ferric hydroxide. BS 3681 has a test for sulphate content on finely ground samples. Iron staining requirements are given in ASTM C33, the tests being covered in ASTM C641.

3 Iron in Ferric form (Fe^{+++}) (e.g. Magnetite Fe_3O_4 found in igneous rocks as an accessory mineral) is resistant to atmospheric oxidation, nor does it react with cement constituents; therefore limits on iron in this form in aggregates are not necessary for chemical reasons.

4 Gaynor and Meininger (1984) recommend a microscopical count of mica particles in the fraction of sand between 300 and 150 μm (No. 50–No. 100) sieve sizes, and if less than about 15% of mica as a number of particles is present in that fraction, the properties of concrete are unlikely to be significantly affected. Grieve (2001) also has further information.

5 In contrast to weathering of argillaceous and ferromagnesian rocks, weathering of feldspathic rocks, e.g. granite and arkose, forms kaolin-type clays that are non-expansive. Eg. in South Africa, Cape and other granites weather to kaolin and some illite.

6 ASTM C117 provides a test for presence of fine material or clay (by washing), ASTM C142 for clay lumps and friable particles, ASTM C1137 for attrition of fine aggregate during mixing, and ASTM D2419 for presence of clay.

7 Test for low density or lightweight materials is given in ASTM C123. See also Clause 9 of SANS 1083: 1976. ASTM C142 covers tests for friable particles.

8 Tests for shell content: BS 812: Part 106, for aggregate size >10mm; alternatively, BS 812: Part 119, to determine acid-soluble component for fine aggregate.

9 There are standard colorimetric tests for detecting the presence of organic matter, e.g. ASTM C40, SANS 5832. The colour of a sample of the suspect sand in a sodium hydroxide solution is compared with a reference solution or a reference colour. Note, however, that these tests do not actually show whether the organic matter is deleterious or not. Therefore, aggregates that fail the colorimetric test could still be used provided it can be shown that there is no detrimental effect on, for instance, compressive strength. This can be checked in a standard mortar or concrete – see ASTM C87, BS 882: 1992, SANS 5834. See also Keen (1979).

10 Standard tests for detection of sugars: see e.g. SANS 5833. These tests usually detect a wide range of different sugars. Sugar in damp conditions can tend to ferment and convert into alcohol, and these tests will not detect such alcohols.

Table 3.16 Aggregate chloride contents permitted in various standards

Type or use of concrete	Maximum total chloride ion (%)		
	In aggregates		In concrete
	BS 882: 1992 (by mass of aggregate)	SANS 1083	ACI 201.2R & 222R
Prestressed concrete and heat-cured concrete	0.01	0.01	0.08
Reinforced concrete made with sulphate-resisting portland cement	0.03	0.03	Between 0.15 and 0.2% (without reference to type of binder)
Reinforced concrete made with portland cement or combinations of portland cement with GGBS or FA	0.05	0.03	
Other unreinforced concrete	No limit	0.08	–

Notes
1 Chlorides are more likely to be present in fine aggregate. Therefore, the above figures relate primarily to fine aggregates.
2 For problems in chemical test methods, e.g. method of extraction of chlorides from aggregates, see Eglinton (1987). For example, washing is appropriate for dense non-porous aggregates where chlorides are only on the surface, but is not appropriate for crushed carbonate rocks of evaporite origin or other materials in which chloride is distributed throughout the particles.

Likewise, sulphur compounds should be restricted due to their tendency to oxidize to produce sulphates. Suggested limiting values are 1 per cent S for natural aggregates and 2 per cent S for slags and other artificial aggregates. Table 3.15 gives further information on sulphates in aggregates.

An aggregate containing a deleterious substance may not need to be entirely condemned for use in concrete. This will depend on its nature and concentration, its form and particle size distribution, and very importantly on the conditions of exposure of the structure and its aesthetic importance. For example, it might be possible to include staining aggregates in the interior members of a building structure with no adverse effects. Engineering judgment coupled with prior experience will be necessary in these cases.

Aggregate petrography

Petrology is the science of rocks, while petrography refers to the methods and techniques of examination and analysis of rocks and rock-forming minerals. Petrography is an indispensable part of modern concrete science and technology. It is a specialist field which has the following purposes for concrete aggregates:

1 To describe aggregate constituents and determine their relative amounts, thereby permitting classification.
2 To determine those physical and chemical characteristics, amenable to petrographic examination, that may have a bearing on the performance of the aggregate in concrete, including detection of potentially deleterious constituents.
3 To attempt to establish the likely performance of aggregates from new or untried sources, by comparing their characteristics with aggregates of known performance.
4 To provide a means for interpreting results of other standard aggregates tests and of selecting further tests for determining the likely performance of an aggregate.

(a)

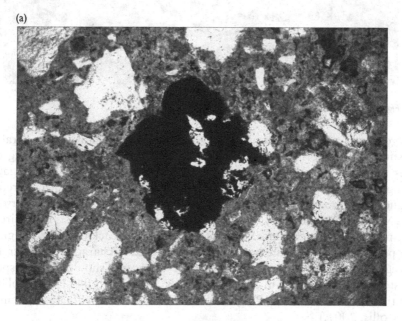

Figure 3.29 (a) Fragment of pyrite introduced as part of the aggregate in the concrete. Pyrite is characterized by its opacity in thin section and this is usually a diagnostic feature. Some silica particles are ingrained in the larger pyrite fragment. Width of the field of view = 1.55 mm. (b) Fragment of pyrite which has begun to oxidize in concrete. Oxidation caused sulphates to be released with consequent sulphate attack on the matrix. The process is associated with a volumetric expansion and this may have caused the fracture visible in some of the sand particles around the pyrite grain. Width of the field of view = 3.1 mm (photomicrographs courtesy of Prof. Ballim, University of the Witwatersrand).

(b)

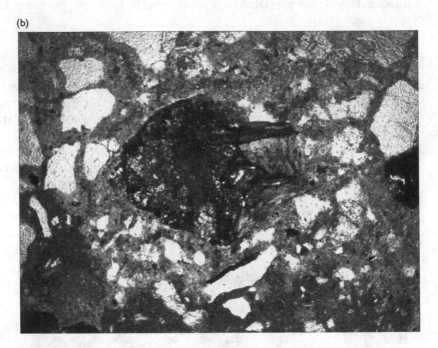

Figure 3.29 (Continued)

A good petrographic analysis does more than simply identify the minerals, constituents, and origin of aggregates. It also considers how the aggregates might perform in concrete. Consequently, the competent aggregate petrographer needs knowledge of the chemistry and physics of concrete. Frequently, petrography may also extend to examination of concrete itself (ASTM C856; St John *et al.*, 1998), particularly when a diagnosis is required to explain poor performance. This section provides background to aggregate petrography, and indicates how it can be used to characterize concrete aggregates and their properties. (The interested reader should consult specialist texts for more detail: ASTM C295; BS 812: Part 104: 1994; Dolar-Mantuani, 1983; Roberts, 1990; Mielenz, 1994; St John *et al.*, 1998; Smith and Collis, 2001.)

Petrographic composition and examination of aggregates

The petrographic composition of an aggregate describes the mineralogical nature of the constituent particles. For instance, granite aggregate will be composed essentially of the minerals quartz, feldspar, and mica. Petrographic examination provides information on (a) the nature and properties (i.e. the composition) of the constituent rocks and minerals, including the

presence and amount of any undesirable constituents, and (b) the degree
of weathering. Particularly when a new or untried source of aggregate is
considered for use in concrete, a thorough petrographic examination to
determine the composition is needed. This is also true for existing sources
when the material is variable or when a distinctly different type of rock
is encountered. Failure to undertake such an examination could result in
the aggregate being used when it is in fact unsuitable for concrete with
consequent poor performance. For example, argillaceous carbonate aggregates from Indiana caused pitting and pop-outs in concrete roads exposed
to winter conditions (West and Shakoor, 1984). Petrographic examination
also indicates the limitations of the aggregate for use in concrete. On the
positive side, petrographic examination can sometimes permit an otherwise
unsuitable material to be beneficiated to avoid or eliminate undesirable
portions or constituents.

A correct description of aggregate composition includes both the overall
composition of the rock and mineral types present, and the nature of the
minerals in the individual grains. Thus, an aggregate deriving from mainly
a quartzitic source with a small amount of interbedded shale would require
a description of the overall proportions of quartzite and shale as well as
the proportion of shale material that might be altered due to weathering
or other causes. The examination also establishes the possible presence of
undesirable minerals and contaminants such as sulphates, depending on
their solubility, as well as the presence of coatings. A very important part
of the examination is to identify and quantify potentially alkali-reactive
constituents and suggest further tests to determine their severity. What
is required therefore is an aggregate evaluation that identifies aspects of
engineering significance and the possible effects of the various constituents
on the properties of concrete, and not merely a petrological description.

The degree of weathering of the aggregate minerals also needs to be
identified. Weathering of certain constituents may give rise to secondary or
tertiary minerals that may be unsuitable in concrete (e.g. active clay minerals
such as smectites), or to additional microporosity due to leaching of constituents which can lead to unsoundness or other unsatisfactory behaviour
in concrete. As an example, aggregates that may be susceptible to freezing
and thawing such as finely porous and highly weathered or altered rocks
must be identified.

The microtexture and microstructure of the aggregate should be described
since this may have a strong influence on aggregate performance. Matters
of relevance are, amongst others, the presence of laminations representing
planes of weakness, and the internal integrity of the material, for example
intergranular strength which can influence friability of the particles. The
examination should also determine the proportions of particles of various
shapes, particularly those that are flat and elongated, since these will have
adverse effects on mixing water requirements.

Figure 3.29 gave an example of the use of petrography to detect a deleterious substance (pyrite) in aggregate. A further example is given in the
micrograph in Figure 3.30 of greywacke aggregate from the Cape Peninsula
in South Africa. This rock is well known for its susceptibility to alkali–
aggregate reaction (see Chapter 6). The problematic constituent is strained
quartz crystals, arrowed in the figure, which are revealed by the use of
crossed polars in a petrographic microscope.

Sampling and analysis techniques

A competent petrographer will be fully acquainted with sampling and analysis techniques, and the concrete engineer or technologist would be well
advised to retain the services of a professional when necessary. (Sampling
of aggregates was covered in Chapter 2.) It is stating the obvious to say that
the reliability of an examination is only as good as the sampling techniques,
analytical methods, and competence and experience of the petrographer.
Variability will always be an important feature of aggregate production, and
regular sampling and examination in both the field and the laboratory are
strongly recommended to ensure quality control of aggregate production.

Examination techniques almost invariably involve optical microscopy of
polished and thin sections, and of grain mounts in immersion oils from
which photomicrographs can be taken. Figures 3.29 and 3.30 give examples of optical photomicrographs. Techniques also include X-ray diffraction

Strained
quartz
crystals

Figure 3.30 Petrographic micrograph of a Cape Peninsula greywacke which is alkali
susceptible. Strained quartz grains are shown enclosed in a fine-grained
biotite. This type of feature is not detectable without petrographic
examination (photomicrograph courtesy of Dr C. Stowe, University of
Cape Town).

(XRD) analysis, differential thermal analysis (DTA), infrared spectroscopy, scanning electron microscopy (SEM) with energy-dispersive X-ray analysis (EDX), and back-scattered electron imaging (BSE) of polished sections. Figure 3.31 shows a back-scattered image, in this case of an andesite aggregate with silica-fume bearing paste cast against it, to carry out fracture studies on the interface (Diamond *et al.*, 1992). The dense character of the paste close to the uncracked interface, even around projecting feldspar crystals at 'A' and 'B', can be seen. Conventional petrographic equipment comprises hand lenses and a stereoscopic polarizing microscope, a facility to prepare polished and thin sections, and a camera for taking photomicrographs. Finally, a comprehensive and professional report is required, which seeks to quantify as many of the important aspects as possible. Guides to good reporting are given in ASTM C295 and in Mielenz (1994) and Smith and Collis (2001). ASTM C295 indicates that the report should contain at least the following elements deriving from the examination: means to identify the sample as to its source and proposed use; test procedures employed; essential data on composition and properties of the aggregates such as particle shape, texture, and possible coatings, as well as a description of the nature and features of each important constituent of the sample; description of any unfavourable effects in concrete of deleterious constituents should they

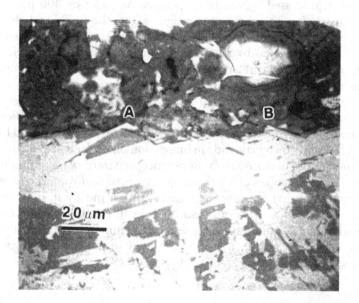

Figure 3.31 BSE image of andesite aggregate with silica-fume bearing paste cast against it. The aggregate is in the lower part of the picture. 'A' and 'B' are projecting feldspar crystals. The paste has a dense character, but contains some shrinkage cracks (from Diamond *et al.*, 1992; reprinted courtesy RILEM).

be found; and recommendations for any additional petrographic, chemical, physical, or geological investigations that may be required to further assess adverse properties indicated by the examination. Tables and photographs should accompany the report as far as possible. (Mielenz [1994] provides considerable detail on the requirements for petrographic examination and reporting.)

Standard petrographic guides

ASTM C295: Standard guide for petrographic examination of aggregates for concrete

ASTM C295 outlines 'the extent to which petrographic techniques should be used, the selection of properties that should be looked for, and the manner in which such techniques may be employed in the examination of samples of aggregates for concrete' (Clause 1.2). It also outlines procedures for the petrographic examination of both unprocessed and processed materials for aggregates. The Standard states that 'the specific procedures employed in the petrographic examination of any sample will depend to a large extent on the purpose of the examination and the nature of the sample'. The minimum sample size for a processed aggregate is specified as 45 kg or 300 pieces, whichever is larger. For processed coarse aggregates, the individual sieve fractions of the aggregate are examined and the proportion of constituent particle types are determined by particle counting, based on petrological composition, condition, and presence of coatings. The examination is done visually, with more complex methods such as thin-section microscopy or X-ray diffraction analysis being applied as necessary. For fine aggregates a quantitative examination is undertaken on each sieve fraction by particle counting under a stereoscopic microscope for fractions over 600 μm, and by examination of samples on a glass slide in immersion oil under a petrological microscope for samples under 600 μm. In practice, separate examination of all the individual size fractions of the fine aggregate is typically unnecessary. It has also been found that for most aggregates and for most purposes, the procedure recommended for the examination of the fine aggregate does not permit adequate identification, and thin-section examination is invariably required. For a full description of the terms used in concrete aggregate petrography, see ASTM C294 or BS 812: Part 104: 1994.

BS 812: Part 104: 1994: Testing aggregates. Procedure for qualitative and quantitative petrographic examination of aggregates

Smith and Collis (2001) provide the following summary for use of BS 812: Part 104: 1994.

The prime use of the method is for the examination of processed aggregates. The method was subjected to precision trials and a proficiency scheme is now administered by the Geological Society Engineering Group. The minimum size of sample for dispatch to the laboratory is shown in *Table 2.8* (*this book*). All aggregates (coarse, fine and all-in) are subjected to an initial qualitative examination to determine the aggregate type and its general characteristics. The sample is then reduced to a representative test portion for quantitative examination. The minimum mass of this test portion should be as shown in *Table 2.9* (*this book*). The quantities given have been calculated to achieve an accuracy of ±10 per cent relative for a constituent present at 20 per cent. A nomogram is given for determining the minimum test portion size required to achieve a ±10 per cent relative error for constituents present at other concentrations (*see Figure 2.11, this book*).

For coarse aggregates the quantitative examination is undertaken on the separate sieve fractions as for ASTM C295, discriminating the particles visually based on significant petrological composition and/or condition; weighing is used rather than particle counting. A more detailed examination may be carried out as necessary; in practice a thin-section examination of selected particles of each significant type is normally required.

For fine aggregate, the method requires that the reduced test portion be split into two size fractions on a 1.18 mm sieve. The coarser fraction retained on the 1.18 mm sieve is further divided into >5 mm, 5–2.36 mm and 2.36–1.18 mm fractions. The quantitative analysis of these individual size fractions is determined by visual hand sorting as for the coarse aggregate, a stereoscopic microscope being used to aid the identification of particles. It will generally be necessary to prepare a thin-section of selected particles for detailed identification. This is undertaken by embedding the particles in a suitable resin. For the fraction passing the 1.18 mm sieve the quantitative analysis is undertaken by embedding the material in resin and preparing a thin-section for point-count analysis under a petrological microscope.

(Smith and Collis, 2001)

Closure

To conclude this chapter, it needs to be stated that the properties of aggregates must be properly understood and their limitations appreciated for their successful use in concrete. Aggregates can be characterized by applying standard and non-standard tests and techniques. This is essential in the case of new or untried sources but is equally important for ongoing quality control of aggregate production. Solid experience with aggregates and an established record of their satisfactory performance in concrete remain important for the engineer who seeks to use aggregates to their full potential in concrete. Knowledge of aggregate properties on their own is insufficient – what is required is a thorough appreciation of how the aggregates behave in concrete as a composite material. This is the subject of the following three chapters.

References

Aïtcin, P.-C. (1998) *High Performance Concrete*, London: E&FN Spon.

Alexander, M.G. (1985) 'Towards standard tests for abrasion resistance of concrete', *Materials and Structures*, 18(106): 297–307.

Alexander, M.G. (1990) Properties of aggregates in concrete. Report on Phase 1 testing of concretes made with aggregates from 13 different quarries, and associated design recommendations, Unpublished Hippo Quarries Report, 147pp.

Alexander, M.G. (1993) 'Two experimental techniques for studying the effects of the interfacial zone between cement paste and rock', *Cement and Concrete Research*, 13(3): 567–575.

Alexander, M.G. and Davis, D.E. (1991) 'Aggregates in concrete – a new assessment of their role', *Concrete Beton*, 59: 10–20.

American Association of State Highway and Transportation Officials (1982) *Standard Method of Test for Soundness of Aggregates by Freezing and Thawing*, AASHTO T103-78, Washington, DC: American Association of State Highway and Transportation Officials.

American Concrete Institute Committee 221 (2001) 'Guide for use of normal weight and heavyweight aggregates in concrete', ACI 221R-96 (Re-approved 2001), *American Concrete Institute Manual of Concrete Practice*, Farmington Hills, MI: American Concrete Institute.

Ballim, Y. and Alexander, M.G. (1988) 'Strength of concrete made with andesite aggregates', *Concrete Beton*, 50: 10–17.

Bazant, Z.P. and Planas, J. (1997) *Fracture and Size Effect in Concrete and Other Quasibrittle Materials*, London & New York: CRC Press.

Caquot, A. (1937) 'Rôle des Matériaux Inertes dans le Béton' (Role of Inert Materials in Concrete), *Mémoire de la Société des Ingénieurs Civils de France* (in French), Societe des Ingenieurs de France.

Collins, R.J. and Bareham, P.D. (1987) 'Alkali–silica reaction: Suppression of expansion using porous aggregate', *Cement and Concrete Research*, 17(1): 89–96.

CRD-C 125-63, *Method of Test for Coefficient of Linear Thermal Expansion of Coarse Aggregate (Strain-Gage Method)*, Vicksburg, MI: US Army, Corps of Engineers, 1963.

CRD-C 124-73, *Method of Test for Specific Heat of Aggregates, Concrete, and Other Materials (Method of Mixtures)*, Vicksburg, MI: US Army, Corps of Engineers, 1973.

CRD-C 119-91, *Standard Test Method for Flat or Elongated Particles in Coarse Aggregate*, Vicksburg, MI: US Army, Corps of Engineers, 1991.

CRD-C 120-94, *Test Method for Flat and Elongated Particles in Fine Aggregate*, Vicksburg, MI: US Army, Corps of Engineers, 1994.

Davis, D.E. (1975) The concrete-making properties of South African aggregates, PhD Thesis, University of the Witwatersrand, Johannesburg, Vol. 1, Ch. 3, Part 1a.

de Larrard, F. (1999) *Concrete Mixture Proportioning: A Scientific Approach*, London: E&FN Spon.

Dewar, J.D. (1986) 'Ready-mixed concrete mix design', *Municipal Engineering*, 3(1): 35–43.

Dewar, J.D. (1999) *Computer Modelling of Concrete Mixtures*, London: E&FN Spon.

Dewar, J.D. and Anderson, R. (1992) *Manual of Ready-Mixed Concrete*, 2nd edn, Glasgow: Blackie & Son.

Diamond, S., Mindess, S., Lie Qu and Alexander, M.G. (1992) 'SEM investigations of the contact zones between rock surfaces and cement paste', *Proceedings RILEM Int. Conference on Interfaces in Cementitious Composites*, Toulouse, France, London: E&FN Spon, 13–22.

Dilek, U. and Leming, M.L. (2004) 'Relationship between particle shape and void content of fine aggregate', *Cement, Concrete and Aggregates*, 26(1): 14–20.

Dolar-Mantuani, L. (1983) *Handbook of Concrete Aggregates*, Park Ridge, NJ: Noyes Publications.

Eglinton, M. (1987) *Concrete and Its Chemical Behaviour*, London: Thomas Telford.

Forster, S.W. (1994) 'Soundness, deleterious substances, and coatings', *Significance of Tests and Properties of Concrete and Concrete Making Materials*, ASTM STP 169C: 411–421, West Conshohocken, PA: American Society for Testing and Materials.

Fulton's Concrete Technology (2001) Eds B.J. Addis and G. Owens, 8th edn, Midrand: Cement and Concrete Institute.

Galloway, J.E. (1994) 'Grading, shape, and surface properties', *Significance of Tests and Properties of Concrete and Concrete Making Materials*, ASTM STP 169C: 401–410, Philadelphia: American Society for Testing and Materials.

Garboczi, E.J. (2002) 'Three-dimensional mathematical analysis of particle shape using X-ray tomography and spherical harmonics: Application to aggregates used in concrete', *Cement and Concrete Research*, 32(10): 1621–1638.

Gaynor, R.D. and Meininger, R.C. (1984) 'Evaluating concrete sands', *Concrete International*, 5(12): 53–60.

Goguel, R. (1994) 'Alkali release by volcanic aggregates in concrete', *Cement and Concrete Research*, 25(4): 841–852.

Grieve, G.R.H. (2001) 'Aggregates for concrete', in B.J. Addis and G. Owens (eds) *Fulton's Concrete Technology*, 8th edn, Midrand: Cement and Concrete Institute.

Helmuth, R.A. (1994) 'The nature of concrete', *Significance of Tests and Properties of Concrete and Concrete Making Materials*, ASTM STP 169C: 5–14, West Conshohocken, PA: American Society for Testing and Materials.

Higgs, N.B. (1987) 'Chlorite: A deleterious constituent with respect to freeze–thaw durability of concrete aggregates', *Cement and Concrete Research*, 17(5): 793–804.

ISRM (1988) 'Suggested methods for determining the fracture toughness of rock', *International Journal of Rock Mechanics and Mining Science*, 25(2): 71–96.

Johansen, V. and Andersen, P.J. (1991) 'Particle packing and concrete properties', in J. Skalny and S. Mindess (eds) *Materials Science of Concrete II*, Westerville, OH: The American Ceramic Society.

Johnston, C.D. (1970a) 'Deformation of concrete and its constituent materials in uni-axial tension', *HRB Record No. 324*, Washington, DC: Highway Research Board.

Johnston, C.D. (1970b) 'Strength and deformation of concrete in uniaxial tension and compression', *Magazine of Concrete Research*, 22(70): 5–16.

Jones, M.R., Zheng, L. and Newlands, M.D. (2002) 'Comparison of particle pack-ing models for proportioning concrete constituents for minimum voids ratio', *Materials and Structures*, 35(249): 301–309.

Kaneuji, M., Winslow, D. and Dolch, W.L. (1980) 'The relationship between an aggregate's p.s.d. and its freeze thaw durability in concrete', *Cement and Concrete Research*, 10(3): 433–442.

Kaplan, M.F. (1959) 'Flexural and compressive strength of concrete as affected by the properties of coarse aggregates', *Journal American Concrete Institute*, 55: 1193–208.

Keen, R.A. (1979) 'Impurities in aggregates for concrete', *C & CA Advisory Note No. 18*, Slough, UK: Cement and Concrete Association.

Kosmatka, S.H., Panarese, W.C., Gissing, K.D. and Macleod, N.F. (1995) *Design and Control of Concrete Mixtures*, 6th Canadian edn, Ottawa, ON: Canadian Portland Cement Association.

Kosmatka, S.H., Kerchoff, B., Panarese, W.C., Macleod, N.F. and McGrath, R.J. (2002) *Design and Control of Concrete Mixtures*, Engineering Bulletin 101, 7th Canadian edn, Ottawa, ON: Cement Association of Canada.

Kumar V, S. and Santhanam, M. (2003) 'Particle packing theories and their appli-cation in concrete mixture proportioning: A review', *Indian Concrete Journal*, 77(9): 1324–1331.

Landgren, R. (1994) 'Unit weight, specific gravity, absorption, and surface mois-ture', *Significance of Tests and Properties of Concrete and Concrete Making Materials*, ASTM STP 169C: 421–428, West Conshohocken, PA: American Soci-ety for Testing and Materials.

Lane, D.S. (1994) 'Thermal properties of aggregates', *Significance of Tests and Prop-erties of Concrete and Concrete Making Materials*, ASTM STP 169C: 438–445, West Conshohocken, PA: American Society for Testing and Materials.

Li, L., Chan, P., Zollinger, D.G. and Lytton, R.L. (1993) 'Quantitative analysis of aggregate shape based on fractals', *ACI Mater. J.*, 90(4): 357–365.

Meininger, R.C. (1994) 'Degradation resistance, strength, and related properties', *Significance of Tests and Properties of Concrete and Concrete Making Materi-als*, ASTM STP 169C: 388–400, West Conshohocken, PA: American Society for Testing and Materials.

Mielenz, R.C. (1994) 'Petrographic evaluation of concrete aggregates', *Significance of Tests and Properties of Concrete and Concrete Making Materials*, ASTM STP 169C: 341–364, West Conshohocken, PA: American Society for Testing and Materials.

Mindess, S. and Alexander, M.G. (1995) 'Mechanical phenomena at cement/aggregate interfaces', in J. Skalny and S. Mindess (eds) *Materials Science of Con-crete IV*, Westerville, OH: The American Ceramic Society.

Mindess, S., Young, J.F. and Darwin, D. (2003) *Concrete*, NJ: Prentice Hall.

Neville, A.M. (1995) 'Properties of aggregates' Chapter 3, *Properties of Concrete*, 4th and final edn, Harlow: Longman.

Okkenhaug, K. and Gjorv, O.E. (2003) 'Frost resistance of hardened concrete: Determining the effects fine aggregates have on the air–void system', *Concrete International*, 25(9): 49–54.

Orchard, D.F. (1973) *Concrete Technology*, Volume 3: *Properties and Testing of Aggregates*, 3rd edn, London: Applied Science Publishers.

Orr, C.M. (1979) 'Rapid weathering dolerites', *The Civil Engineer in South Africa*, 21(7): 161–167.

Ozol, M. (1978) 'Shape, surface texture, surface area, and coatings', *Significance of Tests and Properties of Concrete and Concrete Making Materials*, ASTM STP 169C: 605–607, West Conshohocken, PA: American Society for Testing and Materials.

Perrie, B. (1994) 'Flakiness Index', *Commentary on SABS 1083:1994*, Midrand: Cement and Concrete Institute.

Popovics, S. (1979) *Concrete-Making Materials*, Washington, DC: Hemisphere Pub. Corp.

Porter, J.J. (1962) 'Electron microscopy of sand surface texture', *Journal of Sedimentary Petrology*, 32(1): 124–135.

Powers, M.C. (1953) 'A new roundness scale for sedimentary particles', *Journal of Sedimentary Petrology*, 23(2): 117–119.

Powers, T.C. (1958) 'Structure and physical properties of hardened portland cement paste', *Journal American Ceramic Society*, 41(1): 1–6.

Powers, T.C. (1968) *The Properties of Fresh Concrete*, New York: Wiley.

Roberts, C.A. (1990) 'Petrographic examination of aggregates and concrete in Ontario', *Petrography Applied to Concrete and Concrete Aggregates*, ASTM STP 1061: 5–31, West Conshohocken, PA: American Society for Testing and Materials.

Rogers, C.A. and Senior, S.A. (1994) 'Recent developments in physical testing of aggregates to ensure durable concrete', *Proceedings of Engineering Foundation Conference, Advances in Cement and Concrete*, M.W. Grutzeck and S.L. Sarkar (eds), New York: ASCE, 338–361.

Rogers, C.A., Senior, S.A. and Booth, D. (1989) 'Development of an unconfined freeze–thaw test for coarse aggregates', *Ontario Ministry of Transportation and Communications, Engineering Materials Report EM-8*, Toronto: Ontario Ministry of Transportation and Communications.

Rogers, C.A., Bailey, M. and Price, B. (1991) 'Micro-Deval test for evaluating the quality of fine aggregate for concrete and asphalt', *Ontario Ministry of Transportation, Engineering Materials Office, Report EM-96*, Toronto: Ontario Ministry of Transportation.

Senior, S.A. and Rogers, C.A. (1991) 'Laboratory tests for predicting coarse aggregate performance in Ontario', *Record No. 1301*, Washington: Transportation Research Board, 97.

Shacklock, B.W. (1959) 'Comparison of gap- and continuously-graded concrete mixes', *C & CA Technical Report TRA/240*, Slough, UK: Cement & Concrete Association.

Shilstone, J.M. Sr (1990) 'Concrete mixture optimisation', *Concrete International*, 12(6): 33–39.

Sims, I. and Brown, B. (1998) 'Concrete aggregates', in P.C. Hewlett (ed.) *Lea's Chemistry of Cement and Concrete*, 4th edn, London: Arnold.

Smith, M.R. and Collis, L. (2001) *Aggregates: Sand, Gravel and Crushed Rock Aggregates for Construction Purposes*, Engineering Geology Special Publications 17, London: Geological Society.

South African Council for Scientific and Industrial Research (SA-CSIR) (1959) 'Corrosion of concrete sewers', *CSIR Research Report 163*, Pretoria: South African Council for Scientific and Industrial Research.

Stark, D. (1976) 'Characteristics and utilisation of coarse aggregates associated with D-cracking', *Living with Marginal Aggregates*, ASTM STP 597: 45–58, West Conshohocken, PA: American Society for Testing and Materials.

Stark, D. and Bhatty, M.S.Y. (1986) 'Alkali silica reactivity: Effect of alkali in aggregate on expansion', *Alkalis in Concrete*, ASTM STP 930: 16–30, West Conshohocken, PA: American Society for Testing and Materials.

St John, D.A., Poole, A.W. and Sims, I. (1998) *Concrete Petrography*, London: Arnold.

Teychenné, D.C. (1978) 'The use of crushed rock aggregates in concrete', Garston, UK: Building Research Establishment.

Toufar, W., Born, M. and Klose, E. (1976) 'Contribution of optimisation of components of different density in polydispersed particle systems', *Freiberger Booklet A 558*: 29-44, VEB Deutscher Verlag für Grundstoffindustrie (in German).

Verbeck, G.J. and Haas, W.E. (1950) 'Dilatometer method for determination of thermal coefficient of expansion of fine and coarse aggregate', *Proceedings Highway Research Board*, 30: 187.

Waddell, J.J. (ed.) (1974) *Concrete Construction Handbook*, 2nd edn, New York: McGraw-Hill.

Weinert, H.H. (1980) *The Natural Road Construction Materials of Southern Africa*, Pretoria: Academica.

West, T.R. and Shakoor, A. (1984) 'Influence of petrography of argillaceous carbonates on their frost resistance in concrete', *Cement, Concrete, Aggregates*, 6(2): 84–9.

Wills, M.H. (1967) 'How aggregate particle shape influences concrete mixing water requirements and strength', *Journal of Materials*, 2(4): 843–865.

Winslow, D. (1994) 'The pore system of coarse aggregates', *Significance of Tests and Properties of Concrete and Concrete Making Materials*, ASTM STP 169C: 429–437, West Conshohocken, PA: American Society for Testing and Materials.

Wright, P.G.F. (1955) 'A method of measuring the surface texture of aggregate', *Magazine of Concrete Research*, 7(21): 151–160.

Young, J.F., Mindess, S., Gray, R.J. and Bentur, A. (1998) *The Science and Technology of Civil Engineering Materials*, NJ: Prentice Hall.

Further reading

American Concrete Institute Education Bulletin E1-99 (1999) *Aggregates for Concrete*, Farmington Hills, MI: American Concrete Institute.

American Society for Testing and Materials (1994) *Significance of Tests and Properties of Concrete and Concrete-Making Materials*, ASTM, STP-169A (1966), STP 169B (1978), STP 169C (1994) West Conshohocken, PA: American Society for Testing and Materials.

Bloem, D.L. (1966) 'Concrete aggregates – soundness and deleterious substances', *Significance of Tests and Properties of Concrete and Concrete-Making Materials*, ASTM, STP-169A, West Conshohocken, PA: American Society for Testing and Materials.

Brown, E.T. (ed.) (1981) *Rock Characterisation – Testing and Monitoring – ISRM Suggested Methods*, International Society for Rock Mechanics (ISRM), Commission on Testing Methods, Oxford: Pergamon Press.

Duncan, N. (1969) *Engineering Geology and Rock Mechanics*, Vol. 1, London: Leonard Hill.

Erlin, B. and Stark, D. (1990) 'Petrography applied to concrete and concrete aggregates', *Symposium on Petrography of Concrete and Concrete Aggregates*, ASTM STP 1061, West Conshohocken, PA: American Society for Testing and Materials.

Fookes, P.G. and Revie, W.A. (1982) 'Mica in concrete – a case history from Eastern Nepal', *Concrete*, 16(3): 12–16.

ISRM have a series of 'Suggested Methods' for determining properties of rock samples. See *International Journal of Rock Mechanics and Mining Science* for relevant methods, e.g. 9(10) (Uniaxial Compressive Strength and Deformability – 1972; Water Content, Porosity, Density, Absorption – 1972), 14(3) (Tensile Strength – 1977), 32(1) (Mode I Fracture Toughness – 1995).

Mather, K. and Mather, B. (1950) 'Method of petrographic examination of aggregates for concrete', *Proceedings ASTM, ASTEA*, 50: 1288–1312.

Norbury, D.R. (1984) 'The point-load test, site investigation practice: Assessing BS 5930', in A.B. Hawkins (ed.), *Proceedings of the 20th Regional Meeting of the Engineering Group*, UK: Guildford.

Chapter 4

Aggregates in plastic concrete

It would be no exaggeration to state that aggregates have their greatest influence on the performance of plastic or fresh concrete. Aggregate properties such as shape, surface texture, and grading largely govern the way in which fresh concrete is proportioned, and determine how easily the concrete can be mixed, transported, placed, compacted, and given a suitable finish. Aggregates form the 'link' between the intent of the mix design (i.e. the proportioning of the mix constituents) and the fulfilment of that intent in the hardened state. For, no matter how sophisticated the mix formulation or the binders used, unless the plastic mix can be thoroughly compacted without segregation or excessive bleeding to produce a dense homogeneous material, the concrete will not perform successfully. Aggregates strongly influence the outcome of this process. An understanding of the role of aggregates in plastic concrete is therefore fundamental to the production of good concrete.

Plastic concrete is characterized by several distinct, yet interrelated aspects. Workability is probably the most important, but this is influenced by the water requirement, and both of these are functions of the aggregates. Cohesiveness of the mix is an aspect of workability, but cohesiveness also affects segregation, of which bleeding is a special case. Since the properties of the aggregates influence these aspects, the problem is an interrelated one. This precludes the discussion of any one property in isolation of another, and of necessity there will be overlap. A holistic view of the subject must be taken.

Chapter 3 focused on aggregate properties and characterization *per se*. This chapter deals with the influence of aggregate properties on the performance of plastic concrete. Thus, the focus is sharpened to look at *aggregates in concrete*, and to consider how they influence the plastic properties of the material.

Characterizing the plastic properties of concrete

To outline the role of aggregates in plastic concrete, it is necessary to discuss first how the plastic properties of concrete are characterized, using various approaches and tests to measure them. Most tests are empirical, but there are newer approaches that deal more fundamentally with the rheology of the material. Some definitions applied to plastic concrete are appropriate.

Plastic State of concrete refers to the state in the period between mixing and initial set when it begins to stiffen appreciably. In the plastic state, the concrete can flow and be moulded to a desired shape.

Rheology is defined as the science of the deformation and flow of matter. It covers matter in both solid and non-solid (plastic) forms, and deals with the relationships between stress, deformation, and time, with allowance for other factors such as temperature. For plastic concrete, rheology refers primarily to the flow properties.

Flow refers to the deformation of plastic concrete whereby it can be moved and moulded and made to conform to the formwork shape, without segregation or the use of excessive mechanical effort. A 'flowing' concrete is one in which the force of gravity will largely achieve the required deformation and compaction, and work done is mainly of an internal nature. In a 'non-flowing' concrete, external mechanical effort or work is required to achieve the task.

Workability refers to the ease of mixing, transporting, placing, compacting, and finishing a concrete without segregation of its constituents. It has been defined as the amount of useful internal work required to produce full compaction (Glanville *et al.*, 1947; Neville, 1995).

Consistence (or consistency) describes mobility or ease of flow, and is related to the wetness or dryness of the mix (*Fulton*, 2001). However, concretes of the same consistence may vary in workability. It is sometimes used as an alternative for 'workability', particularly for mortars (Sims and Brown, 1998), and is usually measured by the slump test (see p. 169). (Note that in North American terminology, it is the workability that is most commonly measured by the slump test.)

Segregation is the separation of coarse particles from the mortar phase of the concrete, and the collection of these particles at the perimeter or base of a concrete placement.

Bleeding can be considered as a special case of segregation, in which the solid particles settle in a mix, displacing the water upwards and leaving a layer of water on top of the concrete. Bleeding also results in trapping mix water under large aggregate particles or reinforcing bars.

Thixotropy refers to the property of a material whereby it 'flows' when external work is performed on it, but remains static and retains its shape when undisturbed. For plastic concrete, this property is manifested as the concrete flowing when external compactive effort (e.g. vibration) is applied, but not flowing without this effort.

Two other definitions are also necessary.

Water Requirement of a mix is the quantity of water (in litres per cubic metre (l/m^3)) required to produce concrete of a desired slump, with the given aggregates and binder, without use of admixtures.

Water Demand, or *Standard Water Requirement* (*SWR*) of a mix is the quantity of water (in l/m^3) required to produce concrete with a slump of 75 mm, using aggregates of a nominal maximum size of 19 mm and ordinary portland cement at a w/c ratio of 0.6, and without the use of admixtures (Grieve, 2001).

In conventional concretes, the water requirement is substantially controlled by the aggregates, particularly the fine aggregate. Standard water requirement (SWR) of a mix is essentially constant over a wide range of portland cement contents (up to about $400 \, kg/m^3$). Water requirement depends on the maximum size of the coarse aggregate, particle shape and texture of the aggregates, and the aggregate grading. The SWR is used in South African concrete practice as an index of aggregate quality. Since mix-water requirement is primarily a function of the aggregate properties, determining water requirement under a set of standard conditions permits various aggregate types and combinations to be compared and rated.

Binders other than portland cement also influence water requirement. In general, concretes with slag blends up to about 50 per cent ground granulated blastfurnace slag (GGBS) show little if any difference in slump compared with an equivalent Ordinary Portland Cement (OPC) mix, although they may compact more easily. However, fly ash (FA) blends at FA contents above about 20 per cent show substantial improvement in workability and slump. For constant slump, this translates into appreciable water reductions (as much as 10–15 per cent) for FA mixes compared with OPC mixes. Blends of microsilica (also known as condensed silica fume, CSF) on the other hand tend to make a mix very sticky and it is imperative that a superplasticizing admixture is used to offset this effect, as well as to ensure thorough dispersion of the silica fume in the mix.

Empirical approaches

Empirical approaches and their accompanying test methods have arisen mainly because of the practical need to characterize the plastic properties of concrete on the construction site, where the primary focus is on quality control (Tattersall, 1991; Bartos, 1992). For example, the water content of a mix is frequently controlled by measuring the slump of the concrete at the mixer. For a mix of given cement content, the water content effectively determines the strength of the concrete, and therefore proper control using a reliable site test is very important. Empirical approaches have been necessary simply because it is impossible (or at least highly impractical) to attempt to measure all the various facets of plastic behaviour discussed earlier. Simple but robust methods are required to allow practical concreting operations to be performed with confidence and consistency.

Mixes of 'normal' consistence, or workability

Suitable for vibration or moderate manual compaction; low to high workability.

SLUMP TEST

(BS 1881: Part 102: 1983; ASTM C143; SANS 5862-1)
The slump test is the oldest and still the most widely used workability test, having first appeared as an ASTM Standard in 1922. It employs a mould in the shape of a truncated cone, 300-mm high, 200 mm in diameter at the base and 100 mm in diameter at the top, into which the concrete is compacted by rodding. The apparatus is shown in Figure 4.1a. The difference in height of fresh concrete before and after removal of the mould is measured, and called the 'slump' of the concrete. The test is limited to ordinary fresh concrete, generally within the slump range of 20–180 mm, which covers low to high workability in terms of construction requirements. It is not suitable for non-cohesive mixes such as lean and no-fines mixes. The slump test is extensively used as a means of rapid and continual checking of uniformity of fresh concrete supply. It is *not* a measure of workability as such. Its major disadvantages are that it has no basis in fundamental rheology and is operator sensitive. However, there is hardly a well-run construction site in the world where the slump test may not be seen in regular operation.

COMPACTING (OR COMPACTION) FACTOR TEST

(BS 1881: Part 103: 1983; SANS 5862-4)
The test is meant to measure the degree of compaction achieved by applying a standard amount of work to a sample of fresh concrete, and is suitable for mixes of low to high workability (i.e. similar range to the slump test). A sample of concrete drops progressively through two upper conical hoppers into a steel cylinder, whereafter it is struck off, weighed, and the mass compared with that of a fully compacted sample of the same concrete in the cylinder. The test is seldom if at all in use now (indeed it has been deleted by ASTM), and suffers from the disadvantage that the energy used in compaction is very different from that applied in normal vibration compaction, thus not truly characterizing the practical compactibility of modern mixes.

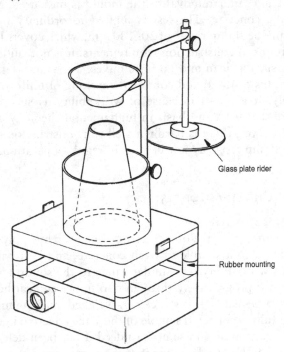

Glass plate rider

Rubber mounting

Figure 4.1 Apparatuses for various tests for plastic properties of concrete: (a) Slump test; (b) Vebe test; (c) Flow-table test.

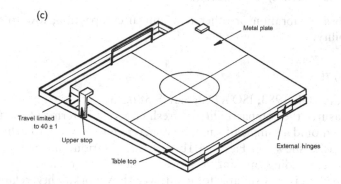

(c)

Metal plate

Travel limited
to 40 ± 1

Upper stop

External hinges

Table top

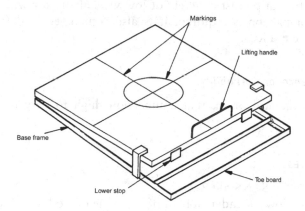

Markings

Lifting handle

Base frame

Lower stop

Toe board

Typical flow table

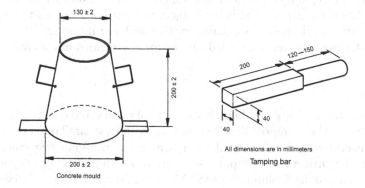

130 ± 2

200 ± 2

200 ± 2

Concrete mould

200

120—150

40

40

All dimensions are in millimeters

Tamping bar

Figure 4.1 (Continued).

Mixes of low consistence, or workability

Suitable only for vibration or other mechanical compaction; low to very low workability.

VeBe (OR VEBE) TEST

(BS 1881: Part 104: 1983; ISO 4110; SANS 5862-3)
The test measures the compactibility of fresh concrete in terms of the time required to remould a sample of slumped concrete into a standard cylinder. The apparatus is shown in Figure 4.1b. It uses vibration and pressure to achieve this remoulding, making it more applicable to modern vibrated mixes. However, it is only suitable for low to very low workability, relatively dry and stiff mixes which have zero or very low slump. It is applied most widely in the production of precast concrete of low workability for which heavy mechanical compaction is required. It is also sometimes used for fibre-reinforced concrete (FRC).

Mixes of high consistence, or workability

Suitable for light mechanical or manual compaction; high to very high workability.

FLOW-TABLE (FLOW) TEST

(BS 1881: Part 105: 1983; SANS 5862-2)
The test measures the 'flow' or radial spread of a sample of fresh concrete which has first been moulded in a truncated cone mould and allowed to slump. After slumping, the sample receives a controlled amount of jolting. The apparatus is shown in Figure 4.1c. It is suitable for mixes of high to very high workability, and is now used for superplasticized and flowing mixes. The test has been criticized as being subject to operator error and insensitivity to different degrees of workability (Dimond and Bloomer, 1977).

Table 4.1 provides recommended ranges for the various test methods.

Other tests

There are several other workability tests, some of them standardized, but not in general use. They comprise the Compaction (Walz) test, similar to the Compaction Factor test, the Flow (LCL) test, which like the Vebe test relies on vibration, the Orimet test for rapid assessment of very highly workable, flowing concrete, and the K-Slump test (ASTM C1362) suitable only for mixes of medium to very high workability. Other tests have been developed for special mixes and concretes, such as FRC and self-compacting concrete (SCC).

Table 4.1 Recommended ranges for workability tests

Workability	Test method
Very low	Vebe
Low	Vebe; (Compacting Factor)*
Medium	(Compacting Factor)*; Slump
High	(Compacting Factor)*; Slump; Flow
Very high	Flow

Note
* Compacting Factor test is placed in parentheses, in view of the fact that it is seldom if ever used today. Nevertheless, it remains a standard test, and is included for completeness.

For the sake of completeness, some of these tests are covered here although aggregate requirements for such concretes are covered in Chapter 7.

INVERTED SLUMP CONE TEST FOR FLOW OF FIBRE-REINFORCED CONCRETE (FRC) (ASTM C995)

This test provides a measure of the consistency and workability of FRC. The test is a better indicator than the standard slump test of workability for FRC placed by vibration because although such concrete can exhibit very low slump due to the presence of fibres, it can still be easily consolidated. The test involves passing a sample of FRC through an inverted standard slump cone suspended within a steel bucket. The sample flows through the slump cone under the action of a vibrator of standard dimensions. The time to empty the slump cone is noted as the test parameter.

FLOW TESTS FOR SELF-COMPACTING CONCRETE (SCC)

Self-compacting concrete relies on its ability to flow freely around obstacles such as reinforcing bars and to fill formwork without external compaction effort. It has special rheological properties that ensure it flows without segregating, and compacts under gravity without entrapping air voids. Workability tests for normal slump concretes clearly will not be adequate for SCC. Thus, a range of tests have been developed specifically to assess the workability and rheological properties of SCC (EFNARC, 2002). Such tests are ideally required to assess three distinct though related properties of fresh SCC: its filling ability (flowability), its passing ability (free from blocking in reinforcement), and its resistance to segregation (stability). No single test yet devised can measure all these properties. None of the test methods have yet been standardized, nor are the tests yet perfected or definitive.

Two tests in common use are the Slump Flow test (JSCE, 1992) and the L Box test (Petersson *et al.*, 1996), both originally developed in Japan. The

Slump Flow test assesses the horizontal free flow of SCC in the absence of obstructions. It is the most commonly used test and gives a good assessment of filling ability but no indication of passing ability, while giving some indication of mix stability. The test involves filling a standard slump cone with SCC, and then allowing the material to flow and spread horizontally on a base plate, the final diameter of the concrete being measured. In addition, the time taken for the concrete to reach the 50-cm spread circle is usually measured. The test is simple to perform but if the 50-cm spread time is required, two people must conduct the test.

The L Box test requires filling the vertical leg of an L-shaped box with SCC, then removing a movable gate to allow concrete to flow into the horizontal section through vertical lengths of reinforcement bar. The slope of the concrete in the horizontal section is measured at the end of the test, which is an indication of its filling and passing ability. Flow times to reach certain distances can also be recorded. A sketch of the apparatus is given in Figure 4.2. This is a widely used test, suitable for laboratory work and possibly site use. Serious stability problems can be detected visually. Further flow tests for SCC are reviewed in EFNARC (2002).

All of the tests discussed above are empirical and suffer from the deficiency that they do not measure a fundamental rheological property of the plastic concrete, nor do they measure 'workability'. A further defect is that they can classify as identical two concretes that may have very different properties or workabilities in practice (Tattersall, 1991). No single test (empirical or otherwise) can properly measure or characterize all of the properties that are subsumed in the term 'workability' (or 'consistence'). Nevertheless, these tests have served the concrete industry well, mainly because they have been used as practical measures for control of concrete quality and consistency.

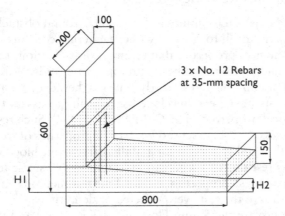

Figure 4.2 Schematic of the apparatus for the L Box flow test for SCC.

Fundamental approaches

More fundamental approaches have been developed to assist in better understanding the plastic behaviour of concrete and to derive fundamental rheological parameters. Research has required the development of these approaches in order that the materials science of concrete might progress. Also, as modern concretes have become increasingly sophisticated, their plastic behaviour has changed. For example, fibre concretes cannot be characterized by some of the empirical tests discussed earlier, while flowing concretes have specific rheological properties that can be assigned to them.

It should be possible to characterize plastic concrete in fundamental rheological terms. The most common approach is to regard it as a 'Bingham Material', in which there are two characteristic material parameters: the Yield Value C' and the Mobility C''. This is illustrated in Figure 4.3. The equation for a Bingham Material is

$$T = C' + C''N \qquad (4.1)$$

where T is torque at angular speed N from a test in a coaxial-cylinders type of viscometer. T can be interpreted more broadly as a shear force term, while N can be interpreted as a rate of shear strain (or deformation) term. The reciprocal of Mobility, that is $1/C''$, is termed the 'Plastic Viscosity'.

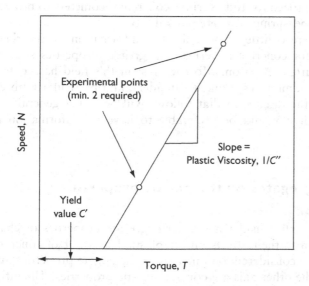

Figure 4.3 Relationship between torque and angular speed for Bingham Material in a coaxial-cylinders viscometer (after Tattersall, 1991).

Note: Two experimental points (shown) are needed to fix the line.

In concrete rheology research, a more practical approach is to use a concrete 'rheometer' which consists of an impeller turning in a mass of concrete inside a large bowl. The torque is measured at different impeller speeds, to construct an experimental line such as that given in Figure 4.3.

The relationship in Equation (4.1) describes the plastic properties of concrete fairly well, certainly in the conventional range of mixes. First, normal concrete will not move when placed in a pile until some shear stress (or force) is applied to it, which means that it has a 'yield value'. Second, a continuous shear needs to be applied to the material in order to keep it moving, implying a 'plastic viscosity'. Equation (4.1) also indicates that at least two points are needed to define the rheological constants. This has given rise to the so-called Two-Point test suggested by Tattersall (1991).

A number of other concrete rheometers are commercially available, developed to characterize the rheology of particular mixes. In a series of tests to evaluate the performance of five such rotational rheometers, either coaxial cylinder, parallel plate, or rotating vane devices, it was found that the yield stress correlated with slump, but correlations differed with the type of rheometer (Brower and Ferraris, 2003). All the rheometers ranked the mixtures in the same order for both yield stress and plastic viscosity. Differences in absolute values between the various rheometers could be attributed to effects such as slip at the instrument-wall interface or confinement of the concrete between moving parts of the rheometer. ACI sub-committee 236-A 'Workability of Fresh Concrete' is pursuing a materials science-based approach to study various concrete rheometers to provide better methods for measuring concrete workability.

Unfortunately, there is little or no published information on the rheological parameters for concrete in terms of aggregate properties such as shape, surface texture, and so on. Most interest in this field has centred on the influence of admixtures, cement extenders, and particularly fibres. Thus, of necessity, the discussion that follows will be rather general and descriptive, although it would be preferable to have the information in quantitative terms.

Influence of aggregates on the plastic properties of concrete

In this section, we shall consider particular aggregate properties or characteristics in relation to their effects on the plastic behaviour of concrete. A matrix needs to be considered: on one axis are aggregate properties and characteristics, on the other axis are concrete plastic properties. The influences of the former upon the latter, and the links between them, need to be explored. This approach is followed in Table 4.2 which summarizes the main points discussed in the text.

Table 4.2 Effect of aggregates on plastic properties of concrete

Aggregate property or characteristic	Aggregate effects on plastic properties of concrete
Particle shape	Shape has a major influence on workability and water requirement of a mix. Rounded or chunky particles roll or slide over each other easily; flaky and angular particles do not. Shape also affects particle packing and particle interlock, and therefore aggregate void content. Differences in standard water requirement (SWR) in excess of 50 l/m^3 can result from use of different aggregates, in particular different sands.
Particle surface texture	Surface texture influences the surface area and the inter-particle friction of aggregates, and thus the water requirement and workability of a mix, but less so than for shape.
Grading	Grading, particularly grading of the fine aggregate, has a very important influence on workability of a mix. It influences the total aggregate surface area (to be wetted), and the relative aggregate volume in a mix. In general, workability is best served by conforming to 'standard' gradings which ensure that voids of any one particle size are overfilled by particles of the next smaller size. Particular attention should be paid to the quantity and nature of the minus 300-μm material. The finer fractions (minus 150 μm and minus 75 μm) have a greater influence on cohesiveness and bleeding of the mix; quantities required in a mix will depend on the nature of the sand, with higher quantities generally being preferable in crushed sands. In HPC, grading of all constituents (including binders) becomes far more important, and optimum packing is sought.
Maximum aggregate size	Larger maximum aggregate size leads to lower water requirement, and generally improves concrete properties, up to a limit. Smaller aggregate sizes are usually required in HSC. For maximum aggregate sizes above 26.5 mm, blending of smaller sizes is required.
Fines content (minus 75-μm material)	In addition to the comments above, caution should be exercised over the nature of the fines (e.g. whether they comprise active clays), and tighter control should be exercised over fines content of natural sands. For crusher fines, improved concrete properties can result from fines contents sometimes in excess of 10%, but freeze–thaw resistance must also be considered.

Particle shape

Particle shape is the single most important factor that influences workability and water requirement of a concrete mix. It impacts strongly on the mix proportions needed to achieve dense, well-compacted concrete. Rounded, less angular particles are able to roll or slide over each other in the plastic mix with the minimum of resistance. Large amounts of flat, elongated, or highly angular particles render the concrete harsh, resulting in voids and honeycombing. As is often the case, the fine aggregate has the greater effect, particularly in gap-graded mixes where particle interference or interaction amongst the coarse aggregate particles is reduced. In those areas where fine aggregates are derived mainly from natural sands of alluvial or fluvial origin, fine aggregates tend to have well-rounded shape. However, in other areas where crusher sands are the norm, fine aggregates have shapes that vary enormously. This depends largely on the origin of the crusher sand, that is the nature of the parent rock, and on the crushing techniques used (see Chapter 2). The shapes of crusher sands can have a large influence on mix proportioning and economy. The influence of a characteristic such as particle shape on workability is difficult to describe accurately or objectively because 'workability' has a very broad definition. There is a close relationship between workability and water requirement, and any effect that alters one will obviously change the other.

While quantitative measures exist for aggregate shape (see Chapter 3), these are not very useful as a basis for assessing the water requirement of an aggregate. Observation and experience are the best guides, and an experienced concrete technologist can often estimate the water requirement to within a few litres per cubic metre.

Particle shape affects workability (and water requirement) by the influence it has on particle packing and particle interlock. This is best illustrated by photographs. In Figure 4.4a, a rounded aggregate is shown which is able to pack in a dense fashion, minimizing the interstitial voids. By contrast, Figure 4.4b reveals poorly shaped, angular particles which 'interlock' with their neighbours thereby inducing internal friction and resisting compaction, and which also do not pack well, giving a large volume of voids that needs to be filled with mortar or paste. The sample in Figure 4.4a will have a lower water requirement than that in Figure 4.4b.

A qualitative indication of the influence of particle shape on SWR is given in the series of photographs in Figure 4.5a–f. These are all fine aggregates used in gap-graded mixes in South Africa, usually with 19-mm nominally single-sized stone as the coarse aggregate. Some are crushed sands while others are natural sands from pit or river sources. The grading of the sands is not identical, but this has a secondary influence to that of particle shape. In a study by Davis on a broad set of South African fine aggregates, the range in SWRs amounted to $70\,l/m^3$. The range in void content for the same set of sands (tested by SANS 5845, which requires a consolidated

(a)

(b)

Figure 4.4 (a) Well-rounded aggregate showing good packing. Void content ≈33%.
(b) Angular and poorly shaped aggregate showing poor packing and particle interlock. Void content ≈45%. (Photo Courtesy of Gill Owens,
Cement and Concrete Institute). Maximum aggregate size for both aggregates approximately 25 mm.

(a)

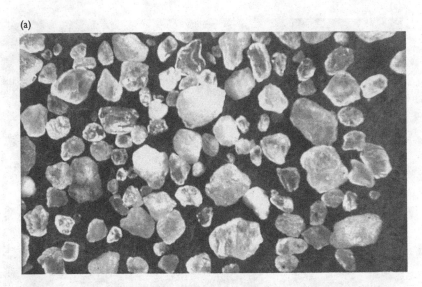

(b)

Figure 4.5 Particle shapes of sands and corresponding standard water requirements (SWRs) (scale indicated by matchsticks in Figures 4.5b–f). (a) Clean rounded quartz sand. SWR = 170–180 l/m³. (b) Rounded pit sand, good particle shape. SWR = 180–185 l/m³. (c) Crushed quartzite sand, reasonable shape. SWR = 200–205 l/m³. (d) Crushed quartzite sand, poor shape. SWR = 220–225 l/m³. (e) Pit sand, good shape, heavily clay-encrusted. SWR = 220–230 l/m³. (f) Weathered granite pit sand. Quartz, decomposed feldspar, and mica. SWR = 220–235 l/m³ (Photos above Courtesy of Gill Owens, Cement and Concrete Institute).

(c)

(d)

Figure 4.5 (Continued).

(e)

(f)

Figure 4.5 (Continued).

(not loose) aggregate sample) was from 28 to 44 per cent, with a mean value of 36 per cent. Similar tests on a range of nominally single-sized 19-mm crushed stone samples gave an average void content (in consolidated samples) of 46 per cent with a range from 34 to 50 per cent (Davis, 1975). The corresponding change in SWR over this range was about $30 \, l/m^3$. Thus, while the range of void contents was the same for the sands and stones (i.e. 16 per cent), the influence on water requirement was far more exaggerated in the case of the fine aggregates. On the basis that concretes made with the different sands all required the same compressive strength, that is the same w/c ratio, the range in mixing water requirement translates into a very large range in cement content between the 'best' and the 'worst' sands. This has major implications for mix economy.

In another study which drew on fine aggregates from different geographic locations in the US, the water requirements of comparable concrete mixes differed by up to $48 \, l/m^3$ (Blanks, 1952). The difference was ascribed mainly to differences in particle shape, the 'best' aggregate being smooth and rounded, while the 'worst' was angular and rough.

In other work on natural sands and gravels (Wills, 1967), the influence of shape was assessed indirectly using the 'loose void content' of the materials (tested using Method A, ASTM C1252). The following was found:

- Water requirement correlated well with results from loose void tests and orifice flow tests, for both sand and gravel.
- For nine fine aggregates with loose void contents ranging from 39 to 50 per cent, the corresponding range in water requirement for concretes made with these sands and control gravel was about $30 \, l/m^3$ (for these mixes, cement content was about $307 \, kg/m^3$).
- For nine gravels, void content ranged from 33 to 42 per cent, and the corresponding water requirements for concretes made with these gravels and a control sand had a range of about $20 \, l/m^3$.
- When the sands and gravels from the same sources were used in combination, water requirement had a range of $45 \, l/m^3$.

These results all indicate a large range in water requirement due to different aggregates. The differences can be ascribed primarily to the fine aggregate, and help to illustrate the influence of particle shape on water requirement and therefore on workability.

This sensitivity of workability and water requirement to particle shape can be used to good effect at the mixing plant to assess the quality of the aggregates. Changes in aggregate shape can arise due to changes in the source of an aggregate or changes within a single source over time. These changes are invariably reflected as differences in water requirement. Once detected at the mixer, adjustments to the mix can be made to ensure a consistent quality of concrete, or the aggregate can be rejected in favour of a better source.

The influence of particle shape on plastic properties of concrete such as cohesiveness, segregation, and bleeding is less clear, and certainly less marked. More angular particles have higher surface area, which should hold water in the mix, but this is usually more than offset by the additional mixing water required to achieve adequate workability. Grading and fines content (material smaller than 75 µm) have a far more powerful influence on these aspects than shape, and will be covered later.

Shell content in aggregate

Chapter 2 indicated that increasing use is likely to be made of marine-dredged aggregates due to growing environmental pressures on land-based deposits. Concerns about marine aggregates relate to the presence of chlorides, which can normally be removed by washing, and shell content. Generally, provided the shell content does not exceed about 25 per cent, there should be no adverse effects on the concrete. Concerns about shell fragments relate to their shape – often flaky and hollow – and not their composition, which is essentially calcium carbonate. Complete hollow shells may render concrete frost susceptible if they occur close to an exposed concrete surface. The question of the influence of shell content on the workability, strength, and elastic modulus of concrete was extensively investigated in the UK by Chapman and Roeder (1970) who found that, even with large shell contents (i.e. up to 10 per cent or more for 20-mm aggregates, and up to 30 per cent or more for 10-mm aggregates or smaller), structural properties of concrete were unimpaired. Hollow shells were found to perform as well as flat shells. The only negative effect was a modest reduction in workability which in practice is usually more than offset by the benefits deriving from the improved texture and shape of marine aggregates. Despite the generally favourable results, BS 882: 1992 imposes fairly strict limits on shell content: not exceeding 20 per cent for 5–10-mm size fractions, and not exceeding 8 per cent for size fractions larger than 10 mm. There is no restriction for aggregate size fractions less than 5 mm.

Particle surface texture

Definitions of surface texture were given in Chapter 3. As roughness increases, the water requirement will rise; concomitantly, for constant water content, the workability will reduce. While surface texture as a property can be distinguished from particle shape, its effects on the plastic properties of a mix often cannot. This is because it is very difficult to separate the two effects when comparing different batches of aggregates. For example, the results for standard water requirement of different aggregates quoted in the section on shape would have included a surface texture effect. The consensus is, however, that the effect of surface texture on water requirement, and thus workability, is less than that of shape (Grieve, 2001). In keeping with the comments on shape, fine aggregates again have the greater influence in respect of surface texture.

Surface texture influences the plastic properties of a mix in two ways: the surface area effect and the interparticle friction effect. Rough particles have a larger surface area than smooth particles of equivalent size and shape. Depending on the geometry of the surface texture, the increase in surface area can amount to 50–100 per cent compared with a completely smooth particle. Further, rough particles induce a higher interparticle friction, requiring greater external effort to move the particles over each other in a mix. These two effects combine to increase the water requirement for rough-textured aggregates.

The photographs in Figures 4.5b and e illustrate different surface textures of two similarly shaped sands. The standard water requirements (SWRs) of these sands are about 180 and 230 l/m^3 respectively. Both are natural sands, but that in Figure 4.5e has heavy clay encrustations giving rougher texture and greater absorption. In general, clean natural sands have smoother texture than crushed sands, since processes of abrasion and attrition generally render them smooth. However, it is possible for natural sands to have rough texture, for example if they are derived from coarse-grained rocks or where rock minerals weather at varying rates. Another factor affecting surface texture is the presence of surface coatings such as adhering dust, or surfaces naturally 'roughened' by weathering. For example, granite pit sands to the north of Johannesburg, which contain weathered orthoclase feldspars to the extent of about 50 per cent, increase the water requirement of a concrete mix by between 5 and 10 l/m^3 (Davis, 1975).

Aggregate grading

Grading of aggregates is important for creating a workable and cohesive concrete mix which can be thoroughly compacted without excessive mechanical effort. This leads to a dense mix which, provided other mix proportions (e.g. w/c) are properly chosen and applied and the concrete is allowed to mature by suitable curing, should provide sufficient strength and durability. The primary importance of grading lies in its influence on the plastic properties of concrete. This is particularly true in lean mixes, or in mixes with high workability.

Chapter 3 indicated that there are two practices in general use for aggregate grading: continuous-grading or gap grading. The adoption of one or the other depends on local practice and is usually governed by the nature of the available aggregates. Continuously graded mixes are preferred when all particle sizes are readily available, usually as alluvial or marine deposits, whereas gap grading may be used where aggregates are produced by crushing, in which case the effects of interparticle friction become more important.

The influence of aggregate grading on plastic properties of concrete will be considered with reference to the aggregate surface area and the relative volume of the aggregates in the mix.

Aggregate surface area

Aggregate surface area determines the amount of water necessary to wet all the solids. The total surface area of a set of aggregates is governed predominantly by the fine aggregate fraction, since for a given mass of aggregate, the surface area increases with reducing particle size. Consequently, the finer the aggregate grading, the greater the total surface area for a given mass, and the water requirement should therefore increase in order to wet all the aggregate surfaces. However, this is not necessarily true as will be discussed later. Table 4.3 illustrates the relationship between surface area and particle size, in relative terms.

The surface area of aggregates is often expressed in terms of the 'specific surface' which is the ratio of the total surface area to the volume of the particles (also sometimes expressed as the ratio of total surface area to the mass of the sample), provided the relative density of all the particles is the same. If all particles were spherical in shape, this parameter would be easy to compute simply from knowledge of the grading. However, particle shapes are normally anything but spherical and this is particularly true for crushed aggregates. The less spherical a particle, the greater is its specific surface. This leads to practical difficulties in measuring or estimating the 'true' specific surface for an aggregate sample comprising particles of varying shapes.

Table 4.3 Relative values of surface area

Particle size fraction	Relative surface area
75–40 mm	$^1/_2$
40–20 mm	1
20–10 mm	2
10–5 mm	4
5–2.5 mm	8
2.5–1.2 mm	16
1.2 mm–600 μm	32
600–300 μm	64
300–150 μm	128

Source: After Neville, 1995.

There have been attempts to devise mix-proportioning methods using the total surface area of the aggregates (Edwards, 1918; Loudon, 1952). At first glance, this would seem to make sense, since sufficient water (or in effect paste) must be provided to coat all the particles adequately, and it should be possible to estimate this paste content from knowledge of the total aggregate surface area. However, there are reasons why these attempts have not been particularly successful and have not achieved common usage. The first is a practical reason, already discussed, that it is actually very difficult to estimate the true specific surface of aggregates. Second, and more importantly, particles smaller than about 150 μm tend to lubricate a mix by facilitating the rolling and sliding of the larger particles over each other. (This effect is very obvious in fly ash mixes where not only are the particle sizes of the fly ash very small, but being of spherical shape they virtually act as 'ball bearings' in the mix.) Consequently, these particle sizes if present in sufficient (but not excessive) quantity improve workability and can actually reduce the water requirement of a mix. This effect was recognized by Murdock (1960) who suggested a 'surface index' of the aggregate, similar to the relative specific surface given in Table 4.3. In contrast to the relative specific surface, the 'surface index' shows reducing numerical values for aggregate sizes smaller than about 600 μm. The surface index can also be modified by an angularity index.

Relative aggregate volume

Aggregate grading also determines the relative volume occupied by the aggregate in concrete. It is desirable both economically and technically to maximize the total aggregate volume in a mix.

A grading that aims to provide the densest possible packing by ensuring that the void content between particles of a given size is completely filled by particles of a smaller size would result in a continuous sequence of sizes, but with large 'steps' between the successive particle sizes (see Chapter 3). Continuously graded aggregates which contain particles of all sizes are usually a better solution, but may suffer from a high degree of internal friction and particle interference particularly if the aggregate shape is poor. Gap-graded mixes in which the 'gap' exists between the continuously graded sand and nominally single-sized stone can often minimize the problem of high internal friction and particle interference, at the same time exploiting the advantages of continuous grading in the sand sizes. This is particularly true for mixes of crushed coarse aggregates or crushed coarse and fine aggregates.

As mentioned, an 'ideal' grading allowing for maximum density by optimally filling the void spaces would usually produce an unworkable concrete. Such impractical gradings are based on 'Fuller Curves' which are parabolic or part parabolic–part linear when plotted on a natural scale. In practice, it is essential to overfill the voids of any one particle size by the next smaller size, and so on down the sequence of sizes, in order to ensure that particles can move freely in the plastic mix. This is achieved by conforming to well-accepted 'standard' gradings, or by using particle packing algorithms for high performance mixes, such as those discussed in Chapter 3.

STANDARD GRADINGS

There are many examples of 'standard' gradings for concrete mixes. An early contribution based on the principle of total specific surface area of the aggregates was that of the UK Road Research Laboratory, Road Note No. 4 (RRL, 1950). Figure 4.6 illustrates four grading curves, numbered 1–4, which correspond to specific surfaces of 1.6, 2.0, 2.5, and 3.3 m^2/kg, respectively. Similar sets of curves are available for other maximum aggregate sizes (38.1 mm and 9.5 mm) (RRL, 1950; McIntosh and Erntroy, 1955). In practice, an aggregate grading will typically lie in a zone between two lines, and the required water content for a given workability would depend on the zone. For example, Zone A in Figure 4.6 represents relatively coarse gradings which would be suitable for rich mixes, while Zone C will give cohesive mixes but may require a higher water content to be workable. 'Type grading curves' such as those in Figure 4.6 apply essentially only to aggregates of the type for which they were derived (in this case rounded river sands and gravels) and they also require continuous aggregate grading from coarse to fine. Thus, they cannot be used universally, and in areas where for example crushed aggregates and gap grading are normal practice, they are of no value.

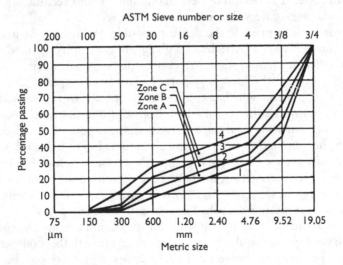

Figure 4.6 Road Note No. 4 type grading curves for 19-mm aggregate, river sand, and gravel (Crown Copyright; RRL, 1950).

Nevertheless, it is desirable particularly from a quality control perspective that aggregates conform to grading limits. These may not be 'standard' gradings in the sense of rigid conformance to a fixed grading curve, but they do help to ensure that the limits on grading are not excessive, thus giving a consistent plastic mix. These grading limits are laid down in the form of 'grading requirements' for aggregates in various national standards. Table 4.4 provides grading requirements for fine aggregate as given in the ASTM, CSA, BS, and SANS documents (ASTM C33; CSA A23.1; BS 882: 1992; SANS 1083). In addition, 'preferred limits' suggested by Grieve (2001) for South African sands are given. These refer more specifically to crushed fine aggregates which usually have somewhat coarser gradings. Limits to fineness modulus (FM) vary in different countries, reflecting local practice. In North America, the limits are from 2.3 to 3.1, while in South Africa the limits are 1.2 to 3.5. The wide limits in the South African standard reflect the fact that a large variety of concrete sands is used, both natural and crushed, occasioned by the relative paucity of good natural concrete sands. The SANS grading requirements are purposely sparse, in order to accommodate the wide range of concrete sands that may be suitable. For the UK, limits in terms of FM are not set. Rather, fine aggregate grading envelopes are specified for Coarse (C), Medium (M), and Fine (F) materials. The median gradings of these envelopes give FM values of approximately 2.9, 2.6, and 1.9 respectively. Clearly, the extremes of the gradings would give a larger range in FM, amounting to about 1.0–3.9, a very wide range.

The limits from the standards should not be interpreted as hard and fast rules, and in certain cases it is possible and even desirable to extend these limits, particularly if alternative sources are unavailable or very costly. ASTM C33 makes such a provision by stating that a fine aggregate is acceptable if it can be shown that concrete in which it is used has properties not inferior to concrete with a reference aggregate that conforms to the standard. It is particularly important with crushed materials to ensure adequate proportions of the 300-μm, 150-μm, and minus 75-μm material to provide cohesiveness and prevent segregation. Table 4.4 shows that the required ranges are very wide, reflecting that aggregates vary widely within countries and between countries, and 'ideal' conditions cannot be prescribed.

Coarse aggregate grading requirements are given in Tables 4.5a and b for various national standards. Both 'graded' and nominally single-sized coarse aggregates are covered. Such requirements help to ensure that gradings are controlled to yield a consistent quality of the final mix. The different national requirements vary, reflecting differences in local materials and practice. Grading of stone has a lesser influence on plastic concrete than that of sand, and as with fine aggregates, perfectly good concrete can be made with very different gradings, even those that may fall outside the grading envelopes.

Table 4.4 BS, CSA, ASTM, and SANS grading requirements for fine aggregates

Sieve size			Cumulative percentage by mass passing sieves							
			BS 882: 1992							
BS (mm)	ASTM, SANS (mm)	CSA (mm)	Overall grading	Coarse grading (C)	Medium grading (M)	Fine grading (F)	ASTM C33	CSA A23.1 (FA1)	SANS 1083: 1994	SA Preferred limits
10.0	9.5	10.0	100				100	100	100	100
5.00	4.75	5.00	89–100				95–100	95–100	90–100	90–100
2.36	2.36	2.50	60–100	60–100	65–100	80–100	80–100	80–100	–	75–100
1.18	1.18	1.25	30–100	30–90	45–100	70–100	50–85	50–90	–	60–90
0.600	0.600	0.630	15–100	15–54	25–80	55–100	25–60	25–65	–	40–60
0.300	0.300	0.315	5–70	5–40	5–48	5–70	10–30	10–35	–	20–40
0.150	0.150	0.160	0–15*				2–10	2–10	5–25	10–20

Notes

* For crushed stone fine aggregate, the permissible limit is increased to 20% except for heavy duty floors.

1 ASTM C33 requires fine aggregate FM to lie between 2.3 and 3.1.

2 ASTM C33 allows reduced percentages passing 300- and 150-μm sieves, when cement content exceeds 297 kg/m^3 (500 lb/yd^3), or if air entrainment is used with a cement content of at least 237 kg/m^3 (400 lb/yd^3).

3 For BS 882: 1992, an aggregate must conform with the overall grading limit, as well as one of the three overlapping grading limits of 'Coarse', 'Medium', or 'Fine'. A given fine aggregate will often actually comply with two of the three gradings 'C', 'M', or 'F'. This ensures that satisfactory materials are not needlessly excluded or fail to comply with an artificially rigid grading envelope. BS 882 also allows the use of sands not complying with the grading limits, provided that it can be shown that such materials produce concrete of the required quality.

4 The SANS grading requirements are purposely sparse, in order to accommodate the wide range of concrete sands that may be suitable. The 'SA Preferred Limits' are taken from Grieve (2001) and are based on extensive experience with a range of crushed and natural sands.

5 CSA Standard A23.1 for fine aggregates differs only slightly from the ASTM C33 limits. CSA allows for a second class of coarser fine aggregate, FA2, to be used in conjunction with FA1 to optimize the particle distribution of the coarse and fine aggregates in a mix.

The ASTM and BS limits in Table 4.5a and b are plotted in Figure 4.7 for both continuous and single-sized 19-mm aggregates. The BS limits are somewhat wider, particularly for single-sized stone.

In practice, maximizing the total aggregate volume in a mix requires both maximizing the coarse aggregate content and ensuring that the fine aggregate has a low water requirement. The former requirement can usually be met by selecting a finer sand, since this allows the stone particles to approach each other more closely in the mix. Figure 4.8 illustrates this effect. The latter requirement involves carefully balancing overall sand fineness (e.g. sand FM) with the proportions of minus 300-μm and minus 75-μm material. The minus 300-μm fraction has a large influence on workability, and generally it is preferable to have 20–40 per cent as this fraction. The minus 75-μm fraction has a larger influence on cohesiveness, but too much of this material can increase the water requirement of a mix unacceptably. Additional useful discussion on aggregate grading can be found in Galloway (1994), who indicates that standard gradings such as those in ASTM C33 may not necessarily lead to the most workable or economical mixes. These gradings may result in a degree of 'gap-grading', whereas it is sometimes preferable to fill space between coarse aggregate particles with smaller coarse particles rather than mortar. Galloway concurs that fine aggregate grading has a much greater effect on concrete workability than does coarse

Table 4.5a Grading requirements for continuously graded coarse aggregate according to BS 882: 1992 and ASTM C33

		Percentage by mass passing sieves					
		Nominal size of graded aggregate					
Sieve size mm	38–5 mm $\frac{1}{2}$–$\frac{3}{16}$ in.		19–5 mm $\frac{3}{4}$–$\frac{3}{16}$ in.		12.5–5 mm $\frac{1}{2}$–$\frac{3}{16}$ in.		
(Nominal)	in.	BS	ASTM	BS	ASTM	BS	ASTM
50	2	100	100	–	–	–	–
38	$1\frac{1}{2}$	90–100	95–100	100	–	–	–
25	1	–	–	–	100	–	–
19	$\frac{3}{4}$	35–70	35–70	90–100	90–100	100	100
13	$\frac{1}{2}$	25–55	–	40–80	–	90–100	90–100
10	$\frac{3}{8}$	10–40	10–30	30–60	20–55	50–85	40–70
5	$\frac{3}{16}$	0–5	0–5	0–10	0–10	0–10	0–15
2.36	No. 8	–	–	–	0–5	–	0–5

Notes
1 Nominal sieve sizes vary slightly for the different standards. The size closest to the nominal size given above applies.
2 Coarse aggregate gradings according to the Canadian Standard CSA A23.1 differ only slightly for certain sizes compared to the ASTM values, taking equivalence of nominal sieve sizes into account.

Table 4.5b Grading requirements for nominally single-sized coarse aggregates according to BS 882: 1992, ASTM C33, and SANS 1083: 1994

Sieve size mm (Nominal)	in.	Percentage by mass passing BS sieves — Nominal size of single-sized aggregate							
		38 mm 1½ in.	19 mm ¾ in.			12.5 mm ½ in.		9.5 mm ⅜ in.	
		BS	BS	SANS	ASTM	BS	SANS	BS	SANS
50	2	100	–	–	–	–	–	–	–
38	1½	85–100	100	–	–	–	–	–	–
25	1	–	–	100	100	–	–	–	–
19	¾	0–25	85–100	85–100	90–100	100	100	–	–
13	½	–	0–70	0–50	–	85–100	85–100	100	100
10	⅜	0–5	0–25	0–25	20–55	0–50	0–55	85–100	85–100
5	3/16	–	0–5	0–5	0–10	0–10	0–5	0–25	0–25
2.36	No. 8	–	–	–	0–5	–	–	0–5	0–5

Notes

1 ASTM C33 contains details of grading requirements for other nominal single-sized aggregates.

2 CSA Standard A23.1 differs only slightly from ASTM requirements.

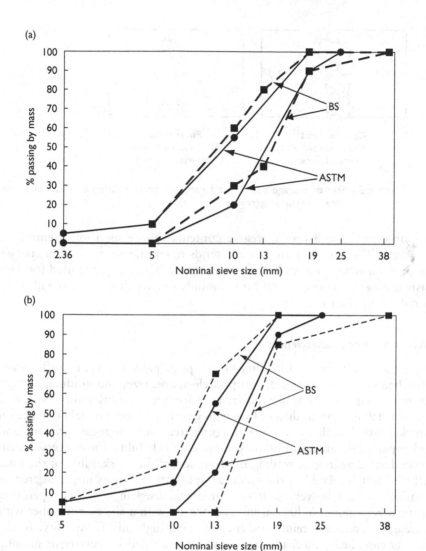

Figure 4.7 ASTM and BS grading limits for (a) 19-mm continuous and (b) single-sized aggregates.

aggregate grading. The fine aggregate should not be too coarse, leading to harshness, or too fine, leading to additional water requirement. For high strength concrete (HSC) in which cement contents are high, coarse sand with an FM of around 3 helps to produce best workability. For pumped concrete, higher fine aggregate contents are required, and attention to a well-graded aggregate with good particle shape and surface texture is very

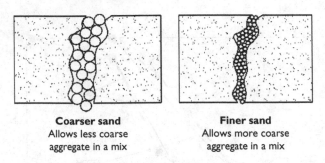

Coarser sand
Allows less coarse
aggregate in a mix

Finer sand
Allows more coarse
aggregate in a mix

Figure 4.8 Effect of average size of sand particles on separation of stone in a mix, hence on coarse aggregate content.

important. Reduced fine aggregate contents can be used in air-entrained concrete. The 150–600-μm fraction tends to entrain more air than coarser or finer particles, and careful control of these fractions is required for consistent air entrainment. Significant amounts smaller than 150 μm will cause a reduction in air content.

Workability and cohesiveness

Aggregates influence workability in two principal ways: (1) by influencing the rheological properties due to particle shape, size, and grading, and (2) by governing the mix-water requirement for a given workability (Sims and Brown, 1998). The grading of fine aggregate has a more crucial influence on workability than that of the coarse aggregate. While aggregate grading (and other particle characteristics) may influence workability, the converse is also true: desired aggregate grading is a function of the workability of the mix, which itself is related to the practical application. For example, aggregate gradings would be very different for a mix used in a precast operation using a zero-slump industrial mix, compared with a site-cast member with extensive secondary reinforcement requiring high mix flowability. In the former case, heavy mechanical compaction is needed and aggregate grading may be less critical, providing full compaction can be achieved. In the latter case, aggregate grading will be more critical requiring low internal friction and adequate cohesiveness.

The proportion of sand retained on the 300-μm sieve determines the coarseness of the sand, and adds 'body' to the sand. The United States Bureau of Reclamation (USBR) recommends that between 15 and 35 per cent be retained as this fraction, although too high a value would imply inadequate finer material.

The minus 300-μm and minus 150-μm fractions critically affect workability and cohesiveness of a mix. International practice allows between

5 and 70 per cent of a sand to pass the 300-μm sieve (refer to Table 4.4). This is a very wide range and local practice will restrict this range substantially. For example, in South Africa where many sands are crushed or are mixed crushed and natural sands, 20–40 per cent passing the 300-μm sieve has been found to be appropriate. The ASTM standard (C33) suggests 10–30 per cent for this size, reflecting North American experience where there is a greater occurrence of natural sands. Certainly, too little of the minus 300-μm and minus 150-μm fractions will produce concretes that will tend to be harsh, to lack cohesion, and to segregate. Conversely, too much of this fine material can lead to 'sticky' mixes, and this is particularly true in high cement content concretes where coarser sand gradings are used.

Excessive quantities of minus 75-μm material will tend to increase the water requirement with concomitant adverse effects on concrete shrinkage. As a guide, this fraction should not exceed 5–10 per cent, less if it is natural sand, possibly more if it is sand crushed from sound fresh rock. On the other hand, less than 5 per cent of this fraction may cause bleeding problems in the concrete. BS 882: 1992 limits the amount of material passing the 75-μm sieve to 4 per cent in fine aggregate, which can be increased to 16 per cent if it is entirely derived from crushed rock. The corresponding respective limits for ASTM C33 are 5 per cent, increased to 7 per cent for crushed material, and for SANS 1083 are 5 per cent, increased to 10 per cent for crushed material. The South African limits may be increased to 10 per cent and 20 per cent respectively, provided that the fine aggregate does not contain deleteriously active clay fines (i.e. it passes the methylene blue test – see later under 'Composition of the particles'), and the aggregate complies with the requirements for clay content, which should not exceed 2 per cent. It is also possible that coarse aggregates may have some minus 75-μm material and this is generally limited to between 1 and 2 per cent. However, ASTM C33 allows considerable modification of these limits, provided the aggregate in question has a demonstrated performance record or that concrete in which it is used shows properties similar to an acceptable concrete.

Coarse aggregate grading has a relatively minor effect on concrete workability. Evidence for this comes from tests on gap-graded crushed aggregate mixes in which the coarse aggregate grading was varied between two extremes, shown in Figure 4.9 (Grieve, 2001). The finer stone grading represents a nominal 19-mm stone with more than normal 6.7- and 9.5-mm fractions, while the coarser grading is a 26.5-mm stone with more than permitted 9.5- and 19-mm sizes. Very acceptable concrete was produced with stone from either extreme of the grading with insignificant effect on water requirement. Experience with such crushed coarse aggregates indicates only small differences in water demand between nominal 19- and 26.5-mm stone (Grieve, 2000, pers.comm.).

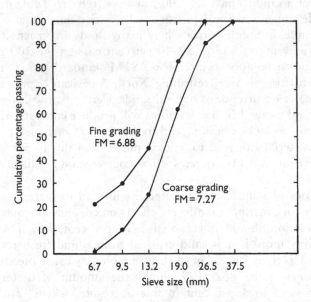

Figure 4.9 Range of stone gradings used to investigate effect of stone grading on concrete workability (from Grieve, 2001).

A note on the workability of gap-graded mixes is appropriate here. The elimination of intermediate particle size fractions reduces the total amount of internal friction in a mix. Consequently, workability of these mixes can be improved, or alternatively lower water contents may be achieved in comparison with equivalent continuously graded mixes. The disadvantage, however, is that for more workable mixes, segregation may become a problem, and therefore gap-graded mixes find best application for lower workabilities (slumps of less than 80–100 mm) and vibratory compaction. The effect of gap grading is to permit such mixes to be readily compacted using vibration, despite low slumps. Pumped concretes and certain paving concretes tend to work better with continuously graded aggregates. However, gap-graded mixes can still be used in these situations provided close attention is paid to aspects such as adequate fines content, rounded particle shape, and smooth surface texture.

Segregation and bleeding

Aggregate grading also affects the tendency for a mix to segregate. If the fine aggregate grading is too coarse, harshness, bleeding, and segregation will occur. Bleeding is a special case of segregation and is particularly influenced by the fines content (the minus 75-μm fraction, including the binder component). Segregation is linked to the cohesiveness of a mix and is an aspect of workability, that is a workable mix is one that does not seg-regate readily. Segregation is aggravated by larger sizes and high relative density of coarse aggregate, high slump mixes, as well as gap-graded mixes. Segregation and cohesiveness are also related to the cement content of a mix: richer mixes are more cohesive and tend not to segregate as easily. Lean mixes will benefit by a greater proportion of aggregate fines. Alter-natively, air entrainment can be used to compensate for a lack of fines, or water-reducing admixtures can permit adequate workability to be achieved without too high a water content.

Bleeding, as mentioned, depends mainly on the fines content. Aspects such as air content, use of chemical admixtures and mineral extenders, and angularity and grading of the aggregate also have an influence. Exces-sive bleeding causes the creation of internal voids and porous zones in the concrete (so-called 'water gain') mainly under coarse aggregate particles, vertical bleed channels, and weak, porous, fines-rich horizontal surfaces in slabs. These effects are illustrated in Figure 4.10. A completely non-bleeding mix is not necessarily an advantage, since this can exacerbate plastic shrink-age cracking particularly in slabs on grade. Bleeding can be controlled by giving attention to grading and angularity of the fine aggregate, use of additional fines, blending of sands, increased cement or extender content, and use of chemical admixtures including air-entrainers. ASTM C33 sets a lower limit of 10 per cent for the 300-μm size, but usually this will be insufficient to avoid workability and bleeding problems. Preferably, at least

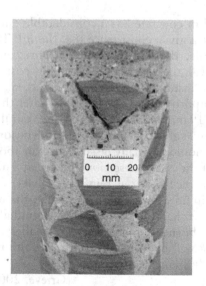 

Figure 4.10 Effects of excessive bleeding near the top surface of a concrete slab. Note the large voids under the aggregate particles. The photos are from opposite faces of the same 70-mm diameter core.

15 per cent should pass the 300-μm sieve and 3 per cent the 150-μm sieve, more if it is a lean mix.

Maximum aggregate size

Maximum aggregate size (also called nominal maximum size) was defined in Chapter 3. It was also pointed out that the differences in definitions can give rise to some confusion for specifiers and contractors. For example, based on Canadian Standard A23.1, a 19-mm nominal maximum stone size must have the entire aggregate pass through the 25-mm sieve. The ASTM C125 definition, on the other hand, would require a 19-mm maximum size stone to all pass through the 19-mm sieve but not all through the next smallest standard sieve size.

The larger the maximum size of aggregate, the less water is required in a mix for a given workability. Consequently, there are advantages to increasing the maximum aggregate size: lower paste content giving better mechanical properties (e.g. lower creep and shrinkage), and improved economy. The explanation for this effect is given in Figure 4.11 which shows schematically that as maximum aggregate size increases, less paste (or mortar) is required to fill between the particles. In the extreme, a solid block of aggregate would require no paste at all!

Referring to gap-graded mixes used in South Africa, the following quote is pertinent:

> The maximum size of stone used in concrete has a considerable effect on the water content required for a given consistence [*i.e. slump*]. The larger the maximum size of stone, the less the water required and consequently the more economical the mix in terms of the cost of cement for a given strength. For instance, a [*gap-graded*] mix made with 37.5 mm stone will require approximately 15 l/m³ less water than a corresponding mix made with 19 mm stone. It is therefore desirable to choose the largest possible stone, subject to practical considerations. For example, nominal stone size should normally be smaller than about one fifth of the least dimension of the concrete section and about 5 mm less than the minimum clear distance between reinforcing bars. For high-quality reinforced concrete in an aggressive environment, the use of a maximum stone size 5 mm less than the cover to steel should also be considered. It should also be borne in mind that the smaller the stone size, the easier the concrete will be to handle and place and the less likely it will be to segregate. (Italic text inserted.)
>
> (Grieve, 2001)

North American practice requires that the maximum aggregate size should be less than one-fifth of the narrowest dimension between sides of

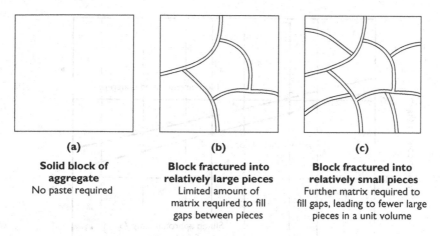

(a)	(b)	(c)
Solid block of aggregate	**Block fractured into relatively large pieces**	**Block fractured into relatively small pieces**
No paste required	Limited amount of matrix required to fill gaps between pieces	Further matrix required to fill gaps, leading to fewer large pieces in a unit volume

Figure 4.11 Effect of maximum aggregate size on matrix (paste) content.

forms, one-third the depth of slabs, and less than three-fourths the clear spacing between bars.

Figure 4.12 illustrates how mixing water content can be reduced as maximum aggregate size increases (Kosmatka *et al.*, 1995), which can result in a sizable reduction in water requirement for a constant workability (Walker *et al.*, 1959). This effect can be ascribed at least partly to a reduction in aggregate specific surface area and thus a lesser area of aggregates to be wetted. *Fulton* (2001) recommends reducing or increasing the water content of a mix by approximately $10 \, l/m^3$ for each standard size increase or decrease in maximum aggregate size respectively. This would apply best to gap-graded mixes of nominally single-sized coarse aggregates.

The reduction in water requirement accompanying larger maximum aggregate size permits either more economical mixes to be made at equal w/c, or stronger mixes if cement contents are not reduced. This assumes of course that there will be no increased cost for the larger aggregate. Aggregate producers may prefer to sell larger aggregates at lower cost due to reduced production costs, which gives a further improvement in mix economy.

However, there are limitations, both practical and technical, to the maximum size of aggregate in a mix. Practically, maximum size is limited by the structural limitations of member cross-section and reinforcement spacing, mentioned above. Technically, the larger the maximum aggregate size, the more heterogeneous the mix becomes and this leads to effects such as enhanced bleeding under large aggregate particles, a very heterogeneous interfacial transition zone (ITZ), and an increased bond stress between matrix and aggregate. This is discussed in Chapter 5.

For gap-graded mixes of nominally single-sized stone up to 26.5 mm, it is not normally necessary to blend the stone with a smaller size. For larger

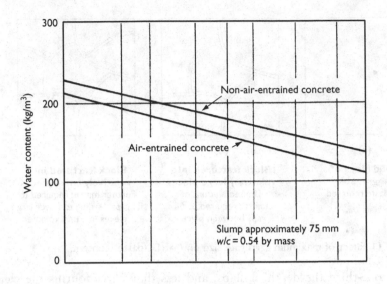

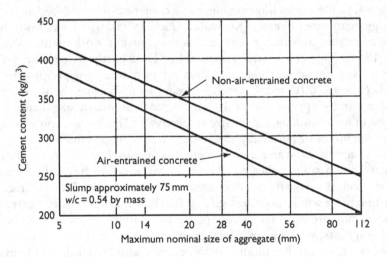

Figure 4.12 Mixing water requirement at constant slump as a function of maximum aggregate size (from Kosmatka *et al.*, 1995).

stone sizes, a wide range of blends may be used and it is unnecessary to be too prescriptive. As the maximum size increases, more intermediate sizes can be used implying that the stone grading will become more continuous. The basis of blending may be to achieve either maximum density or a smooth grading curve (which in itself will also tend to increase the density).

Further notes on fine aggregates and fines content

By now, it should be appreciated that the plastic properties of concrete are more influenced by the characteristics of the fine aggregate than the coarse aggregate. The relative influences of the fine aggregate on the plastic properties of concrete are summarized in Table 4.6. Distinctions are drawn between 'coarser' sands (FM > 2.9), 'finer' sands (FM < 2.0), and 'fines content' (minus 75-μm fraction).

The limit placed on fines content has been the subject of much debate, partly because the fines may be of variable composition. Generally, the limits need to be stricter for natural sands where the likelihood of deleterious clays occurring is greater. Higher levels of fines can often be tolerated in crusher sands, even well in excess of 10 per cent. Such fines will help to control bleeding and reduce the negative effects this might have on reinforcement bond, permeability, and potential durability. These fines will also contribute to the cohesiveness of the mix, particularly for leaner mixes with total binder contents less than about 270 kg/m^3. The presence of adequate fines is also essential for pumped concrete and concrete for water-retaining structures.

On occasions, judicious use of higher non-deleterious fines content can actually give improved hardened concrete properties without requiring excessive water content. For example, Bonavetti and Irassar (1994) used rock fines ('stone dust') as partial replacement for sand in mortars. They found substantial improvements in compressive strength at early ages (<28 days), particularly for limestone dust. The effects were, however, negative for granite dust. The increased strengths were attributed to the 'fine filler effect' and the acceleration of hydration at early ages. Work in South Africa showed that compressive strength of concrete actually increased with increasing aggregate fines content, irrespective of its clay content (*Fulton*, 1994). An extreme case is quoted where a dolomite crusher sand with a fines content in excess of 20 per cent gave adequate workability at a mix-water content of 190 l/m^3 with compressive strengths 15–20 per cent above expected values, and normal shrinkage. Mechanisms for such strength enhancement have not been fully clarified, but the most plausible explanation is the 'fine filler effect', whereby the paste microstructure is made more homogenous by multiple nucleation sites provided by the fines, and the ITZ is improved by reduced micro-bleeding.

A note of caution must be sounded to the above points, however, concerning the resistance of concrete to freezing and thawing and to de-icing salts as well as resistance to abrasion. Neville (1995, p. 157) points out that an excessive amount of ultra-fines may be harmful to these properties. In general, a maximum of 50 kg/m^3 of aggregate fines is recommended in mixes with a maximum aggregate size of 16–63 mm.

In air-entrained concrete, high fines content, particularly if composed of clay, can adversely affect the air content in concrete. This could also affect water requirement, slump, and strength. The causes and effects are,

Table 4.6 Influence of fine aggregate on plastic properties of concrete

Concrete property	'Coarser' sands (FM > 2.9)	'Finer' sands (FM < 2.0)	Fines content* (minus 75-μm fraction)
Water requirement	Tend to lower the water requirement by reducing overall aggregate specific surface	May increase water requirement; however, a certain amount of fines is required to 'lubricate' a mix, and this will not increase the water requirement	Excessive fines content will increase water requirement (this varies for natural or crushed materials)
Workability	Tend to increase workability provided particle interference is not induced	May reduce workability, but see comment above. Minus 300-μm fraction has relatively large influence	This fraction is a portion of the minus 300-μm fraction that is important to workability. Excessive fines will cause mix to be 'sticky', particularly if cement content is also high
Cohesiveness	Will tend to reduce cohesiveness, but coarser sands are suitable in rich mixes	Will influence cohesiveness since they tend to have a larger fines content	Fines content (together with cement content) largely governs mix cohesiveness. Type of fines (clay, silt, etc.) is also important
Segregation	Tend to promote segregation by increasing workability without concomitant increase in cohesiveness	Adequate proportion of minus 300-μm fraction helps control segregation	Helps to control segregation by making mix more cohesive
Bleeding	Tend to promote bleeding by reducing specific surface	Better able to control bleeding provided it has adequate fines content	Bleeding largely controlled by fines content. Inadequate fines leads to bleeding

Note
* Influence of the fines is very dependent on the nature (type) of fines, that is whether they are active clays (smectite, montmorillonite, atapulgite, illite), non-active clays (kaolinite), silts, or fine sand fractions.

however, variable. The committee report, ACI 221R-96, summarizes the effects of fine aggregate on air-entrainment as follows. Increased amounts of minus 75-μm or 150-μm material in fine aggregate generally lead to an increased dosage of air-entraining admixture to obtain the required air content. Conversely, increased amounts of 300–600 μm material will decrease the dosage required for the same air content. Organic materials if present in the fine aggregate may act as air-entrainers, but usually lead to large air bubbles and an undesirable air-void system. Such organics should therefore be avoided in high quality air-entrained mixes.

Other aspects

Composition of the particles

The bulk of aggregates used in concrete comprise clean, hard materials with stable minerals and relatively low absorption. Occasionally, however, aggregates may contain deleterious materials that have adverse effects on mix properties such as workability and water requirement (see Table 3.15). Local experience will usually be the best guide. The composition of the particles will be important knowledge to the concrete engineer.

In the case of new or untried aggregate sources, a petrographic examination is essential in identifying possible problem constituents, particularly in the fines fraction. Trial mixes can also be made up in order to determine plastic properties of a standard mix, and the results compared with a mix using known and proven aggregates of similar type. CSA A23.2-8A has the basis for such a test. A full battery of aggregate tests can be run if necessary, following the outcome of the simpler approaches.

The fines portion of sand can consist of various materials. In the case of natural sands, these would be predominantly fine sand, silts, and clays, while for crushed sands, rock 'flour' (or rock dust) would be dominant. Clays of the montmorillonite, illite, and atapulgite type should be avoided, as they impart dimensional instability (i.e. shrinkage and swelling) to the concrete. These undesirable clays can be identified petrographically or by tests such as the methylene blue indicator test (SANS 6243). A further undesirable feature of clays is that they may also adversely affect setting times and hardening of the concrete. Other dimensionally stable clays can be used up to a point, and their acceptability should be judged more by the effect they have on water requirement.

A particularly troublesome constituent in some aggregates is mica: layered silicate minerals which occur as flaky particles in fine aggregates. This material increases the water requirement, or conversely reduces the workability, if present in excessive amounts depending on the type of mica (Dewar, 1963; Schmitt, 1990). Figure 4.13 indicates that muscovite mica has very substantial effects on both workability and strength of concrete.

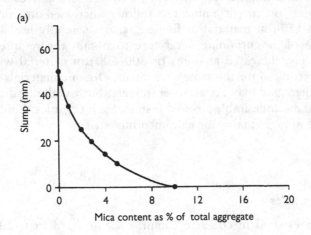

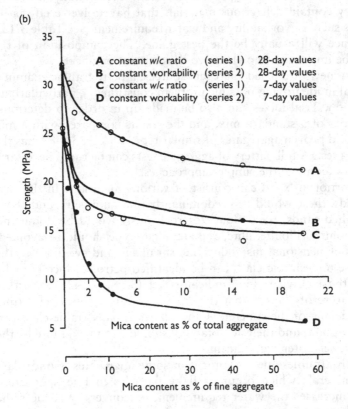

Figure 4.13 (a) Loss in workability with increasing mica content (constant *w/c* ratio).
(b) Effects of muscovite mica content on compressive strength of concrete (from Fookes and Revie, 1982).

Water demand increases (or at constant water content, workability reduces) with increasing mica content. Further, there is a reduction in strength with increasing mica content, keeping other factors such as w/c ratio constant. In constant workability mixes, the combined effects mentioned can lead to very severe loss of strength. An upper limit of 15 per cent mica in the 150–300 μm fraction has been suggested to minimize the deleterious effects of mica (Gaynor and Meininger, 1983), but each case will need to be considered on its own merits.

Absorption

Certain minerals and particles tend to have higher than normal absorption which also increases the water requirement of the concrete. This is true of weathered clays, particularly the smectites, and of microporous materials, for example some fine gravel aggregates in the UK (Sims and Brown, 1998). Aggregates derived from weathered argillaceous rock sources, particularly if un-metamorphosed, should be treated as suspect until trials prove their acceptability or otherwise. Certain basalts that contain secondary alteration minerals may also have higher absorptions and, more seriously, may impart high shrinkage to the concrete (Cole, 1979). (Further detail on shrinking aggregates is provided in Chapter 5.) Of course, if aggregates are in a drier than saturated-surface-dry (SSD) state prior to charging into the mixer, they will absorb water during mixing, leading to slump loss. Stockpiles should be properly wetted in this case, or if this is impractical, additional mixing water should be provided to allow for absorption.

Aggregate attrition during handling and mixing

Aggregates are subjected to substantial surface attrition and abrasion during transport, handling, stockpiling, and mixing. Weaker and softer aggregates can experience breakdown due to these forces of attrition, resulting in a change in the aggregate grading and an increase in fines. In particularly troublesome cases, this can lead to severe loss of workability which might be compensated unwisely by adding extra water during mixing. This phenomenon is aggravated by use of longer mixing times or a period of agitation between mixing and placing which can occur when transport distances to the point of discharge are long. Abrasion and attrition tests for aggregates were discussed in Chapter 3.

ACI Committee 221 (ACI 221R-96) recommends that the susceptibility of a coarse aggregate to degradation during mixing can be assessed by the California durability index test (ASTM D3744). Alternatively, the aggregate can be assessed by increasing the shaking time in a sieve shaker and measuring the additional fines generated. The committee particularly mentions that results using the Los Angeles abrasion test (ASTM C131 and

C535) and the sulphate soundness test (ASTM C88) do not correlate well with degradation of aggregate during mixing, handling, and placing. Use of the Micro-Deval test discussed in Chapter 3 can also be considered for such aggregates.

Finishability of concrete

Finishability of concrete is a property much like workability – difficult to define and measure and dependent on a host of factors, including the particular operating conditions and type of finish required. Some are of the opinion that the amount of material retained on the 300-μm sieve is important. Certainly, most of the important effects discussed earlier such as particle shape, grading, fines, and so on will influence finishability. ACI Committee 221 has a succinct summary of this issue, quoted below.

> The angularity and grading of aggregate, the amount of bleeding, and mixture proportions of the concrete are factors that may influence finishing. When finishing problems occur, the work should be observed very critically, and the material properties and mixture proportions should likewise be reviewed to determine what might be done to improve the situation. Possible remedies to improve finishing of concrete include the use of additional fines in the fine aggregate, the use of a blending sand, more cement, more pozzolan, the use of some chemical admixtures, the use of air entrainment, adjustments to the aggregate grading (both fine and coarse), or changes in mixture proportions. If stickiness is the problem, fewer fines in the fine aggregate, less cement, less pozzolan, adjustments of chemical admixtures, or reduction in air content might help. If the problem is excessive bleeding, its reduction may be accomplished as discussed previously in Section 3.5*. Bleed water can be removed by drags or vacuum mats. If the problem is either fine or coarse aggregate in 9.5–2.36 mm ($^3/_8$ in. – No. 8) sieve sizes 'kicking up' or 'rocking' as the trowel is passed over the concrete, the amount of these sizes may be excessive. Also, this problem may be attributed to a large amount of very flat and elongated particles in the 9.5–4.75 mm ($^3/_8$ in. – No. 4) sieve sizes. Reduction of the amount of these sizes or elimination of these sizes completely can usually improve both the workability and finishing characteristics.
>
> (ACI 221R-96)

* Section 3.5 of ACI 221R-96 recommends that, where bleeding is excessive, attention should be given to the grading and angularity characteristics of the fine aggregate and to the mixture proportions. The use of finer fine aggregates, blending sand, improved control and grading of manufactured fine aggregate, increased cement and/or pozzolan content, use of

some chemical admixtures, and air entrainment are all factors that can reduce bleeding.

Particle packing and water requirement of mixes

Chapter 3 dealt with particle packing theories. It was pointed out that 'ideal' packing in which maximum packing densities (i.e. minimum void ratios) are achieved do not generally lead to workable concretes. Previous discussion indicated that water requirement for adequate plasticity of concrete mixtures will be a complex function of various factors in addition to grading, namely particle shape, surface texture, and maximum aggregate size. Comprehensive and universal relationships covering all these effects are not readily available. Nevertheless, considerable work has been done on the particular aspect of aggregate grading and packing, with applications to water requirement of concrete mixtures for adequate plasticity, that is the amount of water (and air) required to transform an aggregation of dry particles into a plastic mass on mixing. Powers (1968) has reviewed this field comprehensively, and showed that none of the water requirement theories and formulae is universally applicable. The factors that are generally accounted for in such theories are specific surface area of the solid particles and the ratio of the solid volume of the aggregates to that of the cement. In experimental work, water requirement was found to be not necessarily directly proportional to the specific surface area of aggregates for all types of mixes.

In general, the water-requirement of a mixture can be regarded as the sum of the water requirements of the individual size fractions. Since this includes all the solid fractions, that is cement (plus filler or extender as appropriate) plus the aggregates, this general statement can be re-phrased as being the water requirement of the cement and filler fraction, and that of the aggregate. The vast majority of the surface area of a mix is contributed by the fines, represented by the cement plus filler material plus aggregate fines (<150 μm). Therefore, the primary water requirement will be determined by the nature, quantity, and grading of this fines component. The balance of the water requirement is then ascribed to the aggregate portion (>150 μm). Empirical work has indicated that the aggregate water requirement can be expressed as an inverse function of the maximum size of the aggregate. Thus, a general form of the water-requirement expression is:

$$u_w = u_{cement+filler} + f[K/d_{max}] \qquad (4.2)$$

where

u_w is the water requirement of the concrete mix
$u_{cement+filler}$ is the water requirement of the cement plus filler component (materials <150 μm)

d_{max} is the maximum size of the aggregate
K represents empirical constant(s) as required for different types of mixes

with each of the terms on the right-hand side of Equation (4.2) being suitably scaled for their volumetric fractions. Several expressions similar to the one above have been derived for a range of mixes and the empirical constants evaluated (Powers, 1968). Equation (4.2) is reflected in the curves in Figure 4.12.

Turning to practical aspects, concrete mixtures can be designed and manufactured with a wide variety of materials and gradings to meet specified plastic and hardened properties within reasonable tolerances. Packing of aggregate particles is only one part of the overall picture, and theories of ideal total grading based on packing usually break down on the grounds that packing of fine particles depends not only on geometrical factors. While great strides have been made in understanding concrete plasticity and the various complex interrelating factors, the practical use of systematic trials to optimize mix proportions for any given set of materials is still the most satisfactory practical method.

Aggregates and concrete mix proportioning

It is appropriate in this chapter to examine concrete mix proportioning in relation to plastic properties from the perspective of the role of the aggregates. These materials play a decisive role in proportioning of a concrete mix for the appropriate workability, cohesiveness, and so on. The vast majority of mix proportioning is still largely carried out using empirical 'trial and error' procedures, although more sophisticated methods do exist, and it will usually be necessary to make up several batches before an acceptable mix is obtained.

Mix proportioning is the process whereby the proportions of the different constituents in a concrete are selected, and the yield of the mix is determined. (The yield is the volumetric quantity that can be delivered per batch of concrete.) Aggregates play their most important role in governing these proportions to satisfy the requirements for the plastic mix. The reader should refer to other standard texts for the details of mix proportioning; what follows here is a resume of different approaches taken in various parts of the world to address the influence of the aggregates on mix proportioning. A few general comments are appropriate first.

Workability, grading, and concrete mix proportioning

Workability in relation to aggregate grading has already been discussed. The desired workability in combination with the chosen aggregates determines the water requirement of a mix. For any given application or method of compaction, aggregates with desirable properties such as good particle shape, smooth surface texture, balanced gradings, and so on will have lower water requirement than poor aggregates. In addition, since the mix-water requirement is essentially constant over a wide range of binder contents in conventional concretes, it can be said that water requirement of a mix is the water requirement of the given set of aggregates.

The water requirement depends on the aggregate properties in a complex way. While it is possible to measure many of the aggregate properties for purposes such as quality control, these same parameters are not usually used in mix proportioning formulations. Thus, there is no rational method whereby the water requirement of a set of aggregates can be calculated from knowledge (or measured values) of the aggregate properties. The normal approach is for the engineer doing the mix proportioning to examine the aggregates, determine their grading, and then to estimate the likely water requirement for the concrete. At the same time the aggregates can be assessed in terms of the requirements for mix cohesiveness, control on bleeding and segregation, and so on. Experience and judgement are very important in this process. Once the initial assessments have been made, a set of mix proportions can be put together on paper, or using a computer algorithm. The next essential step is to take the mix proportions to the laboratory or to the site, and conduct a series of trials until the desired plastic properties are achieved. Inevitably, adjustments will have to be made to the initial design. These may include adjusting the relative proportions of coarse and fine aggregate, adjusting the grading of the aggregate possibly by blending, and adjusting the water content. This procedure would be repeated if an aggregate source were changed. Aggregates will vary with time even from a single and reliable source, making it essential to continually adjust the mix proportions to achieve consistent quality.

Referring specifically to aggregate grading, good concrete can be made with a wide range of gradings. The deciding factor will be the locally available materials, with the possibility of doing some blending if necessary. Further options may be available by the judicious use of admixtures such as water-reducers or air-entrainers to help overcome deficiencies in the aggregates or their grading. However, it is not possible to make good concrete with poor or deficient aggregates simply by use of chemical admixtures. Coarser sand gradings will reduce the quantity of stone in a mix and would be favoured for rich mixes. Additional stone could be included in a mix if heavier compactive effort can be used. In coarse-graded mixes, special attention needs to be given to cohesiveness if there is a deficiency of fines. By contrast, finer gradings may lead to 'sticky' mixes, particularly if cement contents are high. If the fineness of the grading leads to an unacceptable increase in water requirement, consideration should be given to blending in a coarser sand or, if this is not possible, to use of superplasticizers. In the case of water requirement being linked to the presence of unacceptable fines (e.g. unstable clays), the source of aggregate would need to be changed. Lastly, the preponderance of one particle size in an aggregate can lead to harshness of the mix. Maximum size of coarse aggregate has been discussed previously.

Reference has been made to computer algorithms for mix proportioning. Computer methods of mix proportioning are now in common use in concrete batching and ready mix plants. In many cases, these methods reflect conventional approaches to mix proportioning such as those discussed below. In other cases, they represent attempts at more fundamental approaches reviewed in Chapter 3. A particular example of the latter is the work by Dewar (1995, 1999) on a computerized mix proportioning method that accounts for relative density, bulk density (in effect, voids content), and grading of aggregates. Particle size range, shape, and surface texture are accounted for by the void content determination. The method uses the premise that, when two particulate materials of different size are mixed together, the smallest particles will fill the voids in the larger particles, but the resulting arrangement will be influenced by particle interference which is governed by shape and surface texture. Using mathematical modelling, Dewar's computer simulation provides a full range of possible mixtures from which the appropriate concretes can be selected for the specified mix requirement, based on minimizing the void content for adequate cohesiveness. A typical output for water and cement contents is shown in Figure 4.14. Computerized methods such as those of Dewar, and

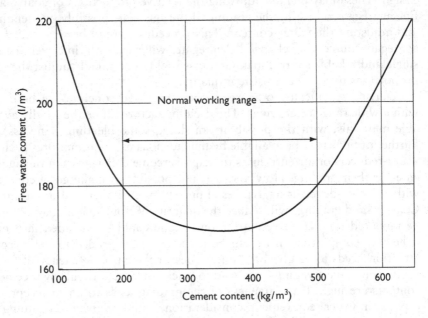

Figure 4.14 Computer simulation of relation between water content and cement content of concrete mixtures (after Dewar, 1995).

also Day (1999) and de Larrard (1999) are useful for production plants, but suffer from the problem that excessive detail is sometimes required and they may not be universally applicable, in view of the wide range in aggregate properties and types.

Approaches to mix proportioning

Mix proportioning procedures in various parts of the world, as they relate to aggregates, are summarized in Table 4.7. Approaches vary widely as might be expected, since a vast range of aggregate materials are represented. These approaches reflect local practice and have been developed with the constraints of economy in mind. Thus, poorer aggregates may well be used in some areas despite the resulting higher paste contents because the cost of importing better aggregates more than outweighs the additional cement cost.

North American approach

The approach in North America is based on the ACI Standard (ACI 211.1-91) 'Standard Practice for Selecting Proportions for Normal, Heavyweight and Mass Concrete', and is used in Canada as well as the USA. (The method in Canada is covered in Chapter 9, 'Designing and Proportioning Normal Concrete Mixtures', Kosmatka et al., 2002.)

Regarding aggregates, the method is based on the premise that workability is strongly influenced by (a) the grading of the aggregates (particle size and distribution), and (b) the nature of the particles (shape, porosity, and surface texture). It is assumed that the optimum content of coarse aggregate depends only on the maximum size of aggregate and the grading of the fine aggregate (i.e. its fineness modulus). In general, aggregates continuously graded over both fine and coarse sizes are used. Water requirement depends on slump and on maximum size, shape, and amount of coarse aggregate. The fine aggregate grading will depend on the cement content, the size of the coarse aggregate, and the type of concreting work to be done.

South African approach

The South African method called the 'C&CI Method of Mix Design' (Fulton, 2001) derives from ACI 211.1-91. As far as the aggregates are concerned it is based on the premise that (a) the water requirement for a given consistence (slump) and with given materials is substantially constant regardless of cement content, and (b) for any particular mix and set of materials, there is an optimum stone content which depends on size, shape, and compacted bulk density of the stone, fineness modulus of the sand, and

Table 4.7 Approaches to mix proportioning in various countries, in relation to aggregates

	North America	South Africa	UK
Basis of mix design method	Method is given in ACI 211.1-91. Aggregate properties (grading, maximum size, shape, porosity, and surface texture) influence workability, and thus water requirement. Coarse aggregate volume depends on maximum size of stone and on average fineness of sand. 'Absolute Volume Method', whereby the sum of the absolute volumes of all mix ingredients is required to be 1 m³, is the preferred method.	Method derives from ACI 211.1-91. Effectively uses 'Absolute Volume Method', based on the 'constancy of water requirement' premise. Mix design aims to achieve an 'optimum stone content', which depends on stone properties (including size) and sand fineness, for a given consistence.	Described in 'Design of Normal Concrete Mixes' (Teychenné et al., 1988, 1997). Distinction is made between crushed and uncrushed aggregates, and a 'mass–density' basis is used for calculation.
Required starting information	Bulk density of dry-rodded coarse aggregate; fineness modulus (FM) of sand; maximum size of coarse aggregate. Approximate mix-water requirement based on slump and maximum size of aggregate. Other: desired slump, air content, cementitious type and content, w/c.	Compacted Bulk Density (CBD) of stone (kg/m³); FM of sand (generally within the range 1.8–3.5); maximum size of stone. Estimation of water requirement based on sand quality. Other: Consistence (slump) of concrete, cement type, w/c.	Type of aggregate: crushed or uncrushed; maximum size of aggregate; relative density of aggregates; percentage of fine aggregate passing the 600-µm sieve. Other: w/c ratio, workability (slump).
Selection of coarse aggregate content	Obtained as a first trial from a table, giving volume of coarse aggregate per unit volume of concrete, on the basis of the dry-rodded coarse aggregate bulk density. This volume depends on the FM of the sand and the nominal maximum size of the stone.	Obtained as first estimate from formula: $M_{st} = CBD\ (K - 0.1FM)$ where M_{st} = mass of stone in 1 m³ of concrete (kg), CBD and FM are as previously defined, and K = a factor which depends on the maximum size of the stone and the workability of the stone.	Total aggregate content found by subtracting cement and water contents from fresh density of fully compacted concrete (from a chart). Coarse aggregate content is the difference between total aggregate content and fine aggregate content.

			Coarse aggregate is divided into size fractions depending on aggregate shape and maximum size. Determined from a chart for 20-mm or 40-mm coarse aggregate. Fine aggregate content depends on: maximum size of coarse aggregate, workability, w/c ratio, and percentage of fine aggregate passing 600-μm sieve.
		the concrete. In addition: the recommended max. M_{st} = CBD × 0.8 (approx.). Follows selection of the contents of all other mix constituents. Sand content is the volume required to make up 1 m³ of concrete, multiplied by the sand RD. The absolute volumes of the other constituents must therefore be determined.	
Selection of fine aggregate content	Fine aggregate content is determined once the amounts of all other ingredients are known. In the 'Absolute Volume Method', fine aggregate volume is found by subtracting the sum of the other absolute volumes of the other ingredients from 1 m³. Fine aggregate grading is constrained within a desirable grading envelope.		
Allowance for adjustments	1 Maximum size of stone is allowed for by adjusting the water requirement. 2 Water requirement can be reduced for better-shaped aggregates: up to 10 l/m³ for subangular aggregates, 20 l/m³ for gravel with some crushed particles, and 25 l/m³ for rounded gravel. 3 For lean mixes, a finer grading of sand is desired, and vice versa. No guidance is given, however.	1 Adjustment of water requirement for maximum size of stone. 2 For Fly Ash blends, stone content may be increased due to enhanced workability. 3 Use of sand blends and/or adjustment of sand/stone ratio, in case of harshness.	1 Water content of mix is adjusted depending on whether crushed or uncrushed aggregates are used. 2 Relative density of aggregates allowed for, in the range 2.4–2.9.

Table 4.7 (Continued)

	North America	South Africa	UK
Comments and comparisons	The North American approach explicitly considers the fineness modulus of the sand and the maximum size of the stone. Fine aggregate properties, while acknowledged to affect workability, are not explicitly accounted for. Coarse aggregate properties are covered indirectly by measuring their dry-rodded bulk density. North American practice is based usually on continuously graded aggregates over all sizes and use of relatively well-shaped gravels. However, crushed aggregates with their generally poorer shapes are acknowledged and allowed for by higher water requirements. The absorption of the aggregates is also explicitly allowed for.	The South African approach explicitly allows for average fineness of the sand (i.e. FM) and maximum size of the stone. Shape and surface texture of the sand are accounted for by judging the sand quality and increasing or decreasing the water requirement accordingly. Stone properties are indirectly accounted for by measuring the CBD of the stone. Stone used is usually single-sized (nominally), typically 19 mm. Sand grading is constrained within desirable grading limits, but a wide range of gradings (and FMs) are permitted.	The UK approach explicitly allows for maximum size of stone and sand fineness (by means of the proportion finer than 600 μm). The effect of allowing for workability is to increase the stone content for low slump mixes. Type of aggregate is also allowed for, i.e. whether it is crushed or uncrushed. The method gives no guidance on how to proceed if one of the aggregate portions (e.g. fine aggregates) is natural, and the stone is crushed. The grading of the coarse aggregate is ignored. Continuous grading of total aggregate is implied. Effect of shape of aggregate is only indirectly covered via adjustments to water content (for crushed or uncrushed aggregates).

desired concrete consistence. Thus, both water requirement and optimum stone content are a function of the aggregate properties, maximum size of the stone, and consistence. In addition, use of various binders (e.g. fly ash) is allowed for. The method contemplates the conventional range of concretes, and HSC is not directly covered. The mix constituents are calculated by mass, with the requirement that the absolute volumes should sum to the total volume of the concrete. The mix proportioning method has been derived to suit South African practice which almost invariably involves the use of crushed aggregates and gap-graded mixtures.

UK approach

This approach as far as aggregates are concerned allows for use of crushed or uncrushed aggregates, by (a) permitting higher concrete strength for crushed aggregates and (b) suggesting higher water content for crushed aggregates (Teychenné et al., 1988, 1997). Total aggregate content is found on a mass-density basis (i.e. fresh concrete density less cement content and water content in kg/m^3). It is assumed that continuously graded aggregates will be used with suitable limitations on the gradings.

In order to show differences between the mix proportioning methods discussed in Table 4.7, and how aggregates are accommodated in these methods, a typical concrete mix has been designed using the various approaches. The outcome is shown in Table 4.8. The requirements for the mix proportions were based on the example given in ACI 211.1-91 (Metric Example, Appendix 2), with some modifications outlined below.

- $w/c = 0.60$ – ordinary portland cement (ASTM Type I). Relative density $= 3.15$.
- Workability required in terms of slump: 100 mm.
- Coarse aggregates: Maximum size 25 mm; well-shaped (rounded) stone. Relative density $= 2.68$

  - Continuously graded for ACI and BS method (assumed uncrushed)
  - Nominally single-sized for South African approach (but assumed crushed)
  - Compacted bulk density (assumed equivalent to dry rodded mass) $= 1550$ kg/m^3.

- Fine aggregates: Well shaped, natural sands, relative density $= 2.64$

  - Continuously graded with FM $= 2.8$.

- Standard plasticizer (water-reducing admixture – WRA) at normal dosage (assume 5 per cent water reduction). Volume of water introduced by plasticizer included in total mixing water.

Table 4.8 Mix proportions of typical concrete mixes designed according to different approaches

	North America, ACI 211.1-91	South Africa, C&CI	UK
	(All quantities in kg/m³ unless otherwise noted)		
w/c ratio	0.6	0.6	0.6
Water requirement[1]	180	180	185
Air content[2]	1.5%	–	–
Cement content[3]	300	300	308
Coarse aggregate content[4]	1039	1070	1055
Fine aggregate content[5]	863	860	862
Total mass (kg)	2382	2410	2410
Total volume (l)	1000	1015	1015
Yield, Wet density (kg/m³)	2382	2374	2374

Notes

1 **ACI:** From Table A.1.5.3.3, mixing water requirement for 25 mm maximum size of aggregate is 193 l/m³. Allowing for a 5 per cent reduction due to the WRA and well-shaped aggregates, the estimated water requirement is 180 l/m³.

C&CI: Water requirements for typical South African crusher sands are generally in excess of 200 l/m³. However, in this case, a well-shaped natural sand is used (somewhat akin to certain Western Cape fine aggregates), together with a moderate dose of normal plasticizer. *Fulton* (2001) indicates the water requirement for 'good' aggregates to be 195 l/m³ for 19-mm aggregates, reduced by 10 l/m³ for 25-mm stone. A further 5 per cent reduction can be effected by the WRA. However, the 100-mm slump requirement would increase the water requirement by about 5 l/m³ compared with a 'standard' slump of 75 mm. This gives an estimate of the water required as about 180 l/m³.

UK: Assuming a 20-mm maximum size uncrushed coarse aggregate and 100-mm slump, the water requirement is 195 l/m³. A 5 per cent reduction can be affected by the WRA. It is noteworthy that if a crushed aggregate were to be used, the UK approach would recommend 225 l/m³ water requirement which is very high.

2 **ACI:** Table A.1.5.3.3 suggests 1.5 per cent entrapped air for 25 mm maximum size of aggregate.

C&CI: Air content is not normally explicitly allowed for in the C&CI method. However, most normal concretes would have about 1–2 per cent air after compaction. This would be regarded as a 'benefit' to the producer by increasing the yield of a mix. Thus, the yield of this mix has been increased by 1.5 per cent.

UK: The same comments apply as for the C&CI method.

3 Cement content is computed from w/c ratio and the estimated water requirement of the mix. The UK approach interestingly allows for an increase in concrete strength if crushed aggregates are being used. This would permit a somewhat higher w/c ratio for a given strength.

4 **ACI:** The quantity of coarse aggregate is estimated from Table A.1.5.3.6, using the fine aggregate FM of 2.8 and 25-mm nominal maximum size of coarse aggregate. This is found to be $0.67 \times 1550 = 1039 \, kg/m^3$ (rounded).

 C&CI: Consolidated Bulk Density of coarse aggregate assumed to be $1550 \, kg/m^3$ (based on well-shaped, relatively smooth stone particles). Factor for calculating stone content (K Factor) taken as 0.97.

 UK: The total aggregate content is first estimated, which requires an estimate of the fresh density of fully compacted concrete ($2410 \, kg/m^3$ in this case) and the RD of the aggregate (a weighted average of 2.66 assumed). The total aggregate content is computed as the difference between the fresh density and the sum of the water and cement contents.

5 **ACI:** Fine aggregate computed on Absolute Volume Basis allowing for all other ingredients including entrapped air.

 C&CI: The fine aggregate content computed from the difference between the sum of the absolute volumes of the other constituents, and $1 \, m^3$.

 UK: The fine aggregate content is determined as a proportion of the total aggregate, requiring knowledge of the percentage of fine aggregate passing the 600-μm sieve (assumed to be 44 per cent in this case based on FM = 2.8). Fine aggregate proportion estimated at 45 per cent.

- All aggregates in an SSD state, and all constituents mass-batched.
- Non air-entrained mix; low absorption aggregates.

Table 4.8 shows that despite the differences between the methods, they all give much the same outcome. The UK approach requires slightly higher water content, but all three methods give virtually identical fine aggregate contents. The ACI method specifically allows for air content, but if this is accounted for in the yield of the mix for the South African and UK approaches, the overall outcome is much the same. It is perhaps noteworthy that the South African and UK methods give identical values for total mass and yield wet density, despite the two methods differing in their approach. Table 4.8 indicates that the properties of the aggregates have a large influence on mix proportions, but that there are practical limitations to the various ratios, these being reflected in the various approaches.

General critique of methods of mix proportioning in relation to aggregates

The methods reviewed above only require explicit information on sand fineness, maximum size and bulk density of coarse aggregate, and an estimate of water requirement for the desired slump. Other information that may be used to adjust aggregate quantities comprises binder type and content, use of admixtures, qualitative judgements of the quality of the sand and stone, and w/c ratio. This information is adequate for making up trial mixes, based on which further adjustments can be made to achieve the specifications. Quantified measures of aggregate properties such as shape, texture, and grading are generally only used to ensure that the materials conform to 'accepted norms', so that 'unacceptable' materials are excluded. As stated earlier, the permissible limits for these properties are wide, reflecting the fact that good concrete can be made with a wide variety of aggregates, a point that cannot be emphasized too strongly.

Although the aggregate properties of shape, surface texture, and grading can be quantified, they are seldom if ever explicitly allowed for in concrete mix proportioning methods. This is understandable, and in fact it would seem to be an unnecessary complication for the normal run of mixes to reduce these properties to factors that could be individually accounted for in a mix proportioning formulation. As far as the coarse aggregates are concerned, these properties are adequately accounted for in terms of a measure of the compacted bulk density of the material. Shape and grading in particular, and surface texture to a lesser degree, are accounted for by the bulk density. In the South African approach, stone content depends on the workability required by means of a factor ('K' factor) that is greater for less workable mixes, whereas in the North American approach, adjustments to stone content for different workabilities are made based on guidelines suggested in the method.

For the fine aggregates, acceptable grading envelopes are usually provided, and the 'quality' of the sand is judged subjectively in order to arrive at a first estimate of mix-water requirement. In general, a greater fineness of sand (i.e. a lower FM) permits more stone to be used in the mix, and vice versa. This reflects the balance that is sought in overall particle specific surface area, and the fact that a certain amount of 'fines' actually aids in plasticizing a mix without increasing the water requirement.

While the North American and South African methods have many points of similarity, the UK approach differs in that total aggregate content is determined first, then the proportion of fine aggregate from which the coarse aggregate content follows. The method is less straightforward than the other approaches, and uses a different measure of sand fineness. Nevertheless, it is suitable for use with British aggregates. This last comment is important: the various mix proportioning methods have been derived or developed to suit the particular materials and concreting practice of the countries in which they are used. No one method is 'better' than another, but may be more suitable with certain types of materials.

This section would be incomplete without reference to the economics of concrete mixes in relation to aggregates. The task of the engineer is to produce a concrete with the required properties in both fresh and hardened states, *in the most economical way*. The cost of the concreting operation involves the costs of the materials, batching, mixing and transporting costs, and the costs of placing and compacting the concrete in the formwork, all without segregation of the mix. It is the *total* cost of this operation that must be optimized. Thus, while poorer aggregates might be selected based on their low cost, if these produce a segregating concrete or a concrete that requires excessive compactive effort due to its harshness or lack of workability, then the total cost of the operation may be unacceptably high and the finished product of poor quality.

Concerning materials costs, cement is by far the most expensive constituent of a mix. Aggregates comprise the bulk of a mix and so the mix costs are sensitive to aggregate costs. The challenge is to balance the overall cost of a mix by selecting aggregates that will help to minimize cement content without being too expensive. Such mixes will usually also be technically superior by minimizing the paste content. It is often economical and advantageous to select a set of good quality aggregates with which good quality concrete can be made at reasonably low water content. For equal strength mixes (at equal w/c), the use of poor aggregates that require high water contents for adequate workability will lead to elevated cement contents, and the cost of the extra cement may more than offset any cost reduction related to the aggregates. As an example, for a mix with a w/c ratio of 0.45, a difference in water requirement of about $20\,l/m^3$ for two different sets of aggregates would result in a difference in cement content of about 50 kg/m³. When accumulated over several thousand cubic metres of concrete on a large job, this difference could amount to a substantial sum of money. Thus, aggregate selection is crucial to both the economics and the technical quality of a concrete mix.

Closure

This chapter has reviewed the important properties of aggregates that influence the plastic properties of concrete. It is in this area that aggregates generally have their most visible and measurable effects on concrete. Fine aggregates, due to their greater surface area, tend to have a larger influence than coarse aggregates. It is imperative that the concrete engineer fully understands the nature of the aggregates and their likely influences on the plastic concrete, in order to produce hardened concrete that is fully compacted and delivered at an economic cost. We now turn to the subject of hardened concrete in the next two chapters, in order to examine how aggregates influence the physical, mechanical, durability and transport properties.

References

American Concrete Institute Committee 221 (2001) 'Guide for use of normal weight and heavyweight aggregates in concrete', ACI 221R-96 (Re-approved 2001), *American Concrete Institute Manual of Concrete Practice*, Farmington Hills, MI: American Concrete Institute.
American Concrete Institute Committee 211 (2002) 'Standard practice for selecting proportions for normal, heavyweight, and mass concrete', ACI 211.1-91 (Re-approved 2002), *American Concrete Institute Manual of Concrete Practice*, Farmington Hills, MI: American Concrete Institute.
Bartos, P. (1992) *Fresh Concrete Properties and Tests*, London: Elsevier.
Blanks, R.F. (1952) 'Good concrete depends upon good aggregate', *American Society of Civil Engineers*, 22(9): 651–656, 845.
Bonavetti, V.L. and Irassar, E.F. (1994) 'The effect of stone dust content in sand', *Cement and Concrete Research*, 24(3): 580–590.
Brower, L.E. and Ferraris, C.F. (2003) 'Comparison of concrete rheometers', *Concrete International*, 25(8): 41–47.
Chapman, G.P. and Roeder, A.R. (1970) 'The effects of sea-shells in concrete aggregates', *Concrete*, 4(2): 71–79.
Cole, W.F. (1979) 'Dimensionally unstable grey basalt', *Cement and Concrete Research*, 9(4): 425–430.
Davis, D.E. (1975) 'The concrete-making properties of South African aggregate', PhD thesis, University of the Witwatersrand, Johannesburg, Vol. 1, Ch. 3, Part 1A.
Day, K.W. (1999) *Concrete Mix Design, Quality Control and Specification*, London: E&FN Spon.
de Larrard, F. (1999) *Concrete Mixture Proportioning: A Scientific Approach*, London: E&FN Spon.

Dewar, J.D. (1963) 'Effects of mica in the fine aggregate on the water requirement and strength of concrete', *TRA/370*, London: Cement and Concrete Association.

Dewar, J.D. (1995) 'A concrete laboratory in a computer – case studies of simulation of laboratory test mixes', *Proceedings of 11th European Ready Mixed Concrete (ERMCO) Congress*, Istanbul, 185–193.

Dewar, J.D. (1999) *Computer Modelling of Concrete Mixtures*, London: E&FN Spon.

Dimond, C.R. and Bloomer, S.J. (1977) 'A consideration of the DIN flow table', *Concrete*, 11(12): 29–30.

DIN 1048: Part 1, *Testing Concrete: Testing of Fresh Concrete*, Berlin: Beuth Verlag.

Edwards, L.N. (1918) 'Proportioning the materials of mortars and concretes by surface area of aggregates', *Proceedings ASTM*, 18, Part II, 235–302.

EFNARC (2002) *Specification and Guidelines for Self-Compacting Concrete*, Farnham, Surrey: EFNARC.

Fookes, P.G. and Revie, W.A. (1982) 'Mica in concrete – a case history from Eastern Nepal', *Concrete*, 16(3): 12–16.

Fulton's Concrete Technology (1994) Eds B.J. Addis and G. Owens, 7th edn, Midrand: Cement and Concrete Institute.

Fulton's Concrete Technology (2001) Eds B.J. Addis and G. Owens, 8th edn, Midrand: Cement and Concrete Institute.

Galloway, J.E. (1994) 'Grading, shape, and surface properties', in *Significance of Tests and Properties of Concrete and Concrete Making Materials*, ASTM STP 169C: 401–410, West Conshohocken, PA: American Society for Testing and Materials.

Gaynor, R.D. and Meininger, R.C. (1983) 'Evaluating concrete sands', *Concrete International: Design and Construction*, 5(12): 53–60.

Glanville, W.H., Collins, A.R. and Matthews, D.D. (1947) 'The grading of aggregates and the workability of concrete', *Road Research Technical Paper No. 5*, 2nd edn, Crowthorne: HMSO.

Grieve, G.R.H. (2001) 'Aggregates for concrete', in B.J. Addis and G. Owens (eds) *Fulton's Concrete Technology*, 8th edn, Midrand: Cement and Concrete Institute.

International Standards Organisation, 'Fresh concrete – Determination of the consistency – Vebe test', ISO 4110: 1979, Geneva: International Standards Organisation.

Japan Society of Civil Engineers (1992) *Recommendations for Design and Construction of Anti-Washout Underwater Concrete*, Concrete Library of JSCE, 19: 89.

Kosmatka, S.H., Panarese, W.C., Gissing, K.D. and Macleod, N.F. (1995) *Design and Control of Concrete Mixtures*, 6th Canadian edn, Ottawa, ON: Canadian Portland Cement Association.

Kosmatka, S.H., Kerchoff, B., Panarese, W.C., Macleod, N.F. and McGrath, R.J. (2002) *Design and Control of Concrete Mixtures*, Engineering Bulletin 101, 7th Canadian edn, Ottawa, ON: Cement Association of Canada.

Loudon, A.G. (1952) 'The computation of permeability from simple soil tests', *Géotechnique*, 3(4): 65–83.

McIntosh, J.D. and Erntroy, H.C. (1955) 'The workability of concrete mixes with 3/8 in. aggregates', *Cement & Concrete Association Research Report No. 2*, London: Cement & Concrete Association.

Murdock, L.J. (1960) 'The workability of concrete', *Magazine of Concrete Research*, 12(36): 135–144.

Neville, A.M. (1995) *Properties of Concrete*, 4th and final edn, Harlow: Longman.

Petersson, O., Billberg, P. and Van, B.K. (1996) 'A model for self-compacting concrete', in *Proceedings P. Bartos (ed.) RILEM Conference on Production Methods and Workability of Concrete*, London: Chapman & Hall, 483–490.

Powers, T.C. (1968) *The Properties of Fresh Concrete*, New York: John Wiley & Sons.

Road Research Laboratory (1950) 'Design of concrete mixes', *DSIR Road Note No. 4*, London: HMSO.

Schmitt, J.W. (1990) 'Effects of mica, aggregate coatings and water-soluble impurities on concrete', *Concrete International*, 12(12): 54–57.

Sims, I. and Brown, B. (1998) 'Concrete aggregates', in P.C. Hewlett (ed.) *Lea's Chemistry of Cement and Concrete*, 4th edn, London: Arnold.

Tattersall, G.H. (1991) *Workability and Quality Control of Concrete*, London: E&FN Spon.

Teychenné, D.C., Franklin, R.E., Erntroy, H.C. (Original authors) and Teychenné, D.C., Nicholls, J.C., Franklin, R.E. and Hobbs, D.W. (Revision Panel, 1988 edn) (1988) 'Design of normal concrete mixes', *Building Research Establishment (BRE Report 106)*, Garston: Building Research Establishment, 42pp.

Teychenné, D.C., Franklin, R., Erntroy, H.C. and Marsh, B.K. (1997) *Design of Normal Concrete Mixes*, 2nd edn, Garston: Building Research Establishment.

Walker, S., Bloem, D.L. and Gaynor, R.D. (1959) 'Relationship of concrete strength to maximum size of aggregate', *Proceedings Highway Res. Bd. No. 38*, Washington, DC: Highway Research Board, 367–379.

Wills, M.H. (1967) 'How aggregate particle shape influences concrete mixing water requirement and strength', *Journal of Materials*, 2(4): 843–865.

Further reading

Malhotra, V.M. (1964) 'Correlation between particle shape and surface texture of fine aggregates and their water requirements', *Materials Research and Standards*, 4(12): 656–658.

Pike, D.C. (1989) 'Report to CAB/2 on a project to examine the regulation of fines in sands for concrete and mortar', Private circulation document 89/17125 (Technical Committee CAB/2, Aggregates), Unpublished, London: British Standards Institution.

Chapter 5

Aggregates in hardened concrete

Physical and mechanical properties

Aggregates have long been regarded simply as 'inert fillers' in concrete, there to provide bulk and economy. A 1931 monograph on 'Cements and Aggregates' stated that 'The coarse aggregate in concrete is simply an inert filler used to reduce the cost', and went on to say that the type of coarse aggregate has relatively small effect on the strength of the concrete, provided it is sound (Baker, 1931). This view has unfortunately prevailed even up to the present time among engineers.

We now know that aggregates can have very profound influences on the physical and mechanical properties of hardened concrete. In this chapter, we will outline the concept of the interfacial transitional zone (ITZ), and stress that appropriate 'engineering' of the ITZ has a role to play in controlling the performance of the concrete composite. Even in conventional concretes where no particular attempt is made to engineer the ITZ, there is abundant evidence that different aggregate types influence mechanical properties. Different aggregate types may interact differently with the matrix, and these differences may be technically important depending on the magnitude of the influence. On occasions, improvements in strength induced by use of different aggregates may be economically important, by permitting significant reductions in cement content. Whether engineers can exploit such effects will depend on the geographical proximity of different aggregate sources to a construction site, and their cost. Technical and economic exploitation is possible where aggregates are derived from dedicated quarries producing a rock type of assured consistency. It would be less likely in areas where aggregate sources comprise mixed and variable gravels.

Fundamental aspects

The properties and performance of cement-based materials are governed largely by their microstructure, that is the nature of the solid and pore phases at a scale of microns and smaller. The microstructure contains distinct phases which interact with larger scale constituents such as aggregates.

Such large-scale inclusions in the cement matrix impart their own characteristics to the overall composite, so that the engineering properties are an integration of the characteristics of the various phases at their respective scales. In this context, we are concerned mostly with the influence of large-scale aggregates which can be regarded as generally hard inclusions in a relatively homogeneous and softer matrix (which of course at a nanometer scale is itself a very heterogeneous composite!) Consequently, in concrete we are dealing with a highly complex, inhomogenous material that by its very nature will not be simple to describe or model. It is beyond the scope of this book to deal comprehensively with the complex interaction between microstructure and properties of the composite. Our focus is rather on the particular contribution of the aggregates and how they influence the engineering properties. Nevertheless, a brief introduction to the multi-phase nature of concrete is appropriate, and in particular the nature and role of the ITZ between paste and aggregate.

Phases in concrete and their role

Microstructural observations using SEM techniques reveal that concrete can be represented as a three-phase model: a 'bulk' cement paste phase, an aggregate phase, and a phase linking these two, referred to as the interfacial transition zone. This zone is up to 50-μm thick and surrounds the aggregate particles. It contains phases similar to those in the bulk paste (e.g. calcium hydroxide crystals, ettringite, CSH, and pores), but their volume fractions are significantly different from those in the bulk paste: usually there is less unhydrated cement and higher porosity, with larger pores than those in the bulk paste. Large oriented crystals of CH may often be present in the ITZ with a greater concentration of ettringite and a lower concentration of CSH. A diagrammatic representation of the ITZ showing the essential features is given in Figure 5.1. The ITZ may constitute about 30–60 per cent of the hardened paste phase. The presence of an ITZ with properties significantly different from the bulk paste must influence properties of the composite material.

Formation and characteristics of ITZ

Two major mechanisms contribute to the formation of the ITZ. First, there is the so-called 'wall effect', whereby packing constraints of small cement particles against larger aggregate surfaces cause less efficient geometrical arrangement of the cement. This results in fewer cement particles near the aggregate surface than in the bulk paste. Bleeding tends to exaggerate this effect by forming a layer of water on the aggregate surface which also increases the w/c ratio locally. ITZ microstructure therefore depends on the particle size distribution of the binder and its ability to pack efficiently

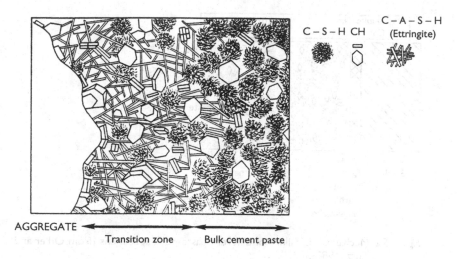

$C-S-H$ CH $C-A-S-H$
(Ettringite)

AGGREGATE
Transition zone   Bulk cement paste

Figure 5.1 Schematic view of the ITZ around an aggregate particle (from Mehta and Monteiro, 1993).

at the aggregate surface. The second mechanism is the one-sided growth effect, whereby reactive growth of the cement particles occurs only from the paste side, rather than from all directions. These two mechanisms increase the porosity of the ITZ relative to the bulk paste. The ITZ around a single particle is itself highly variable due to 'trapped' water below the coarse aggregate particles.

The ITZ is generally weaker than the bulk paste, although this is not always the case. In most conventional concretes, failure and fracture by cracking first takes place predominantly in the ITZ, before branching into the bulk paste (Mindess and Alexander, 1995). Studies of fracture of paste–aggregate interfaces show that, depending on the nature of the rock and paste matrix, failure may occur at the interface or at some distance into the paste matrix. The latter occurs when the interface is strengthened by silica fume for instance, or when a rock bonds particularly well with the paste, as shown in Figure 5.2 (Odler and Zurz, 1988; Alexander *et al.*, 1995).

Study of the ITZ and its role in cement-based materials has been the subject of much research (Maso, 1992; Katz *et al.*, 1998; Alexander *et al.*, 1999). The properties of the ITZ should be viewed not as well-defined material properties, but rather as system properties dependent on the composition as well as the method of fabrication of the cement composite (Bentur and Alexander, 2000). The ITZ adds another level of heterogeneity to the system over and above that inherent in the binder phase. Inverse modelling in which real concretes are tested and the results simulated by numerical meso-level models (micron-scale) provides a means for quantifying the properties of the ITZ.

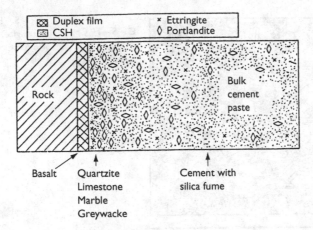

Figure 5.2 Mechanism of failure of different paste–rock specimens (from Odler and Zurz, 1988).

In work on the influence of aggregate size on the ITZ, Elsharief *et al.* (2003) studied two series of mortars with aggregate sizes in the 2.36–4.75-mm and 150–300-μm range, at w/c ratios of 0.55 and 0.40. Backscattered electron imaging was used to characterize the ITZ microstructure in terms of porosity and unhydrated cement content. Porosity results are shown in Figure 5.3. The smaller aggregate sizes gave reduced porosities and increased content of unhydrated particles in the ITZ. Reducing the w/c ratio had a greater influence on the 150–300-μm mortar than on the larger aggregate mortar, such that at $w/c = 0.40$ and 180 days, the porosity beyond about 15 μm from the aggregate particles was virtually indistinguishable from the bulk paste. The development of the ITZ microstructure with time depended on its initial microstructure: the larger the initial content of unhydrated cement in the ITZ, such as occurred for the w/c 0.4 mortar, the greater the reduction in porosity with time. The ITZ densified with time due to additional hydration, which also reduced the thickness of the ITZ. These microstructural observations find expression in the performance of concrete as discussed later in the chapter.

Notwithstanding the above, the ITZ is not necessarily ubiquitous in all concretes, but depends on the nature of the materials, mixture proportions, and manufacture of the concrete. The conventional view of the ITZ as a distinct zone around aggregate particles has been challenged by Diamond (2003), who found from studies on mortars that the hardened cement paste (hcp) consisted of patches of dense, non-porous regions and highly porous patches; such patches indifferently occupying both classical 'ITZ' and 'bulk' locations. Chapter 6 makes further mention of this.

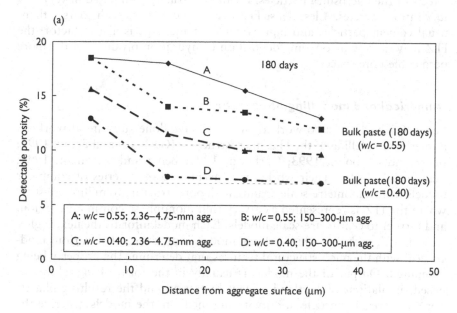

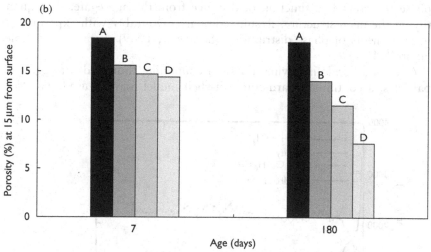

Figure 5.3 Influence of aggregate size on porosity of the ITZ in mortars with *w/c* ratios of 0.55 and 0.40: (a) Porosity with distance from the aggregate surface; (b) Variation with age of porosity at 15 μm (from data in Elsharief *et al.*, 2003).

Possibly the best approach is to regard the ITZ as another phase in concrete which can be 'engineered' to a greater or lesser extent. It can be modified by the addition of very fine fillers such as silica fume, in which case the ITZ is smaller and may in fact be absent due to the densifying effects of the microsilica particles. It can be virtually eliminated in very low w/c ratio concretes. A less dense ITZ occurs in concretes with coarser-than-usual cement particles and high water contents. Age is also a factor: the ITZ may densify by ongoing deposition of hydration products in the more permeable pore space.

Numerical and modelling approaches

Bentz, Garboczi, and co-workers at NIST have done substantial work on numerical modelling of the ITZ microstructure (Garboczi and Bentz, 1991; Bentz and Garboczi, 1993; Garboczi, 1993; Bentz and Stutzman, 1994; Bentz et al., 1995). Their work has concentrated on a series of computer models from nanometre-scale dealing with pore structure to millimetre-scale where the ITZ is represented as a very thin monophase continuous region, and further to centimetre-scale models. Each model informs the next higher scale model, such that a hierarchy of models is produced. The various models deal with the microstructural features that determine the property being computed. Output of the models is usually in the form of digital-image-based simulations of portland cement hydration and the resulting phases. For an aggregate particle in a hydrating medium, the models generate the phase fractions as a function of distance from the aggregate, thus quantifying the microstructure. Results compare favourably with experimental measurements of phase distribution (Scrivener, 1989) – see, for example, Figure 5.4.

At a larger scale involving a system containing thousands of aggregate particles, a continuum hard-core/soft-shell model may be used. The 'soft

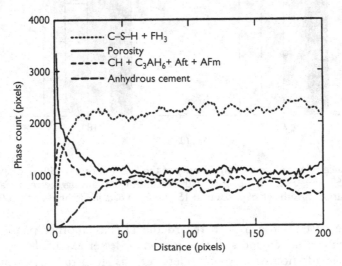

Figure 5.4 Phase volume fractions in simulated ITZ (from Bentz and Stutzman, 1994).

shells' are permitted to overlap during initial packing of the model, while the 'hard core' simulated aggregate particles are not. The soft shells may also be considered as enveloping ITZs which are penetrable and can therefore permit transport of substances through the bulk material – see Chapter 6. From the perspective of concrete mechanical properties, the soft shell ITZs represent zones of weaker strength and lower stiffness.

The suite of computer-based models has been used to study aspects such as ITZ chemistry and the effects of cement type and characteristics, cement extenders, and admixtures. With reference to aggregates, the models can be used to study the effects of, for example, sorptivity and reactivity of aggregates. Absorptive particles such as lightweight aggregates act as filters, drawing in water and pulling cement particles towards their surface, densifying the ITZ, and improving its properties. A reactive aggregate such as cement clinker (while being somewhat unrealistic) is both slightly absorptive and reactive, eliminating the wall effect and the one-sided growth effect, and resulting in the virtual elimination of the ITZ.

A recent development regarding aggregate modelling is a mathematical procedure using spherical harmonic functions to characterize shape and size of aggregate particles. The 3-D particle images are acquired using X-ray tomography (see Figure 3.7). This approach allows comparison of composite performance based on precise morphological aspects of particles, and incorporation of random particles into multi-particle computational models (Garboczi, 2002).

Our interest in microstructure here is largely practical, that is we are interested in its effects on the engineering properties of the composite concrete. Computer-based models can be used not only to simulate microstructure but also to infer the effects of microstructure on concrete properties. Where relevant, these are reviewed later in this chapter.

The following sections will discuss the influence of different aggregate types and properties on several properties of concrete: compressive and tensile strength, fracture behaviour, deformation behaviour, and wear properties.

Concrete strength

Concrete structures must be sufficiently strong and stable to resist applied loads and they must also be stiff enough to provide load resistance without undue deformation. The role of aggregates and their effects on mechanical properties are important due to

- The current development and use of high strength concrete (HSC), where aggregate effects become more pronounced.
- Application of the concepts of materials engineering in which properties of the material are controlled and 'engineered' by selection of con-

stituents including aggregates, and an understanding of their interaction in the composite.

• The need to preserve aggregate resources and develop alternative sources.

Review

Duff Abram's original (1918) w/c ratio law for concrete strength stated that 'For a given cement and conventional aggregates in workable mixtures, under similar conditions of placement, curing and test, the strength of concrete is *solely* a function of the ratio of cement to the free water in the plastic mixture' (Abrams, 1918; italics added for emphasis). This was limited to normal structural concrete, that is graded aggregates with maximum size not exceeding 38 mm, and with w/c ratios greater than about 0.4 (to ensure adequate compaction). This oversimplification led to the incorrect assumption that physical and elastic properties of aggregate have no effect on the compressive strength of concrete – in effect the 'inert filler' concept. The importance of Abram's statement lay in recognizing the key role of w/c ratio, but his 'law' is really a special case of a series of relationships between compressive strength and w/c ratio in which other factors, including aggregate factors, also play a role.

This is particularly true of HSC in which, due to the dense microstructure and strong transition zone, the mechanical properties of the aggregates become very important. The inclusion of aggregates in the failure mechanisms of HSC creates a further 'class' of concretes which no longer rigidly obeys Abram's w/c ratio law. At present, however, no universal or sensible mathematical relationships have been derived to allow general prediction of concrete strength as a function of both paste and aggregate variables.

Early studies on the effects of aggregates on concrete strength were conducted in both the United States and the United Kingdom. In 1961, Gilkey published a review paper on the limitations of the Abram's w/c ratio theory (Gilkey, 1961), in which he contended that the law constituted a serious oversimplification that blocked sound thinking. His work focused attention on the heterogeneity and non-isotropicity of concrete. He proposed that the w/c ratio law be extended to include aspects such as the ratio of cement to aggregate, and the grading, surface texture, stiffness, and maximum size of the aggregate. The w/c ratio law could be generalized into a series of relationships represented by a family of overlapping, approximately parallel strength curves.

The work of Bloem and others focused mainly on the maximum size of coarse aggregate which influenced concrete strength independently of w/c ratio (Walker and Bloem, 1960; Bloem, 1961; Bloem and Gaynor, 1963). Strength decreased as aggregate size increased over the full range of w/c, with typical results shown in Figure 5.5. This effect could be offset

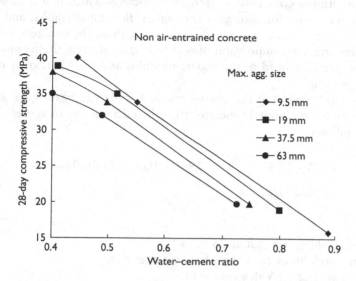

Figure 5.5 Water/cement ratio versus strength for different maximum size of aggregates (based on data in Walker and Bloem, 1960).

in leaner concretes of lower strength by the reduction in mixing water requirement that accompanies an increase in aggregate size. In general, these relationships are of limited significance, since aggregates of the same size but from different sources can give much larger strength differences than the influence of size *per se*.

These effects are additional to, and separate from, the influence of aggregate characteristics such as shape, surface texture, grading, and maximum size on water requirement and thus strength of a mix. It is generally true that, other factors being equal, lower mix-water content yields improved concrete mechanical properties.

The early UK work involved concretes varying in strength from 13 MPa (7 days) to 86 MPa (91 days), using aggregates with dynamic elastic moduli from 23 to 78 GPa (Kaplan, 1959a,b). The average maximum differences in compressive and flexural strength for the same mix proportions at the same age were about 21 and 31 per cent, respectively. For the strongest set of mixes, an increase in compressive strength of 17 MPa (at 91 days) could be obtained by careful selection of aggregates without changing the mix proportions. Aggregate properties affecting compressive strength were surface texture, shape, and modulus of elasticity. Aggregate strength within the limits of the tests had no effect on compressive strength. Regarding surface texture, concretes containing smooth gravels had crack initiation at lower compressive stresses than concretes containing coarser textured aggregates. These differences in crack initiation stresses were greater than the differences in ultimate compressive strength. Concretes which best resisted pre-cracking in compression also gave the highest flexural strengths, and a unique relationship existed between flexural strength of the concrete and crack initiation stress in compression. Regarding flexural strength, the most important properties related to the elastic modulus and surface texture of the coarse aggregate.

Other UK work involving the use of crushed stone aggregate gave an empirical relationship between concrete strength and a number of aggregate variables as follows:

$$\log_{10}(u) = 3.896 - 1.293w/c + 0.0296E + 0.00547i + 0.0168n$$
$$+ 0.0225a/c \tag{5.1}$$

where
u is the cube crushing strength at 28 days (Ψ)
E is the static modulus of elasticity of coarse aggregate
i is the Aggregate Impact Value (per cent)
n is the angularity number (see Chapter 3 for discussion of these two variables)
a/c is the coarse aggregate/cement ratio (Bennett and Khilji, 1964).

The variables in Equation (5.1) reflect aggregate properties that could influence concrete strength, that is stiffness (E), strength (i), a measure of shape (n), and aggregate volume concentration (a/c). The modulus of elasticity of the coarse aggregate had a considerable effect on the strength, exceeded only by the influence of w/c ratio. De Larrard and Belloc (1997) presented a semi-empirical approach to quantify the influence of aggregates on the compressive strength of concrete. The method centred around the concept of 'Maximum Paste Thickness' (MPT), which can be interpreted as the mean distance between aggregate particles, and is described in terms of three important mix variables: the maximum size of aggregate D_{max}, the aggregate volume concentration g, and the granular packing density of the aggregate g^*, thus:

$$\mathrm{MPT} = D_{max} \left(\sqrt[3]{\frac{g^*}{g}} - 1 \right) \tag{5.2}$$

These parameters can be measured or estimated for a mix and provide the 'topological' effects concerned with volume concentration and maximum size of aggregate. De Larrard gave four independent sets of data which indicated that the strength of a particulate composite such as concrete depends on MPT – see Figure 5.6.

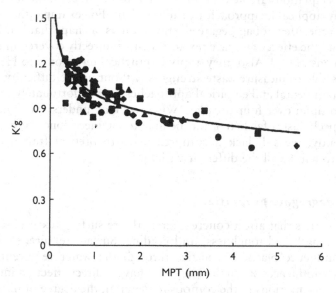

Figure 5.6 Compressive strength (normalized for effect of w/c) as a function of MPT (from de Larrard and Belloc, 1997).

Note
Symbols represent four independent sets of data.

The further effects of bond between paste and aggregate and limitations on compressive strength due to inherent strength of the rock (aggregate) are described by the empirical equation:

$$f'c_c = \frac{a \cdot f'c_m}{b \cdot f'c_m + 1} \qquad (5.3)$$

where
$f'c_c$ = composite (concrete) strength
$f'c_m$ = matrix strength
a, b = empirical constants depending on the type of aggregate (and the age of testing).

The matrix strength $f'c_m$ is found from the paste strength which renders the approach rather awkward for normal concretes where this value is difficult to measure due to the tendency for pastes to segregate at normal w/c ratios. For very high matrix strengths, the composite strength tends to the value a/b so that this ratio is controlled by the intrinsic strength of the rock. Conversely, for low matrix strengths (or alternatively very high aggregate strengths), the strength of the composite is approximately equal to $a \cdot f'c_m$, so that 'a' can be taken to describe the bond between paste and aggregate.

De Larrard's equations are helpful but still fail to provide a fundamental and universally applicable approach to the problem. For example they do not account for the effects of aggregate elastic modulus or shape that Kaplan pointed out, and the effect of surface texture is only indirectly covered in the experimental constant 'a'. Also, they apply in practical terms only to HSCs where it is possible to measure paste strengths without undue difficulty.

All these experimental and empirical approaches have unfortunately been little improved upon over four decades. While we may understand failure and fracture mechanisms far better and be able to engineer concrete mixes more intelligently, we still lack a comprehensive theoretical framework within which to handle all the different variables.

Influence of aggregate properties

Aggregate properties that affect concrete strength are surface texture, stiffness, shape, strength and toughness, and grading. Surface texture, shape, and grading influence plastic properties through the water requirement which indirectly influences strength. They also have a direct effect by influencing stress concentrations in the composite material, the degree of microcracking and crack branching before and during failure, tortuosity of crack paths, macro- and micro-roughness effects at interfaces, and amount of bonded surface area of the aggregates.

For a proper description of concrete strength, we must consider strength of the binder phase, of the aggregate phase, and of the interfacial bond between the phases. The strength of the binder phase is not of direct interest here, whereas strength of the aggregate and of the aggregate-paste bond are. Aggregate *type* will be considered first. While this approach is not altogether useful in that it treats aggregates generically without accounting for their physical or mechanical properties, it helps to illustrate that different aggregate types can have profound influences on strength.

Aggregate type

Aggregate type *per se* is not the important factor. The physical and mechanical properties of the aggregate are really of concern. Within any aggregate type, such properties can vary extremely widely. Irrespective of aggregate type, aggregates with similar properties should perform similarly in concrete, assuming that the important factors can be quantified. Aggregate type is considered here because this is a convenient way of reporting some of the data in the literature.

The work by Kaplan on aggregate type has already been mentioned. Further evidence comes from an experimental study on 23 aggregate types in common use in South Africa, which provides the bulk of the data discussed here. All coarse aggregates were crushed materials and in most cases the fine aggregates were also crushed from the same source as the stone. Rock types represented were andesite (two sources), dolerite (three sources), dolomite (two sources), felsite (two sources), granite (three sources), greywacke (one source), quartzite (eight sources), siltstone (one source), and tillite (one source).

On the basis of the relationship between compressive strength and cement/water ratio, various aggregate types displayed similar behaviour, shown in Figure 5.7. (Cement/water ratio was used instead of the more common water/cement (w/c) ratio due to the roughly linear relationship obtained.) Group 1 aggregates comprised an andesite, two dolerites, two quartzites, and a felsite and siltstone, and gave higher concrete strengths. The unconfined compressive strength (UCS) of these rocks varied from about 100 to 540 MPa, which indicates that rock strength was not a strong influencing factor. These tests were carried out with one type of ordinary portland cement, and strength variations may be attributed primarily to aggregate effects which include the ITZ associated with different aggregates. Differences in concrete strength over the range of aggregates and w/c ratios varied from 30 to 55 per cent on a relative basis (absolute variations were from 13 to 17 MPa). These variations are technically substantial and may often be economically important. Strength depends on many factors, not all of which can be isolated in these results.

In HSC, aggregate type also has an important influence on concrete strength (Aïtcin, 1989). Canadian tests on two concretes of $w/c = 0.275$,

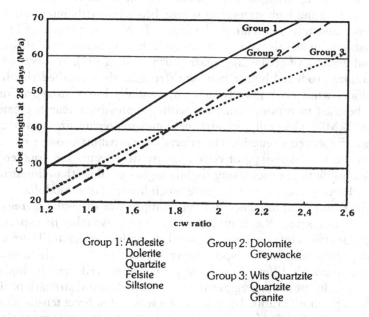

Group 1: Andesite   Group 2: Dolomite
        Dolerite   Greywacke
        Quartzite
        Felsite   Group 3: Wits Quartzite
        Siltstone   Quartzite
                               Granite

Figure 5.7 Relationship between 28-day compressive strength and c/w ratio for different aggregate groups (from Alexander and Davis, 1991).

identical in every respect except that one contained 10-mm fresh diabase and the other 14-mm weathered granite as coarse aggregate, showed strength variations of between 10 and 14 per cent depending on the age of testing. Variations were larger at later ages ($\geq$28 days) amounting in absolute terms to a maximum of about 16 MPa.

Other studies have examined the effects of coarse aggregate type on concrete strength. One study involved three coarse aggregate types – a crushed basalt, crushed limestone, and rounded gravel, all 19-mm nominal maximum size (Özturan and Çeçen, 1997). Information such as surface texture or degree of weathering was not given. Concrete strengths varied from 40 to 90 MPa and it was confirmed that aggregate type had a measurable influence on strength. For example, the two crushed aggregates gave up to 20 per cent higher compressive, flexural, and splitting tensile strength in both low and high strength concretes. For concrete of 40 MPa compressive strength, the absolute differences were smaller, but the relative differences were much the same as for the HSC. These findings confirm the aggregate-specific nature of the problem. Another study found that aggregate types which gave superior strength in normal strength mixes did not necessarily give the best performance in high strength mixes (Sengul *et al.*, 2002). The important factors were elastic mismatch between aggregate and matrix at lower strengths, and strength of the aggregate itself at higher concrete strengths. For all strengths, a strong interfacial zone contributes to enhanced concrete strength. These factors reflect the different modes of failure for normal and high strength concrete, being primarily intergranular in the former and transgranular in the latter. Finally, in a study of HPC using Oklahoma materials, compressive strength of concrete was affected by type and grading of coarse aggregate (Bush *et al.*, 1997). For example, crushed granite produced higher concrete strengths than crushed rhyolite and limestone, while river gravels gave substantially lower strengths and could not be used to produce concrete with compressive strength greater than about 70 MPa. Generally, smaller maximum aggregate sizes (less than 19 mm) gave improved strengths. The effects were attributed to less microcracking in the ITZ. The type of coarse aggregate affected the concrete's elastic modulus, with granites giving slightly higher E values than limestone or rhyolite. River gravels, however, gave much lower elastic moduli.

Aggregate factors that influence concrete compressive strength (aggregate strength, surface texture, elastic modulus, etc.) can reasonably be expected to have similar effects on tensile and flexural strength of concrete. There are considerably fewer data on this aspect, however. Kaplan (1959a), in flexural tests on concrete with different coarse aggregates, showed that the higher the elastic modulus of the aggregates the greater the flexural strength of the concrete for a given w/c ratio. The results of a series of indirect tensile (cube splitting) tests using South African crushed andesite and quartzite aggregates are shown in Figure 5.8. Strength premiums of the order of 20 per cent were

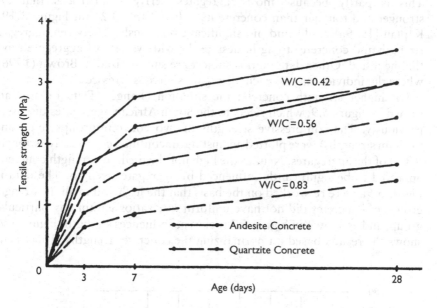

Figure 5.8 Cube splitting tensile strengths of andesite and quartzite concrete (from Alexander and Ballim, 1986).

achieved which were very similar to the compressive strength premiums obtained from use of these aggregates. The andesite aggregates had elastic moduli in the range 81–105 GPa, while the quartzite aggregates had values in the range 67–79 GPa. The crushed andesite had a significantly rougher micro-texture than the quartzite, giving better bond with the cement matrix.

Aggregate shape is also an important factor when designing concrete mixes for enhanced tensile or flexural strength. Angular crushed aggregates, provided they are competent and sound, impart better tensile properties than rounded smooth-textured aggregates.

Aggregate strength

For conventional concretes with compressive strengths less than about 60 MPa, aggregate strength does not have any strong influence on concrete strength. This has already been alluded to. In a book on concrete roads (1955) the UK Road Research Laboratory stated,

> Provided that aggregates are stronger than the concrete of which they form part, their inherent strength is not likely to influence the strength of the concrete, either in crushing or in flexure.

(RRL, 1955)

This is partly because most aggregates derive from rocks that are stronger and tougher than concrete itself (see Table 3.2 and Figure 3.19). Kaplan (1959a) could find 'no significant relationship' between aggregate strength and concrete strength, despite the wide variety of aggregate types that he tested. Other data on this subject are summarized in Brown (1996), where the individuality of each aggregate source is stressed.

For higher strength concrete, the situation changes. Data on this are shown in Figure 5.9, which refers to the South African aggregates discussed previously. The compressive strengths of two concrete groups (high and medium strengths) were plotted against the unconfined compressive strength (UCS) of the aggregates. Neither the high nor the medium strength concretes appeared to be significantly influenced by aggregate strength. The data in Figure 5.9a were re-worked on the basis that the high and medium strength groups of concretes did not have uniform w/c ratios within any particular group, and as shown in Figure 5.7 rock type influences strength. Figure 5.9b shows the results based on normalizing the concrete strengths. There is no

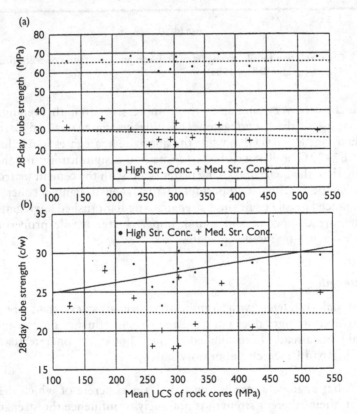

Figure 5.9 Influence of aggregate strength on 28-day concrete cube strength: (a) measured values and (b) normalized values (after Alexander and Davis, 1991).

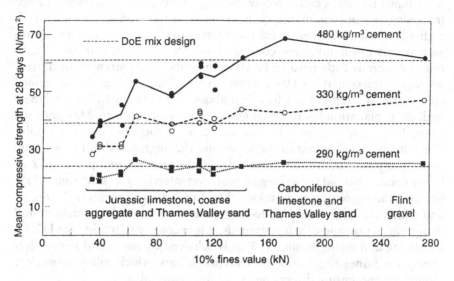

Figure 5.10 Effect of 10% fines value on concrete strength at 28 days, with three types of UK coarse aggregate (from Collins, 1983).

influence of aggregate strength on the medium strength concretes, but a positive trend exists now for the HSCs.

Similar results using three UK coarse aggregates (with Thames Valley Sand in all cases) are shown in Figure 5.10. Aggregate strength is represented by the 10 per cent fines value. For low to medium strength concretes, 10% FACT had little effect on concrete strength whereas for higher strength concrete (>50 MPa) higher strength aggregates had an important influence. From Figure 5.10, 10 per cent fines values in excess of about 150 kN are required for higher strength concretes. BS 882: 1992 gives quantitative requirements for coarse aggregate in terms of the 10 per cent fines value. For example, heavy duty concrete floor finishes require a 10 per cent fines value greater than 150 kN, other pavement wearing surfaces greater than 100 kN, and in general no aggregate should have a 10 per cent fines value less than 50 kN.

Work on HSC (Aïtcin and Mehta, 1990) indicates the importance of proper aggregate selection. The aggregates should preferably be fine grained and very strong, with rough surface textures and somewhat angular shape. The need to produce a homogeneous paste microstructure and a dense strong ITZ is also important. All components of the concrete – paste, aggregate, and the ITZ – should be optimized in terms of strength, avoiding the 'weak-link' problem.

These observations can be explained first in terms of the different modes of failure of low and high strength concrete. Low strength concretes fail predominantly by cracks following aggregate–paste interfaces before traversing the mortar matrix (intergranular failure); HSCs have a much higher incidence of aggregate fractures across the failure plane, when aggregate strength will influence concrete strength (transgranular failure). In practice there is little need to be concerned about the strength of normal aggregate unless making HSC, where an aggregate having an unconfined compressive strength of not less than about 200 MPa should be used. Second, the explanation lies in the composite nature of concrete. Many normal strength concretes and even high cement content concretes may not exhibit true composite behaviour in the sense that the aggregates do not act fully compositely with the matrix. This is due to the presence of a porous ITZ of lower density around the aggregate particles, effectively a 'soft' zone which prevents the aggregates from taking their full share of the internal stresses, and which also reduces the paste–aggregate bond strength. To achieve full composite action and enhance strength, it is necessary to modify the ITZ by densifying and strengthening it. This is achieved by the use of microfillers (e.g. silica fume) together with superplasticizers which allow mix-water reductions and ensure dispersion of the fine material.

Arguments about the effect of the ITZ and its importance in HSC are ratified by other data which indicate a significant influence of aggregate strength on concrete strength, shown in Figure 5.11 (Goldman and Bentur, 1993). The matrix comprised portland cement with silica fume and superplasticizer additions, all at equal w/c. The increase in concrete strength for mixes of equal w/c was occasioned by an increase in aggregate strength (indicated by a reducing aggregate crushing value).

Aggregate–paste bond strength and influence of the ITZ

Aggregate–paste bond strength is influenced by surface texture, shape, and the nature of the aggregate, and these will be covered later. The bond itself depends on the ITZ between paste and aggregate. Strengthening of the paste–aggregate bond may result in improved strengths, but these are usually limited to the order of only 20–40 per cent. A regression relationship between bond strength and concrete strength has been developed in the form

$$\sigma = b_0 + b_1 M_1 + b_2 M_2 \qquad\qquad (5.4)$$

where
σ is concrete strength in compression or in flexure
b_0, b_1, b_2 are linear regression coefficients
M_1, M_2 are moduli of rupture of the paste and the aggregate–cement bond, respectively (Alexander and Taplin, 1962, 1964).

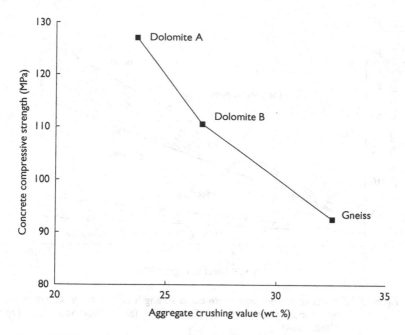

Figure 5.11 Correlation between aggregate compressive value and concrete crushing strength (from Goldman and Bentur, 1993; ©1993, reprinted with permission from Elsevier).

Figure 5.12, which is a plot of Equation (5.4) where $M_2 = 0$ for the 'no bond' case and $M_2 = M_1$ for the 'perfect bond' case, shows that flexural strength of the paste has a much greater effect on concrete strength than change in bond strength.

Strengthening and improving ITZ microstructure can increase the aggregate–paste bond strength. This was shown in experimental work in which the ITZ was densified and strengthened by microfillers in the cement paste (Goldman and Bentur, 1993). The microfillers were silica fume and carbon black, both with very fine particle sizes ($<0.5\,\mu m$). Silica fume contrasts with carbon black in that it also has pozzolanic properties and is able to provide additional hydration products. Despite this fundamental difference between the two microfillers, silica fume concrete and carbon black concrete achieved similar, substantially higher strengths, compared with an equivalent reference concrete – see Figure 5.13. The microfiller effect provided the bulk of the additional strength, presumably through the mechanism of increasing the paste–aggregate bond strength by modifying the microstructure of the ITZ thus making it denser and stronger.

Similar effects can be obtained in other ways. Very fine quartz powder fillers can be used to improve particle packing in silica fume-superplasticized

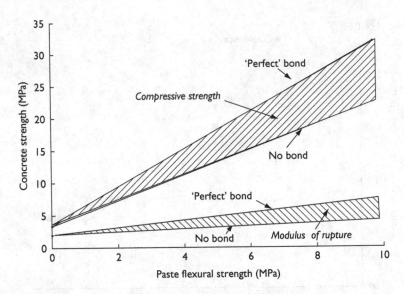

Figure 5.12 Effect of cement–aggregate bond strength on the compressive strength and modulus of rupture of concrete (after Alexander and Taplin, 1962, 1964).

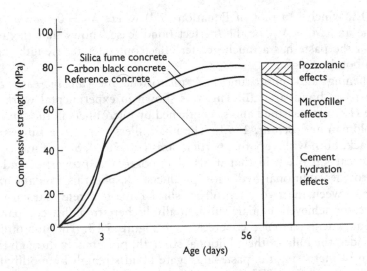

Figure 5.13 Strength-time curves for concretes to illustrate the contribution of pozzolanic and microfiller effects ($w/c = 0.46$) (after Goldman and Bentur, 1993).

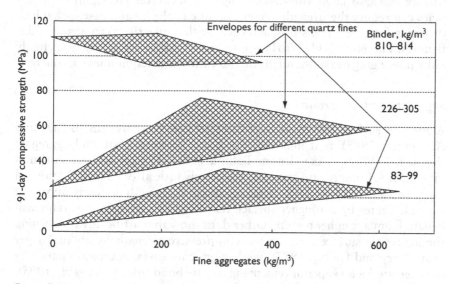

Figure 5.14 Concrete strength at 91 days versus dosage of fine aggregate (adapted from Kronlof, 1994).

concrete mixtures (Kronlof, 1994). Non-reactive fine fillers alter and refine microstructure, particularly that of the ITZ, giving improved strength. Figure 5.14 shows the possibilities of this approach. Much of the improvement can be ascribed to a decrease in the water/binder ratio, but microstructural effects and improved interaction between paste and aggregate are also important. The potential of aggregates to contribute directly to concrete strength are fully realized only in concretes in which the microstructure of the paste and of the ITZ has been modified, making them true composite materials. Notwithstanding the above, a RILEM committee dealing with the ITZ concluded that ITZ characteristics may have significant, if only moderate, influences on compressive strength, whereas they may have a far more dramatic effect on the properties of fibre-reinforced cement composites (Bentur and Alexander, 2000).

Aggregate shape

The weight of experimental evidence points to aggregate shape having an effect on concrete strength, with flexural strength more affected than compressive strength. Much of the evidence is anecdotal, with angular particles being considered to improve concrete strength over rounded particles. A quantitative assessment of the influence of shape was attempted by Kaplan (1959a), who also found that flexural strength was more improved by angular aggregates than compressive strength. These improvements can be explained by the greater degree of mechanical interlock, internal friction, and increased surface area associated with angular aggregates. Conversely, highly flaky particles can reduce the strength of concrete due to the greater ease with which they may fracture under stress, and because they give rise to compaction difficulties. This effect is likely to be pronounced with coarse aggregates. Highly flaky fine aggregates substantially increase the water requirement of concrete.

Aggregate surface texture

Aggregate surface texture is important to concrete strength (Ballim and Alexander, 1988) as it improves the bond between paste and aggregate (Kaplan, 1959a,b; Alexander and Ballim, 1986). Typical data are shown in Figure 5.15. Compressive strength is relatively little affected over the normal range of values (15–55 MPa), whereas flexural strength is clearly improved over all values by a rougher surface texture. High strength concretes will benefit from a rougher texture. Other data show that artificially roughening the aggregate surface can increase compressive strength by about 10 per cent (Perry and Gillott, 1977), while aggregate surfaces that are naturally rougher produce a superior cement–aggregate bond (Alexander et al., 1995).

Table 5.1 gives results from a study using crushed quartzite and andesite aggregate. Andesite aggregate concretes were up to about 28 per cent stronger in both tension and compression than equivalent quartzite aggregate mixes. This increased strength was attributed at least partly to the fact that the fractured andesite had a considerably more contorted and irregular surface texture than the quartzite and did not exhibit surface disintegration, as shown in the micrographs in Figure 5.16. This improved surface texture improved the aggregate bond strength. The results in Table 5.1 are consistent with the predictions of Figure 5.12.

Aggregate elastic modulus

Aggregate modulus of elasticity also plays a role in affecting concrete strength, with data given in Figure 5.17 (Kaplan, 1959b). There is a continuous increase in concrete strength with increasing aggregate elastic modulus, the likely mechanism being that stiffer aggregates attract a greater share of the load in the composite.

It is usually very difficult to separate the effects of texture and elastic modulus experimentally. The elastic modulus effect was examined in the South African study mentioned previously but no clear correlations emerged.

Relative effects of shape, surface texture, and elastic modulus of the aggregate

Kaplan's seminal work on the effects of aggregates on concrete strength provides one of the few quantitative studies that are available (Jones and

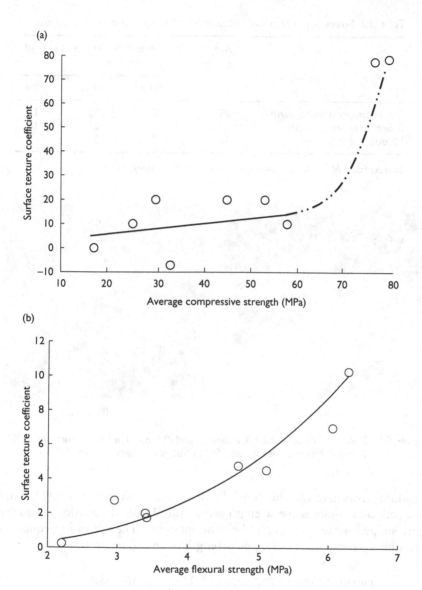

Figure 5.15 Experimental relationships between particle surface texture of aggregates, and (a) compressive and (b) flexural strength of concretes using similar aggregate materials (after Kaplan, 1959a).

Kaplan, 1957; Kaplan, 1959a,b). He carried out a statistical analysis of variance of the results to ascertain to what extent the variation in concrete strength could be attributed to variations in shape, surface texture, and elastic modulus of the coarse aggregates. After making allowance for

Table 5.1 Strength premiums of andesite concrete over quartzite concrete

	Age (days)	Percentage increase in strength for w/c ratio		
		0.83	*0.56*	*0.42*
Cube compressive strength	28	28	23	17
Indirect tensile strength	28	19	24	18
Modulus of rupture	35	16	27	9

Source: From Alexander and Ballim, 1987; reprinted, courtesy RILEM.

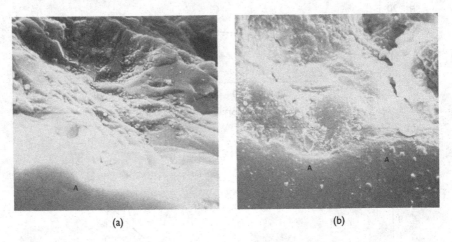

(a)   (b)

Figure 5.16 Photomicrographs of (a) andesite and (b) quartzite rock surfaces, 2000×
(from Ballim and Alexander, 1988; photos courtesy of Prof. Ballim).

random errors in testing, he showed that variations in both concrete flexural strength and compressive strength were indeed due to variations in aggregate shape, surface texture, and elastic modulus. The regression equations relating aggregate properties and strength of the concrete were:

$$\text{Flexural strength } f'_r (\text{p.s.i}) = 18.1 E_a + 9.1 AN + 3.8 S + 375 \qquad (5.5)$$

$$\text{Compressive strength } f'_c (\text{p.s.i}) = 64 E_a + 35 AN + 25 S + 5400 \qquad (5.6)$$

where
E_a = aggregate elastic modulus (p.s.i)
AN = angularity number (i.e. percentage voids − 33)
S = surface texture roughness factor (according to the method of Wright, 1955).

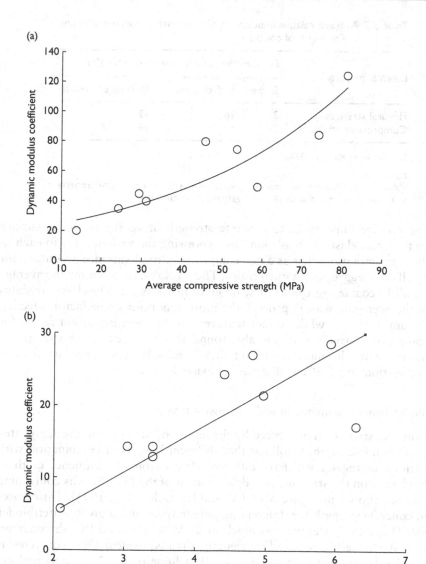

Figure 5.17 Experimental relationships between dynamic modulus coefficient of aggregates, and (a) compressive and (b) flexural strengths of concretes (made using similar aggregates) (after Kaplan, 1959b).

These best-fit equations are from Kaplan's investigation and indicate that concrete strength is related sensibly to measurable aggregate properties. The equations are not meant to indicate that other aggregate factors, for example strength or absorptivity, will have no influence on concrete strength.

Table 5.2 Average relative importance of aggregate properties affecting the strength of concrete

Concrete property	Relative effect of aggregate properties (%)		
	Shape	Surface texture	Modulus of elasticity
Flexural strength	31	26	43
Compressive strength	22	44	34

Source: After Kaplan, 1959b.

Note
Values in the table are the ratio of the variance due to each aggregate property to the total variance due to all three aggregate characteristics.

The relative importance to concrete strength of aggregate shape, surface texture, and elasticity was found by expressing the variance due to each of the aggregate properties as a percentage of the total variance accounted for by all three aggregate characteristics. These results, representing the average for all 13 coarse aggregate types, are given in Table 5.2. The elastic modulus of the aggregate was, in general, the most important single factor affecting flexural strength, while surface texture was the most important factor for compressive strength. It was also found that the greater the strength of the concrete, the more important these effects became. This has obvious implications for HSC, as discussed previously.

Binder type in combination with aggregate type

Concrete strength is influenced by the nature of the ITZ and the aggregate–paste bond strength. It follows that different binders in combination with various aggregates will have different effects from the influence of these binders upon the structure and development of the ITZ. Results confirming this are shown in Figure 5.18 (Alexander and Milne, 1995). Differences in concrete strength for different aggregate types and a given water/binder ratio (*w/c* in figure) were as much as 20 MPa. Slag and fly ash concretes usually showed lower 28-day strengths than equivalent OPC concretes at equal *w/c*. However, at equal *w/c*, silica fume paste (7 per cent replacement) had similar strength to an OPC paste at 28 days, and consequently the enhancement of compressive strength in silica fume concretes can be interpreted as primarily due to an improvement in interfacial bond and the nature of the ITZ.

Aggregate grading, maximum size, and coarse aggregate packing

Aggregate grading primarily influences concrete water content and thus *w/c* ratio which is the main determinant of strength. Fines content can

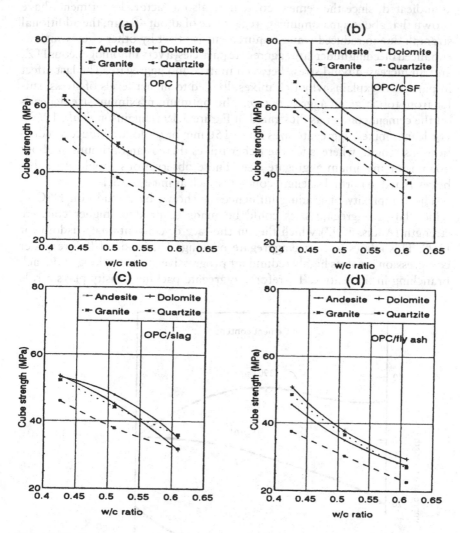

Figure 5.18 Concrete cube strength at 28 days for different aggregate and cement types (c is cementitious content). Binder types: 100 OPC; 93:7 OPC:CSF; 70:30 OPC:Fly Ash; 50:50 OPC:Slag (GGBS) (from Alexander and Milne, 1995) (reprinted with permission from American Concrete Institute, Materials Journal, 92(3); 227–35).

have an indirect influence on strength by affecting the compactibility of concrete, and as we have seen earlier, larger amounts of very fine material may increase strength by improving the nature of the ITZ.

The early work of Bloem and Gaynor (1963) on maximum aggregate size has been mentioned – see Figure 5.5. However, the effects are not as simple as indicated, since the cement content is also a factor. Experiments have shown that above a maximum aggregate size of about 40 mm, the additional strength due to reduced water requirement is offset by other effects such as enhanced bleeding under large aggregate particles, a very inhomogeneous ITZ, and an increased bond stress between matrix and aggregate. The last effect in particular explains why rich mixes, loaded to higher levels of stress, suffer from too large an aggregate size. The optimum maximum size depends on the cement content as illustrated in Figure 5.19 (Higginson *et al.*, 1963). For leaner mixes, aggregate sizes up to 150 mm or so do not induce a reduction in strength, whereas for the richer mixes there is an optimum of about 40 mm for maximum aggregate size. These observations should probably be restricted to normal strength concretes such as in Figure 5.19.

The complexity of grading influences is shown by a study on HSC in which larger aggregate sizes could be made to produce higher concrete strength (Addis, 1992) which flies in the face of conventional wisdom for HSC. This relates to coarse aggregate packing. The mode of failure under compression involves highly redundant progressive microcracking and crack branching in the matrix. Therefore, aggregate packing density plays a role

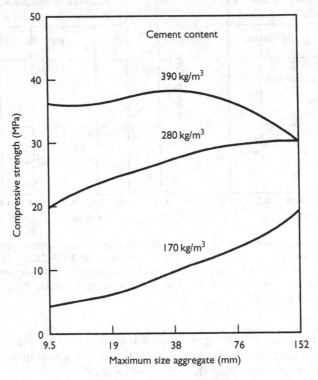

Figure 5.19 28-day compressive strength of concrete as influenced by maximum aggregate size and cement content (from Higginson *et al.*, 1963).

in affecting strength. In the study mentioned, packing was characterized by the consolidated bulk density (CBD). Two concretes were tested in which the fineness modulus (FM) of the stone fraction was similar, but in which one had a nominal 19-mm stone (CBD = 1560 kg/m³), while the other had a blend of 26.5- and 6.7-mm stone (CBD = 1750 kg/m³). Strength improvements of the order of 10 per cent for the blended stone were found. In contrast with accepted practice, the blended mix with the larger size stone did not give reduced compressive strengths in HSC. This finding is probably limited to the use of coarse, angular crushed aggregates where packing density is improved by blending of stone sizes. In North American practice where continuously graded aggregates are the norm, shape, surface texture, and inherent strength of the particles will all be very important for concrete strength.

The complexity of the failure modes and mechanisms for plain concrete in compression has hindered the development of a coherent microstructural failure model to date. Even in tensile or flexural failure, multiple micro-cracking occurs, and the effects of aggregate shape, surface texture, and grading are important. The effects discussed above may be worth considering in individual situations, but a full explanation of them in respect of concrete strength still eludes us. Many of these effects are also inter-related. For example, aggregate packing can affect aggregate–paste bond strength, and densification of the ITZ by microfillers will likewise influence bond strength. Different aggregate types will influence concrete strength due to their physical and mechanical characteristics (e.g. strength, particle shape, surface texture, elasticity), their packing density (which is a function of shape and texture), and their interface characteristics (which includes micro-roughness). It is thus impossible to separate out different effects intrinsically, which further complicates the search for a comprehensive and composite model for concrete strength. Nevertheless, an understanding of the important influences will allow concrete designers to make the best use of available aggregates and binders to optimize the concrete properties.

Concrete fracture properties

Concrete can be characterized by suitable fracture parameters. For brittle materials, these parameters derive from linear elastic fracture mechanics (LEFM) and describe the strength of the material in the presence of stress-raising flaws and cracks. While concrete is not truly a brittle material, for present purposes it is sufficient to consider the simplified LEFM approach based on Mode I (crack-opening mode) fracture.

Two suitable parameters are fracture toughness K_c and fracture energy (or specific work of fracture) R_c. Fracture toughness is a parameter relating to the critical stress field surrounding a crack, obtained from the expression $K_I = Y\sigma\sqrt{(\pi c)}$, where c is the crack half-length, σ is the global field stress in the vicinity of the crack, and Y is a factor that takes into account the geometry of the specimen. When the crack propagates unstably in the material, K_I is considered to reach a critical value K_c (strictly K_{Ic}) which is the fracture toughness. Fracture energy R_c refers to the energy required to cause a crack to grow, this energy usually being provided by the external load. It is a measure of the total amount of work required to propagate a crack through a material. For fracture of a homogenous linear elastic brittle material, $R_c = (K_c^2/E)$ (plane stress state), where E is the elastic modulus. Further detail on these parameters in relation to concrete can be found in Mindess et al. (2003).

There is little published information on the influence of aggregate type on concrete fracture properties. Differences in the shape and texture of aggregates will affect fracture surfaces and the nature of interface (bond) cracks. In a study on fracture energy of HSC, natural gravels gave lower fracture energies (Giaccio et al., 1993). Differences between these aggregates and a crushed basalt aggregate amounted to as much as 30–40 per cent, which correlated closely with differences in concrete compressive strength.

The influence of aggregates on fracture properties depends on the nature of the ITZ. This zone may be amenable to microstructural study in real concretes using techniques such as SEM and image analysis (Scrivener, 1999), but it is much more difficult to measure the mechanical properties of the ITZ directly. Problems associated with such studies are:

1   The bond between the hcp and the aggregate is neither perfect nor continuous around the entire aggregate particle. Thus, the state of stress within the transition zone will vary with location and age.
2   Different surface finishes of the aggregate, ranging from very smooth polished surfaces to rough fractured surfaces, will provide different bond properties and there is no agreement on which method of surface preparation should be used in experimental work.
3   The interfacial zone itself is variable in composition and its mechanical properties vary spatially. There are therefore no unambiguous values that can be regarded as 'constants'.
4   Bleeding leads to an uneven distribution of water around the aggregate particles, which exacerbates the variability in the transition zone.

There have nevertheless been attempts to measure mechanical properties of the ITZ (Mindess, 1996). Failure in cementitious composites invariably involves crack propagation in the interfacial region, and so parameters such as fracture toughness and fracture energy are probably more appropriate to measure. Such measurements are more sensitive to different aggregate types and to modifications to the interfacial zone than simple bond strength measurements.

In one study involving Mode II shear fracture, the interfacial fracture toughness approximately doubled in going from normal strength mortar to high strength mortar cast against granite (Buyukozturk and Lee, 1993). In other work, using a pushout specimen, the interfacial fracture energy more than doubled by incorporating silica fume in the matrix (Mitsui *et al.*, 1994). However, as seen later, this depends upon the type of aggregate.

Even more dramatic changes in interfacial properties can be obtained if the aggregate surface is pre-treated or coated, generally by coatings containing silica fume. For instance, Ping and Beaudoin (1992) found increased densification of the interface when the aggregate particles were coated with silica fume. The practicality of using these techniques in real concretes can be questioned.

Concrete fracture properties will also depend on the influence that aggregates have on the so-called 'fracture process zone' – the zone of non-linearity that develops ahead of primary cracks. For this reason, concretes show greater toughness than their equivalent mortars or pure pastes, since the aggregates increase the size of the process zone leading to greater fracture energy and fracture toughness. A study on the influence of aggregate size showed that fracture parameters increased with aggregate size, the value of R_c doubling as size increased from 5 to 20 mm. As to the effects of aggregate volume, the same study showed that, for medium strength concrete ($w/c = 0.44$), fracture parameters increased as total aggregate volume increased from 40 to 80 per cent, while for HSC ($w/c = 0.26$), maximum fracture values occurred at a total aggregate volume of about 60 per cent. These effects were ascribed primarily to the influence of the aggregates on the fracture process zone (Chen and Liu, 2004).

Fracture mechanics tests of the interfacial zone

Chapter 3 indicated that a practical way of measuring fracture parameters is using the ISRM Test Method I, with drilled rock cores (ISRM, 1988). In this method a chevron-notched cylindrical beam specimen is tested in centre-point loading (Figure 3.24).

'Artificial' interfaces can be created by casting cement paste against a naturally fractured rock surface. This limits the direct use of the results, but permits comparisons (Mindess and Alexander, 1995). Interface values can also be compared with values measured in rock cores, cast paste, or mortar cylinders. The fracture toughnesses (K_c) and fracture energies (R_c) at an age of 28 days (paste-related specimens) for various materials incorporating andesite, dolomite, and granite rocks are given in Table 5.3. The rock specimens had substantially higher fracture properties than the paste-related specimens. For paste specimens, the replacement of 15 per cent of the cement by silica fume reduced values for both K_c and R_c for the same water/binder ratio, implying embrittlement of the material. For the dolomite specimens, fracture usually occurred at the rock/paste interface. For the andesite specimens, fracture occurred at the interface in specimens without

Table 5.3 Fracture toughness K_c and work of fracture (fracture energy) R_c measured at 28 days (for paste-related specimens), $w/c = 0.3$; CSF at 15% replacement

	K_c (MN/m$^{1.5}$)	R_c (J/m^2)
Paste		
OPC	0.54	14.9
OPC + SF	0.50	9.1
Rock		
andesite	3.49	147.1
dolomite	2.23	59.9
granite	2.22	162.6
Paste/Rock interfaces		
andesite/OPC	0.80	24.6
andesite/OPC + SF	0.71	14.5
dolomite/OPC	0.57	7.5
dolomite/OPC + SF	0.64	9.9
		(65 days)
granite/OPC*	0.63	25.2

Source: From Mindess and Alexander, 1995; reprinted with permission of the American Ceramic Society, www.ceramics.org. All rights reserved.

Note
* These tests were done with an OPC different from the other interface tests.

silica fume, but in the paste away from the interface when silica fume was added. The measured fracture toughness is then not that of the interface directly but is representative of the paste. Because of its very fine particle size, silica fume densifies the paste and the interfacial zone usually leading to embrittlement. Figure 5.2 also showed that interfaces were sensitive to the type of aggregate, with fracture occurring in the paste rather than the interface for certain rock types, indicating an interface tougher than the paste.

The tests discussed above represent interfaces between paste and rough fractured rock surfaces (Diamond and Mindess, 1992; Diamond *et al.*, 1992; Mindess and Diamond, 1992, 1994). Different rock types produced different roughnesses at different levels of scale. Therefore, the measured interface mechanical properties included the contribution from the mechanical interlock effect which was different for each aggregate type.

Computer simulations of concrete fracture

The computer simulation work of Bentz, Garboczi, and co-workers has been extended to study concrete properties influenced by the presence of aggregates and the ITZ (Bentz *et al.*, 1995). A finite element lattice model developed at Delft University was used for plane stress simulations of concrete fracture (Schlangen and Van Mier, 1992a). The model is loaded and internal forces are computed in the lattice network. In each step of the analysis, the load is adjusted so that only one element exceeds its prescribed tensile strength. Cracks are then propagated by removing this element, the process being repeated until fracture is complete. The lattice network is superimposed on a three-phase representation of concrete (aggregates, cement paste matrix, and ITZ), and lattice elements are given characteristic mechanical properties (strength and stiffness) depending on which particular phase they represent – see Figures 5.20a and b (Schlangen and Van Mier, 1992b). This arrangement was used to simulate tensile loading of

(a)

(b)

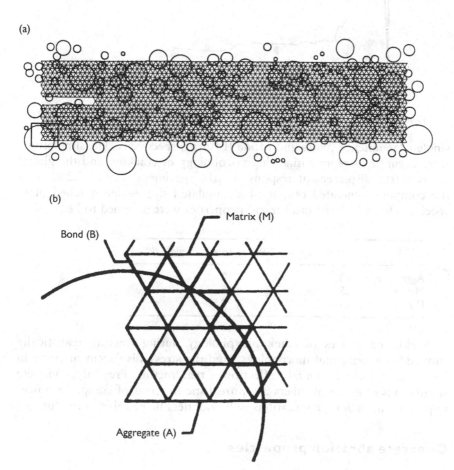

Figure 5.20 (a) Triangular lattice projected on generated concrete microstructure. (b) Definition of aggregate, matrix, and ITZ (bond) elements. (c) Stress–displacement response for simulation I (material constants given below) (from Schlangen and Van Mier, 1992b; ©1995, reprinted with permission of the American Ceramic Society).

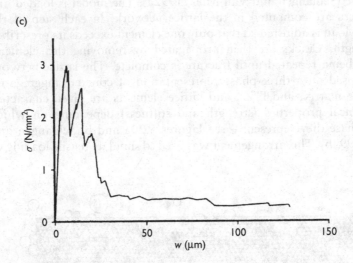

Figure 5.20 (Continued).

single-edge notched plates and Mode II (shear) specimens. The simulations were accurate in representing the morphology of cracking and the global stress versus displacement response of the specimens. Figure 5.20c gives the computer-generated output of a simulated single-edge notched plate specimen for which the mechanical properties were assumed to be:

	Elastic modulus (GPa)	Tensile strength (MPa)
Aggregate	70	10
Matrix	25	5
ITZ	25	0.5

Pixel-based images of crack morphology during fracture realistically simulated experimental morphology, and the stress–displacement curve in Figure 5.20c likewise is a fair reflection of real fracture. Presently, complete simulations of entire members are unrealistic in terms of computer power required, but such representations will doubtless be possible in the future.

Concrete abrasion properties

Surface abrasion and erosion

Surface abrasion of concrete is a complex phenomenon since abrasive actions can many and varied: friction, attrition, grinding, rolling, impact, high local stresses, or the action of fluids containing abrasive media. Therefore, concrete abrasion tests vary widely in their action and effects. Some tests tend to preferentially abrade the softer, weaker matrix, for example the sandblast test (ASTM C418), while others in effect measure the abrasion resistance of the harder aggregates. Among the latter are tests such as the so-called 'Bohme test' (DIN 52108) in which the entire surface is abraded by the action of a horizontally rotating steel grinding wheel, or ASTM C779, Procedure A.

For most practical cases of concrete abrasion, aggregates play a role by representing hard inclusions in a softer matrix. Two most common abrasive actions are impact – where hard particles impinge on the concrete surface causing shattering or debonding of aggregate particles – and grinding, rubbing, or scratching in which particles or objects wear down the concrete surface by relative movement causing gouging and indentation. For both types of action, the harder and tougher the aggregate and the greater the aggregate volume concentration, the more abrasion resistant is the concrete. Hardness and toughness are different properties and will affect abrasion resistance differently depending on the type of abrasive action. Hard aggregates resist grinding and sliding abrasion, while tough aggregates (which may in fact be quite soft) resist crushing and impact better. Brittle aggregates, even if strong and hard, should be avoided in concrete exposed to impact action, since they can fracture and chip under such forces.

There is not necessarily a direct correlation between the abrasion resistance of concrete and the quality of the coarse aggregate determined by the Los Angeles Abrasion Test (ASTM C131). This is because the LA test is more a measure of the impact resistance of aggregates than their abrasion resistance.

Aggregates also play a role in concrete abrasion by virtue of the aggregate–paste bond. Aggregates that bind strongly with the matrix will resist abrasion better than weakly bonding particles that can be easily plucked out. This bonding effect is primarily mechanical, relating to the shape and surface texture of the particles. Angular, chunky particles that can 'lock in' to the matrix are preferred, particularly if they also have rough macro-or-microtextures that can develop strong bond with the matrix. Sound, crushed hard rock aggregates of the preferred shape will be more desirable than rounded, smooth particles if abrasion resistance is important.

Coarse aggregates are embedded below the surface, and so they come into play only when sufficient wear has occurred to expose them. Provided the aggregates are sufficiently strong and hard, their relative influence reduces as the strength of the concrete (and hence abrasion resistance of the matrix *per se*) increases.

Sims and Brown (1998) reviewed research by the UK Road Research Laboratory into the resistance of concrete road surfaces to military tank traffic. This research highlighted the importance of concrete compressive strength in producing abrasion-resistant surfaces, with aggregate type playing an important role in the lower strength concretes (<40 MPa), shown

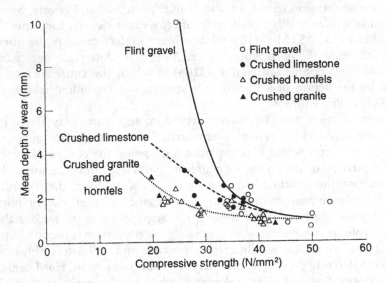

Figure 5.21 Wear resistance of concrete road surfaces influenced by different aggregate types and concrete compressive strength (after RRL, 1955).

in Figure 5.21. Flint aggregate concrete suffered most severely because of debonding, while the harder and stronger granite and hornfels aggregates were superior.

Aggregates influence concrete resistance to surface erosion which generally arises from the action of flowing water. This is often combined with abrasive particulate loads so that erosion and abrasion occur simultaneously, and the important aggregate requirements for abrasion resistance apply equally to erosion resistance. Other than the obvious requirement that the concrete matrix be strong and dense, the aggregates should be hard and tough and able to bond strongly with the matrix.

Aggregates and skid-resistant concrete pavements

An important property of any concrete pavement or trafficked wearing surface is skid-resistance. This may be necessary on high-speed freeways as much as on sidewalks for foot traffic – in both cases, slippery conditions are to be avoided, particularly in wet weather. On concrete pavements, the skid-resistance derives from a combination of two effects: the macrotexture which is governed by the mean depth of the surface asperities and which controls the drainage, and the microtexture which relates to the texture of the surface at any point. Aggregate particles play a role in this either by being purposely exposed during construction to form a rough macrotexture, or during the life of the pavement by the softer matrix being worn away leaving the coarse aggregate exposed. A wear-resistant aggregate is therefore required, such as a diabase. Multi-mineral aggregate types that contain a proportion of silica or quartz are preferred to single-mineral types such as limestone which tend to wear uniformly, thus becoming smooth. The multi-mineral types have minerals that wear at differential rates resulting in a naturally rough microtexture.

Concrete deformation properties

Short-term ('elastic') and long-term (shrinkage and creep) deformations of concrete may be profoundly affected by type of aggregate. In general, the influence of aggregates on elastic modulus and long-term deformation is more pronounced than the effects on strength. The role of aggregates is important for the following reasons:

- The development of HSC where aggregate effects become more pronounced and where increase in strength is not necessarily matched by a corresponding increase in stiffness.
- The need to 'engineer' the material by selection of appropriate constituents and an understanding of interaction in the composite.
- Use of partial prestressing and unbonded tendons where deformations become more critical.
- A greater mix of cast-in-situ and precast elements in modern structures.
- Exploitation of new cementitious materials and alternative aggregate sources.

Early work in this area (La Rue, 1946; Kaplan, 1959b) established the following conclusions: concretes having the same compressive strength may have different dynamic moduli of elasticity if different aggregates are used; conversely, concretes of different mix proportions may have different compressive strengths for the same dynamic modulus; for equal compressive strength, the elastic modulus increases with increasing aggregate/cement ratio; changes with age in parameters such as dynamic modulus, ultrasonic pulse velocity, and Poisson's ratio of concrete may be ascribed chiefly to changes in the mortar phase.

Mathematical expressions to predict elastic modulus have been developed based on concrete being modelled as a two-phase material – a matrix phase and an aggregate phase (Hashin, 1962; Hirsch, 1962; Counto, 1964; Hansen, 1965; Hobbs, 1971). Different geometrical arrangements of the phases give rise to the models shown in Figure 5.22. The Parallel model is based on uniform strain giving an upper bound, whereas the Series model is based on uniform stress and gives a lower-bound solution. Other intermediate models (Hirsch, Counto, Hashin/Hobbs) usually give reasonable

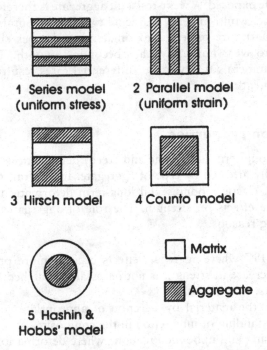

Figure 5.22 Phase models for concrete and mortar elastic modulus (from *Fulton's Concrete Technology*, 1994, based on Hashin, 1962; Hirsch, 1962; Counto, 1964; Hansen, 1965; Hobbs, 1971).

predictions of concrete elastic modulus provided phase properties and relative volume concentrations are known. The performance of the models is compared in Figure 5.23, while Table 5.4 provides a summary of the models.

Models 1 and 2 represent two extremes of phase arrangement. Neither model is correct, since the matrix and aggregate phases of loaded concrete experience neither uniform stress nor uniform strain. A drawback of the Series and Hirsch models is that E_c becomes zero for the case $E_a = 0$, which is not true. The Counto and Hashin/Hobbs models overcome this limitation. The latter model is based on a realistic phase arrangement in which spherical aggregate particles are embedded in a continuous matrix, and is derived on the additional assumptions that there is no interaction between aggregate particles and no local bond failure occurs. The first assumption is reasonable but the second assumption is only partially true. The three intermediate models give similar results for the normal range of aggregate volume concentration (see Figure 5.23) and experimental results tend to lie close to these curves.

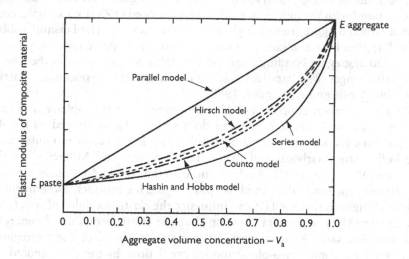

Figure 5.23 Comparison of phase models for concrete and mortar (from *Fulton's Concrete Technology*, 1994).

Table 5.4 Elastic modulus of concrete according to two-phase model formulae

Model	Characteristics	Formula
1. Parallel model (uniform strain)	Upper-bound solution. Applies to soft particles embedded in a hard matrix, i.e. low-density aggregate concrete	$E_c = E_m(1 - V_a) + E_a V_a$
2. Series model (uniform stress)	Lower-bound solution. Applies best to hard particles embedded in a soft matrix, i.e. normal-density aggregate concrete	$\dfrac{1}{E_c} = \dfrac{1 - V_a}{E_m} + \dfrac{V_a}{E_a}$
3. Hirsch model	Combination of models 1 and 2	$\dfrac{1}{E_c} = 0.5\left(\dfrac{1}{(1 - V_a)E_m + V_a E_a}\right)$ $+ 0.5\left(\dfrac{1 - V_a}{E_m} + \dfrac{V_a}{E_a}\right)$
4. Counto model	Generally applies for full range of aggregate stiffness	$\dfrac{1}{E_c} = \dfrac{1 - \sqrt{V_a}}{E_m}$ $+ \dfrac{1}{\left((1 - \sqrt{V_a})/\sqrt{V_a}\right)E_m + E_a}$
5. Hashin and Hobbs model	Assumes equal Poisson's ratio for two phases, and no particle interaction	$E_c = E_m\left[\dfrac{(1 - V_a)E_m + (1 + V_a)E_a}{(1 + V_a)E_m + (1 - V_a)E_a}\right]$

Source: From *Fulton's Concrete Technology*, 1994, based on Hashin, 1962; Hirsch, 1962; Counto, 1964; Hansen, 1965; Hobbs, 1971.

Notes

E_c, E_m, and E_a are the elastic moduli of the concrete, matrix, and aggregate respectively. V_a is the volume concentration of the aggregate.

The Series model seems to apply best with aggregates that do not develop a good bond with the matrix or exhibit very marked ITZs (for example, certain weathered aggregates or high w/c ratio concrete). The Hashin/Hobbs model applies best in mixes with more homogenous paste microstructures and good aggregate bonding, making it suitable for concretes in the lower w/c ratio ranges. The Parallel model generally greatly overestimates elastic modulus (Grills and Alexander, 1989).

As stated earlier, concrete can also be treated as a three-phase material and it is probably more correct to do so. The ITZ is treated as a soft porous zone around the aggregate particles with lower elastic modulus than the bulk paste (Garboczi and Bentz, 1991; Nilson and Monteiro, 1993; Lutz and Monteiro, 1995). Analyses indicate that the ITZ modulus may be 15–50 per cent lower than the bulk paste value for mortars. Experimental evidence suggests that the ITZ can influence the elastic modulus of concrete. Meininger (1994) cites data showing that the elastic modulus of concrete did not necessarily increase with an increase in the E of the aggregate, contrary to a simple two-phase model prediction. In the data quoted, a natural rounded aggregate with an E of about 41 GPa produced concrete with E ranging from 28 to 34 GPa after 90 days, compared with a similar concrete with a quarried aggregate with an E of 62 GPa giving concrete E of 21–28 GPa. Such anomalies can best be explained by postulating a different nature of the ITZ that may develop with the two rock types.

There is less experimental evidence on the influence of aggregates on deformation in HSC. Data indicate that concrete elastic moduli may vary as much as 40 per cent for equal compressive strength, while concrete shrinkage differences with different aggregates may amount to 20–25 per cent, with similar differences for specific creep (Leming, 1990). As was the case for strength, the properties (in this case the elastic properties) of the aggregates become more crucial in HSC. Simplified code approaches for prediction may no longer be appropriate, and while two-phase models may be quite accurate (<10 per cent error) for low porosity, fine-grained, high E_a aggregates, they may be inaccurate for poor quality aggregates such as compressible sandstones (Baalbaki et al., 1991, 1992). The aggregate becomes a more 'active' component in HSC rendering it a more truly composite material. However, as will be seen later, even in normal strength concretes aggregate elastic modulus has a very important influence on concrete deformation (Bennett and Khilji, 1964; Nimityongskul and Smith, 1970).

Stress–strain relationships

The stress–strain relationship for short-term loading is neither truly elastic nor linear. However, for most concretes tested according to standards such as ASTM C469 or BS 1881, the stress–strain relationship is usually sufficiently linear for an elastic modulus to be defined and measured. The

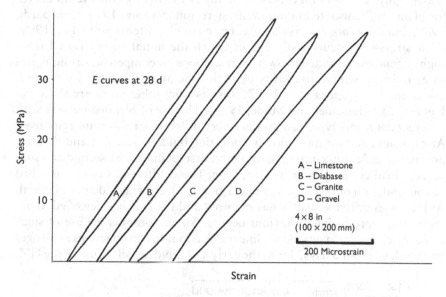

Figure 5.24 Elastic modulus measurements of various high performance concretes (after Aïtcin and Mehta, 1990).

modulus of elasticity and shape of the stress–strain curve are influenced by the nature of the aggregate (for example, whether it is hard and fresh or soft and weathered) and the nature of the ITZ.

Stress–strain curves for concretes with four different coarse aggregate types are shown in Figure 5.24 (Aïtcin, 1989; Aïtcin and Mehta, 1990). The shapes of the stress–strain curves and the hysteresis loops differ, as do the elastic moduli. Crushed fine-grained aggregates (in this case diabase and limestone) gave superior strength in HSC compared with smooth river gravel or crushed granite that contained a soft mineral inclusion. The characteristics of the more competent aggregates are observable in the figure (note the hysteresis loops), and may be attributed in part to a stronger transition zone. In the same sequence of studies, the influence of coarse aggregate preparation on properties of HSC was investigated (Aïtcin, 1989). The three aggregates studied were similar, comprising mostly diabase but were produced in different ways, either by crushing from a natural gravel, by blasting and crushing from rock, or used in natural pea gravel form. The highest strength and elastic modulus and narrowest hysteresis loops occurred for pea gravel diabase. Poorer performance of concrete containing diabase that had been blasted was ascribed to weakening and fracturing of the aggregate during blasting. Hysteresis loops are also affected by microstructure with HSCs having considerably smaller hysteresis loops than normal strength concretes (Sengul *et al.*, 2002).

Generally, stress–strain curves show linear or slightly concave-up curves for plain OPC concrete in the low stress region (Swamy, 1970; Collepardi, 1989), but this is not always the case. For example, Aïtcin and Mehta (1990) show stress–strain curves of HSC in which the initial curves varied from slightly concave-up with narrow hysteresis loops for competent, strong aggregates and a dense ITZ, to convex-up with wide hysteresis loops for weaker, less competent aggregates and ITZ. Results from other work are shown in Figure 5.25 (Alexander and Milne, 1995). The use of blended binders with different aggregate types can produce very different stress–strain responses. At early ages and for more slowly hydrating binders (e.g. slag and fly ash), non-linear and convex-up curves occur on first loading, with second and subsequent load cycles showing the opposite. These differences can be ascribed to consolidation of young pastes under first load, creating a denser material. At later ages once the matrix has matured and the ITZ has densified, more linear and reproducible behaviour occurs. An exception is the use of silica fume which results in practically linear behaviour for all concretes at all ages greater than about 3 days, due to the early and rapid densification of the ITZ.

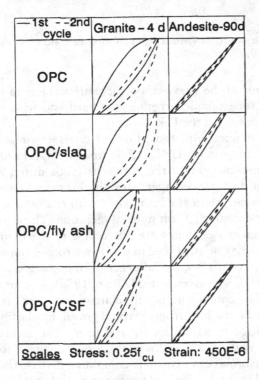

Figure 5.25 Schematic of stress–strain behaviour for granite and andesite concrete at different ages (from Alexander and Milne, 1995) (reprinted with permission from American Concrete Institute, Materials Journal, 92(3): 227–35).

Aggregates and elastic deformations

The elastic modulus of concrete (E_c) is influenced by the stiffness of the paste and aggregate phases, their volume concentrations, and interface characteristics. The stiffer the individual phases, the higher the elastic modulus and the lower the long-term movement of the concrete. Concrete elastic modulus also increases with increasing volume concentration of the stiffer phase (usually the aggregate). For a given aggregate, E_c increases with strength of the concrete since stiffness of the matrix increases with concrete strength.

The major contribution to E_c comes from stiffness of the aggregate and its volume concentration. This is reflected in a linear expression for concrete elastic modulus at a given age of the general form

$$E_c = K_0 + \beta f_{cu} \tag{5.7}$$

where
E_c is concrete elastic modulus
f_{cu} is characteristic concrete strength
β is a coefficient
K_0 is an aggregate stiffness factor related to the aggregate elastic modulus and volume concentration (Teychenné, 1978).

Equation (5.7) is a linear approximation of a more general power curve. For any given aggregate/cement (a/c) ratio, a reasonably linear relationship exists between strength and elastic modulus – see Figure 5.26. Concretes

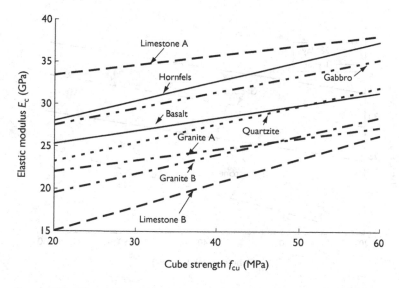

Figure 5.26 Elastic modulus–cube strength relation at 28 days, UK aggregates (based on Teychenné, 1978).

with similar compressive strengths but with aggregates of differing stiffness may differ substantially in E_c value, in fact by as much as 100 per cent. Figure 5.26 also indicates that both K_0 and β (slope of the line) can vary, depending on aggregate type.

E_c can increase substantially with age, the increase being greater than would be predicted based only on increase in compressive strength with age. This is shown in Figure 5.27 for two aggregates. The E_c–f_{cu} relationship for the limestone becomes less sensitive to concrete strength with time, whereas that for the gabbro aggregate shows an increase in the K_0 value only. These observations give rise to interesting mechanistic questions regarding the nature of the interaction between aggregate and matrix. Time-dependent changes in the ITZ (e.g. additional pore-filling and densification) are the most plausible explanation for these effects.

The powerful influence of aggregates on concrete elastic modulus was confirmed in the study of 23 South African aggregate types, which showed there is no unique E_c–f_{cu} relationship for concrete (Alexander and Davis, 1991, 1992). Each aggregate type produces a different age-dependent relationship, illustrated in Figure 5.28. Slopes of the regression lines become flatter with increased age, and at later ages concretes achieve higher E values than similar concretes of equal strength at earlier ages. This is attributed to densification of the ITZ with continued hydration giving composite behaviour of the material.

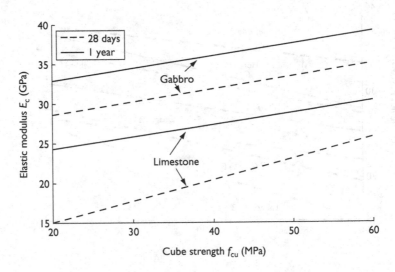

Figure 5.27 Increase in concrete E from 28 days to one year, UK aggregates (based on Teychenné, 1978).

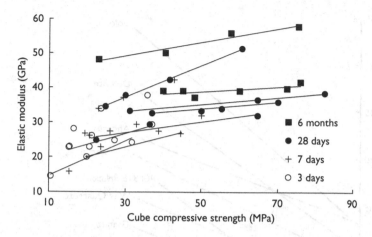

Figure 5.28 Composite plot of elastic modulus at different ages for three aggregate types (from Alexander, 1998; © 1998, reprinted with permission of the American Ceramic Society).

Prediction of concrete elastic modulus incorporating aggregate effects

It is often necessary to predict E_c at two important ages: earlier ages (say up to 28 days), to account for dead load effects such as removal of false-work (formwork) and early prestressing; and later ages (say from 3 months onwards), to account for long-term and live-load effects. The expression for E_c given in Equation (5.7) can be applied to these two time periods with constants determined experimentally. Figures 5.29 and 5.30 show trend lines for various aggregate types and indicate the aggregate-specific nature of the problem as well as the influence of age. The ACI prediction for E_c is also shown on Figure 5.29 for comparison, based on a concrete density of 2400 kg/m³ (ACI 318, 2002).

Figure 5.31 shows a typical regression relationship for dolomite concrete for ages from 3 to 28 days, with $K_0 = 23.3$ GPa and $\beta = 0.47$ GPa/MPa. Experimentally determined constants for South African aggregate types are given in Table 5.5 with the regression values rounded to the nearest 0.05 for β and an integer value for K_0. In general, the β values reduce with age while the K_0 values increase, often quite dramatically. This is consistent in mechanistic terms with strengthening and densification of the ITZ, permitting the aggregate to act more compositely with the matrix.

Aggregates and blended cements

Table 5.5 applies to concretes made with ordinary portland cement and will not necessarily be valid for other cement types. For blended cements, the

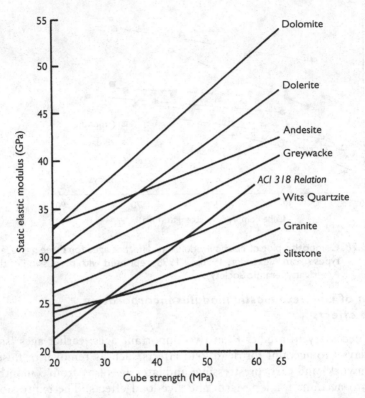

Figure 5.29 Relationship between static elastic modulus and cube strength of concrete for ages from 3 to 28 days (from Alexander, 1998; © 1998,
reprinted with permission of the American Ceramic Society).

same approach can be used but with different K_0 and β values. Figure 5.32
illustrates results for an OPC concrete and an OPC/CSF blend with granite
aggregates (Alexander and Milne, 1995). Cement type influences the elastic
modulus–strength relationships at different ages. Silica fume concretes generally yield trend lines consistent at all ages, and are as stiff as, or slightly
stiffer than, OPC concretes. Slag and fly ash concretes reflect lower K_0
values and higher β values than OPC concrete, implying a slower rate of
strength gain and poorer interfacial properties at early ages.

Rocks of the same type but from different sources

Variations occur between performance characteristics of rocks of the same
type but from different sources, and even within a single quarry there
are often differences. The quartzites shown in Table 5.5 provide a good
example. This may pose some uncertainty regarding design values for an

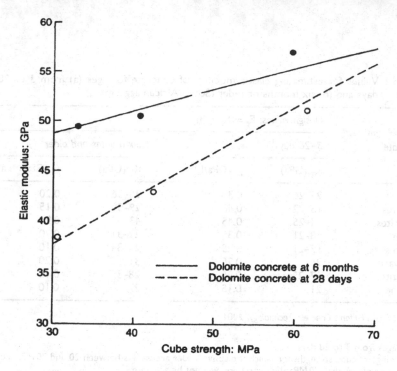

Figure 5.30 Increase in elastic modulus of dolomite concrete from 28 days to 6 months (from Alexander, 1998; © 1998, reprinted with permission of the American Ceramic Society).

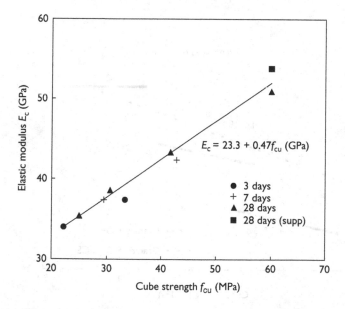

Figure 5.31 Elastic modulus–cube strength relationship for dolomite concrete for ages from 3 to 28 days (from Alexander, 1998; © 1998, reprinted with permission of the American Ceramic Society).

Table 5.5 Values for estimating elastic modulus of concrete for ages (a) from 3 to 28 days and (b) six months or older (South African aggregates)

| South African aggregate type | Design values: $E_c = K_0 + \beta f_{cu}$ | | | |
| | 3–28 days[1] | | Six months and older[2] | |
	K_0 (GPa)	β (GPa/MPa)	K_0 (GPa)	β (GPa/MPa)
Andesites	25–26	0.30	35–36	0.20
Dolerites	15–22	0.40	29–39	0.15
Dolomites	24–25	0.45	43–49	0.20
Felsites	18–21	0.35	29–31	0.20
Granites	17–21	0.25	25–34	0.10
Greywacke	24	0.25	31	0.20
Quartzites	17–23	0.25	28–35	0.15
Siltstone	21	0.15	27	0.10

Source: From *Fulton's Concrete Technology*, 2001.

Notes
1 For ages from 3 to 28 days:
The range of cube strengths for which the above values are valid is between 20 and 70 MPa. For strengths of less than 20 MPa the expression will not be accurate.
2 For ages of six months or older:
The range of cube strengths for which the above values are valid is between 30 and 90 MPa.

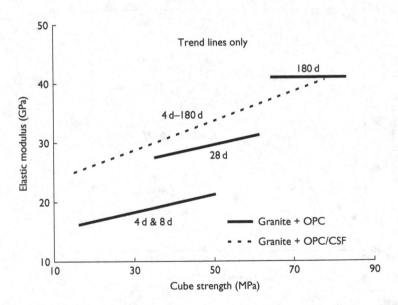

Figure 5.32 Elastic modulus versus cube strength relationships for OPC and OPC/CSF blends with granite aggregate (adapted from Alexander and Milne, 1995).

aggregate from a source not investigated. A first approach would be to use values from an equivalent rock type, provided there is evidence that the two are reasonably similar. For example, Table 5.5 indicates that for 40-MPa concrete made with a range of quartzites, E_c at 28 days has a comparatively small range from 27 to 33 GPa.

Relative contributions of sand or stone to concrete elastic modulus

Table 5.6 shows results for elastic modulus tests on concrete and the equivalent mortar matrix, in terms of K_0 and β values (Alexander and Davis, 1991). Results refer to concretes in which the crushed sand fraction was from the identical source as the stone fraction. The matrix phase (i.e. mortar) makes an important contribution to concrete elastic modulus and β factors are very similar for a given concrete and its corresponding mortar. Some coarse aggregates such as the granites used in this study have very little stiffness-enhancing effect, while others such as the andesite and dolerite markedly increase E_c when added to the mortar. The sensitivity of the aggregate to the matrix (the β value) is governed by the fine aggregate/paste fraction, while the effect of the coarse aggregate is to increase K_0. The aggregate-specific nature of the problem is well illustrated by these results.

Relationship between aggregate stiffness factor K_0 and aggregate elastic modulus E_a

A relationship might be expected between the aggregate stiffness factor K_0 and the aggregate elastic modulus E_a. Work on UK aggregates (Teychenné,

Table 5.6 Linear regression of elastic modulus on strength, for concrete and mortar, at 28 days

Aggregate type	Concrete		Mortar	
	K_0 (GPa)	β (GPa/MPa)	K_0 (GPa)	β (GPa/MPa)
Andesite	30.7	0.21	21.5	0.23
Dolerite	31.4	0.22	21.6	0.22
Dolomite	25.5	0.41	21.2	0.47
Felsite	24.5	0.28	17.0	0.25
Granite 1	21.5	0.19	18.7	0.21
Granite 2	27.7	0.15	24.1	0.15
Greywacke	25.4	0.23	23.7	0.22
Wits Quartzite 1	29.4	0.10	25.8	0.09
Quartzite 2	27.9	0.18	25.6	0.17
Quartzite 3	26.5	0.21	25.9	0.15

Source: From Alexander and Davis, 1991.

1978), in which the average values for K_0 and β were 20 GPa and 0.2 respectively, gave a tentative correlation as

$$K_0 = 0.38E_a \tag{5.8}$$

The advantage of Equation (5.8), if it were generally applicable, is that K_0 could be readily determined from a knowledge of E_a.

In the South African aggregate study, due to the range of K_0 and β values (see Table 5.5), the probability was low that a confident relationship between K_0 and E_a would exist. Also the mean β value for all the aggregates was 0.28, and not 0.2 as found in the UK study.

The correlation equation (with $\beta = 0.28$) was:

$$K_0 = 8.57 + 0.17E_a \quad \text{(coefficient of correlation} = 0.66) \tag{5.9}$$

This indicates that K_0 is not purely an aggregate stiffness factor, while β is not purely a paste-interaction factor. The relationships depend on the nature of the interaction between aggregate and matrix, that is on the nature of the ITZ. In normal strength concretes, the ITZ is likely to be highly variable and dependent on the aggregate, and the aggregate may consequently not be a fully active component in the mix.

Influence of aggregates on concrete Poisson's ratio

Aggregates influence concrete Poisson's ratio similarly to how they influence elastic modulus. There are considerably fewer data available on this aspect. Much of the data discussed previously relates to the South African study which generally employed crushed, single-sized coarse aggregates and gap-grading. However, the same basic behaviour can be expected to apply to continuously graded aggregates, certainly at a comparative level, even though it will be necessary to derive the specific relationships for any given set of materials. The reason for this is that the total aggregate volumes for equivalent continuously graded and gap-graded mixes are not substantially different and aggregates in both types of mixes tend to act discretely, that is they are not usually in direct contact. The work of Aïtcin (1989) and Aïtcin and Mehta (1990) also corroborates these observations.

Long-term deformations

Long-term time-dependent deformations of concrete comprise shrinkage and creep. Here too, aggregates can strongly influence deformations. Not much work has been done on the influence of aggregate properties on shrinkage and creep of concrete, except for shrinking aggregates (see p. 281).

The influence of aggregates on shrinkage and creep is primarily due to two effects:

- The dilution effect of the aggregate – the more aggregate in a mix, the lower the creep and shrinkage. This will depend on the standard water requirement (SWR) of the aggregates, which together with w/c ratio determines the paste and aggregate content. Aggregates with lower SWR will produce concretes with lower creep and shrinkage properties, other factors being equal.
- The restraint effect, due to the stiffness of the aggregate which restrains the paste movement, provided that the aggregate is stiffer than the paste.

The effect of aggregate volume content on concrete shrinkage is important – see Figure 5.33. For concretes with total aggregate volume contents of 65–75 per cent, concrete shrinkage will be less than 20 per cent of the paste shrinkage. The effect of aggregate stiffness is indirectly shown in Figure 5.34, considering that concrete elastic modulus is closely related to the modulus of the aggregate. Practically, aggregate stiffness is probably important only in concretes with low stiffness aggregates such as low density aggregates or weathered aggregates (Hobbs and Parrott, 1979).

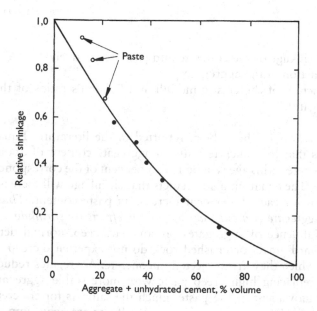

Figure 5.33 Effect of aggregate concentration on shrinkage of concrete (Powers, 1971) (from *Fulton*, 2001).

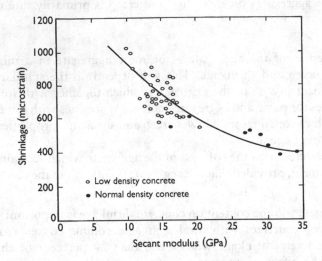

<figure>
Figure 5.34 Relation between drying shrinkage after two years and secant modulus of elasticity of concrete (at a stress:strength ratio of 0.4) at 28 days (Reichard, 1964) (from *Fulton*, 2001).
</figure>

Concrete shrinkage is related to aggregate stiffness and concentration by the equation (Pickett, 1956)

$$S_c = S_p(1 - V_a)^n \tag{5.10}$$

where
S_c and S_p are the shrinkage of the concrete and paste respectively
V_a is the volume fraction of the aggregate
n is a complex function of the elastic moduli and Poisson's ratios of the concrete and aggregate.

Values of n from 1.2 to 1.7 have been reported in the literature. Equation (5.10) predicts that for concrete with an aggregate content of about 70 per cent, the concrete shrinkage will be 10–15 per cent of the corresponding paste shrinkage. The equation also predicts that shrinkage will increase by about 6 per cent for each 1 per cent increase in paste content. *Thus, minor changes in aggregate content may have large effects on shrinkage.*

Turning to the influence of aggregates on concrete creep, normal density aggregates of hard gravel or crushed rock do not experience creep at the stress levels to which they are subjected in concrete. Aggregates reduce concrete creep by replacing high creep paste with more stable aggregate and by restraining movement of the paste, much the same as for concrete shrinkage. Aggregate volume concentration and stiffness are again important factors. Aggregate properties such as grading, maximum particle size, and shape indirectly influence concrete creep through their effect on water requirement and hence volume concentration.

Concrete creep can be expressed in terms of aggregate volume concentration and paste creep as follows (Hobbs, 1971)

$$\varepsilon_c = \frac{1 - V_a}{1 + V_a} \cdot \varepsilon_p \qquad (5.11)$$

where
ε_c and ε_p are the creep of the concrete and the paste, respectively
V_a is the volume fraction of the aggregate.

Creep of normal concretes varies between about 15 and 30 per cent of the creep of their corresponding pastes, depending on the quantity and elastic properties of the aggregate. For paste contents in the range of 25–35 per cent by volume, concrete creep increases by about 5 per cent for each 1 per cent increase in paste content. The higher the elastic modulus of the aggregate, the greater will be the restraint of the paste creep. This is shown in Figure 5.35 where the creep of concretes containing low modulus aggregates may be up to four times that of concretes with stiffer aggregates. Creep appears to be unaffected once the aggregate modulus exceeds about 70 GPa.

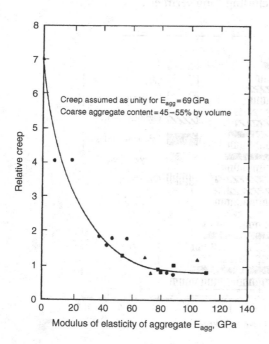

Figure 5.35 The effect of aggregate stiffness on creep of concrete (Evans and Kong, 1966) (from Fulton, 2001).

Effects of aggregate type

It is not possible to generalize on the effects of different aggregate types on concrete creep and shrinkage. This is shown by several historical studies in which the effects of specific generic types of aggregates were not consistent between the studies (Troxell *et al.*, 1958; Rüsch *et al.*, 1962). Particular aggregate types will differ from one locality to another. In general, sandstone aggregates often produce higher creep and shrinkage in concrete.

Data from the South African aggregate study are given in Figures 5.36 and 5.37. Shrinkage and creep were influenced by the factors mentioned above, and are shown as relative deformations to allow comparison over the range of aggregates. A wide range of relative strains due to the use of different aggregates can arise. Even within a generic group of aggregates, large differences occur. Correlations between shrinkage or creep and fundamental parameters, such as concrete water content and aggregate stiffness, were non-existent or tenuous at best. The complexity of the problem will demand trial testing when a suspect or unusual aggregate is being used or the structure is likely to be deflection-sensitive.

Total strains (sum of elastic, shrinkage, and creep strains) plotted against concrete elastic modulus from the same study are shown in Figure 5.38. Concrete elastic modulus may be regarded as a rough index of overall concrete stiffness including long-term effects.

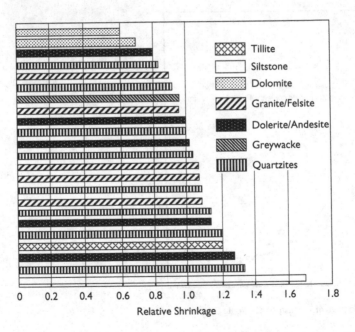

Figure 5.36 Relative shrinkage for a range of aggregate types (from Alexander, 1998; © 1998, reprinted with permission of the American Ceramic Society).

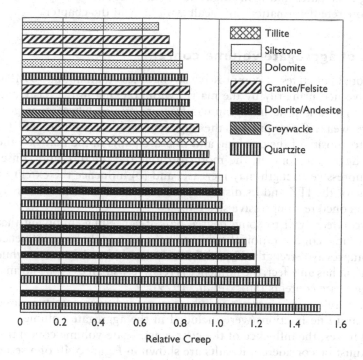

Figure 5.37 Relative creep coefficients for a range of aggregate types (from Davis and Alexander, 1992).

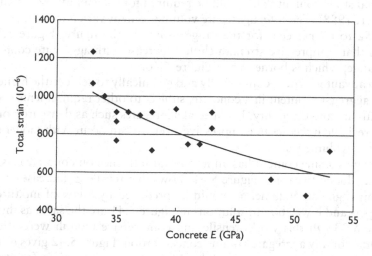

Figure 5.38 Total strains (elastic + shrinkage + creep) under sustained compressive stress of 5 MPa versus concrete elastic modulus (from Davis and Alexander, 1992).

Shrinking aggregates and lightweight aggregates, due to their particular effects on concrete deformation, are dealt with later in the chapter.

Influence of aggregate volume content

In conventional concretes of compressive strength less than about 50 MPa, failure is governed primarily by the matrix with cracking in the paste and ITZ phases. There is little intragranular fracture except where aggregate particles are weak. Consequently, the aggregate volume content will (a) govern the tortuosity of the cracking and fracture paths in the matrix – the greater the aggregate content, the more tortuous the fracture and consequently compressive strength may increase, and (b) influence the extent or total volume of the ITZ and its distribution in the matrix. This effect may decrease the concrete compressive strength.

These two effects will to some degree be self-cancelling, implying that aggregate volume concentration should not have an unduly large influence on, say, compressive strength. Figure 5.39 shows this to be true – while volume content has an effect, the influence is relatively small over the normal range of aggregate contents (Stock *et al.*, 1979).

Figure 5.39 refers to concretes in which a graded aggregate was used and coarse and fine fractions were included in the aggregate volume. For gap-graded mixes, the influence of the coarse aggregate volume concentration alone must be considered. Results are shown in Figure 5.40 of a series of mixes based on a crushed quartzite sand in which the stone content of crushed quartzite or andesite stone was varied from 0 to 100 per cent of the normal stone content that would be required for such a mix (Ballim and Alexander, 1988). The *total* aggregate volume content varied from approximately 52 to 69 per cent for the range shown in the figure. Figure 5.39 indicates that compressive strength should increase with aggregate content in this range, which is borne out in Figure 5.40.

It is an advantage both economically and technically to achieve the highest possible aggregate content in a concrete, subject to other requirements such as workability and durability. For special aggregates such as dense iron ore, very appreciable increases in compressive strength can occur with moderate increases of volume fraction.

Aggregate volume content has an important influence on concrete elastic modulus, alluded to earlier. Figure 5.41 shows that tensile E_c increases with increasing aggregate volume, as would be predicted by a law of mixtures. (The upper- and lower-bound solutions in Figure 5.41 are the same as those in Figure 5.23.) In this work, tensile and compressive moduli were effectively equal for any aggregate volume concentration. Figure 5.42 gives other results for compressive elastic modulus of concrete, where E_c is far more sensitive to addition of coarse aggregate in the form of crushed andesite than crushed granite.

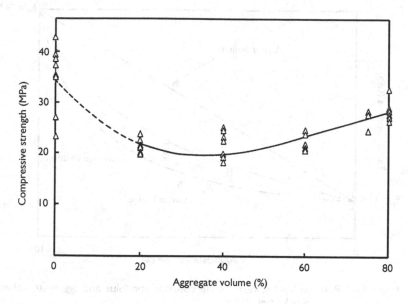

Figure 5.39 Relationship between concrete compressive strength and aggregate volume (from Stock *et al.*, 1979).

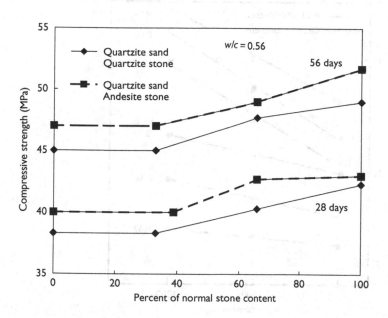

Figure 5.40 Effect of stone content and age on the compressive strength of concretes containing blends of andesite and quartzite aggregates (from Ballim and Alexander, 1988).

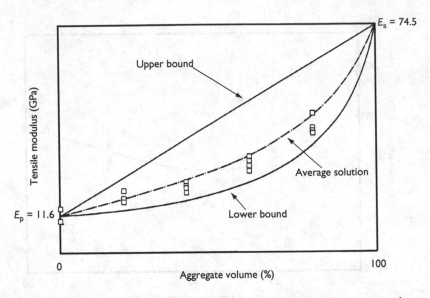

Figure 5.41 Relationship between tensile concrete modulus and aggregate volume (from Stock et al., 1979).

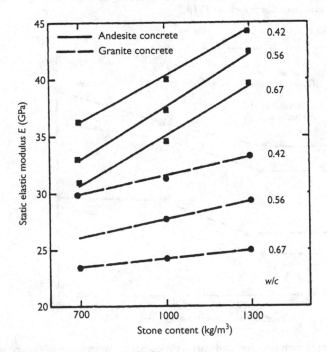

Figure 5.42 Concrete elastic modulus as a function of stone content and w/c ratio for granite and andesite concretes (from Grills and Alexander, 1989).

Aggregate dimensional stability, thermal properties, and corresponding concrete properties

Aggregates have a profound influence on dimensional stability and thermal properties of concrete. These properties include moisture movements (swelling and shrinkage), thermal movements (expansion and contraction), and thermal conductivity, diffusivity, and specific heat. The effects of aggregates on concrete shrinkage were covered earlier; this section will address the particular problem of unusual concrete shrinkage caused by dimensionally unstable shrinking aggregates. Relevant thermal properties of aggregates as well also as concretes in which they are embedded will also be discussed.

Moisture stability

Shrinking aggregates and lightweight aggregates

Shrinking aggregates and lightweight aggregates impart higher shrinkage and creep to concrete. Lightweight aggregates are covered in Chapter 7, but the principles discussed here apply equally to them. Aggregates influence concrete shrinkage in several ways. First, the nature of the aggregates (shape, texture, grading) dictates the water content of the concrete, which is linked to drying shrinkage. Second, aggregates influence concrete shrinkage through the mechanisms of dilution and restraint (see p. 273), reducing the inherent paste shrinkage – see Figure 5.33. This assumes that the aggregates are dimensionally stable and non-shrinking. Aggregates that exhibit dimensional instability on wetting and drying can lead to very large concrete drying shrinkage – typically in excess of 1000×10^{-6}. This can affect structural performance by giving rise to excessive deflections and cracking, which is more severe in harsher drying environments. It might be possible to use shrinking aggregates that do not exhibit too large shrinkage in concrete (say $<1000 \times 10^{-6}$) in moist situations where concrete shrinkage will be reduced, and shrinkage is not a critical factor.

The factors that govern the behaviour of shrinking aggregates are complex and interrelated. One important factor is the nature of any unstable aggregate minerals such as smectites, and rock types that weather to produce such minerals may be susceptible. However, other tests indicate that the presence in the aggregate of swelling clay minerals is not a prerequisite for large shrinkage movement of concrete. Clays such as illite and kaolinite with stable lattices were found to be present in many aggregate samples responsible for large moisture movements of concrete (Grieve, 2001). Other factors relate to high aggregate absorption (see p. 283), large total internal surface area, and aggregates with low elastic moduli (Pike, 1990).

Susceptible rock types are certain basic igneous rocks (altered dolerites and basalts), some types of metamorphosed shales, slates, and greywackes, and certain mudstones and sandstones. Weathering of these rocks plays a role, and if weathered in either tropical or temperate climates, they must be considered potentially highly shrinkable unless shown otherwise. The properties of a given aggregate type can vary considerably from different sources, and any suspect aggregate needs to be evaluated. There are many aggregates deriving from the rock types mentioned that perform quite satisfactorily in concrete.

Problems with shrinking aggregates have been reported from the US (Hansen and Nielson, 1965), the UK with certain Scottish dolerites and other rock types (Neville, 1981), and from South Africa with certain coarse and fine aggregates mainly of sedimentary origin from the Karoo Supergroup (Stutterheim, 1954; Grieve, 2001). A recent survey of New Zealand aggregates revealed that drying shrinkage of concretes made with a wide selection of aggregate types gave highly variable results, with differences of as much as 100 per cent within particular groups of aggregates (Mackechnie, 2003). In the New Zealand study, aggregate type and quality were the most important factors influencing shrinkage rather than water demand. Several greywackes and mixed gravels produced concrete with low water demands (below $160 \, l/m^3$) but with drying shrinkage in excess of 1000×10^{-6} after 56 days. In contrast, concretes made with andesite and basalt aggregates had low to moderate shrinkage despite having fairly high water demands. The best predictor of concrete drying shrinkage was aggregate absorption, with values above 0.8 per cent indicating a substantially greater risk of increased drying shrinkage. This is borne out by earlier work on Scottish aggregates in the UK by Edwards (1966), whose results are given in Figure 5.43. In South Africa, a good relationship between the content of secondary minerals in heavily altered dolerites and their shrinkage potential was demonstrated by Weinert (1968).

Testing for shrinking aggregates is best carried out by evaluating shrinkage of a representative concrete mix and comparing the results against an identical mix with a reference non-shrinking aggregate. There is debate as to whether a temperature-accelerated test is acceptable or if drying of the specimens should be carried out in a standard laboratory environment (20 °C, 60 per cent RH). Temperature-accelerated tests frequently suffer from poor precision, and the end-point of drying is often not clearly defined (Pike, 1990). The Building Research Establishment Digest 35 (BRE, 1968), now withdrawn, included a concrete drying shrinkage test procedure and recommendations for limiting concrete use depending on four levels of concrete shrinkage. The first three levels ranged from low values to a concrete drying shrinkage of 850×10^{-6}, with increasing restrictions being placed on types of structural concrete usage as the upper limit was approached. Concretes with drying shrinkages exceeding 850×10^{-6} were recommended for use only in conditions where complete drying would not occur, for mass concrete encapsulated with air entrained concrete, or for members not exposed to the weather. BRE Digest 35 has been superseded by BRE Digest

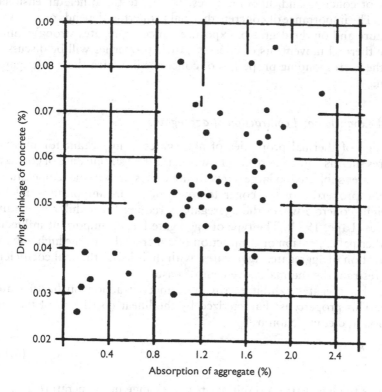

Figure 5.43 Influence of aggregate water absorption on drying shrinkage of concretes in which they were used (from Edwards, 1966).

357 (BRE, 1991) which refers to a shrinkage test for classifying the drying shrinkage of concrete aggregates (BS 812: Part 120: 1989). The drying environment for the 200 × 50 × 50-mm samples is 105 °C for 3 days. The concrete mix proportions are standardized in the test, and the results are therefore a comparative measure of aggregate shrinkage. The test results are used to classify aggregate shrinkages as follows: concrete shrinkage not exceeding 750×10^{-6} – may be used for all concreting purposes; greater than 750×10^{-6} – may be used in conditions similar to those given in BRE Digest 35 for the $>850 \times 10^{-6}$ category (with the additional proviso that members not exposed to the weather should be symmetrically and heavily reinforced).

Thermal movements

Most materials exhibit dimensional changes with change in temperature. Thermal movements should be considered in conjunction with other deformations of concrete and, in some cases, dominate the other dimensional changes. The importance of concrete thermal properties depends on the type of structure and on the degree of exposure. Since aggregates strongly influence the thermal movements of concrete, their properties will be discussed before the corresponding properties of concretes in which these aggregates are used.

Thermal expansion and contraction of aggregates

Knowledge of thermal properties of aggregates is important for massive structures such as dams where heat flow and thermal stability are of concern, for structures subjected to large temperature cycles or extreme temperatures where the properties of the concrete constituents are important, and for lightweight concrete where the aggregate is required to exhibit insulating properties (Lane, 1994). The type of aggregate has an important influence on the thermal expansion or contraction of concrete due to the high volume concentration of aggregate. Aggregates, with their lower thermal coefficient values, restrain the thermal movement of paste.

Mineral aggregates exhibit expansion and contraction on heating and cooling. This property is characterized by the linear coefficient of thermal expansion and contraction α,

$$\alpha = \frac{\varepsilon}{\Delta T} \qquad (5.12)$$

where ε is the linear strain resulting from a change in temperature ΔT.

The α value for a rock is a composite value of the corresponding properties of the constituent minerals, scaled according to a law of mixtures, and depends also on the porosity of the rock. Many rocks and aggregate particles are practically isotropic in regard to thermal movements. Occasionally some rocks may exhibit marked anisotropy, with the difference in α for the three directions amounting to a factor of two or more. For example, calcite can have a linear thermal expansion coefficient as high as $25.8 \times 10^{-6}/°C$ parallel to its C axis, and as low as $-4.7 \times 10^{-6}/°C$ perpendicular to the C axis. Potassium feldspars can also exhibit marked anisotropy. Provided the anisotropy is not extreme, a mean value of linear thermal coefficient for the aggregate in concrete can be assumed.

Table 3.2 gave typical values of thermal coefficients for different rock types. The overall range of values is large, and there can be large variations for any given rock type. New or unknown aggregates should be tested if there is concern as to their thermal expansion behaviour. Generally, silica-rich rocks such as quartzites and metamorphosed sandstones have higher α values than calcite-rich rocks such as dolomites and limestones or those containing high calcium feldspars. The thermal coefficients of rocks therefore vary largely in proportion to their quartz content.

Concrete is a composite based on a binder and embedded aggregate particles, and therefore the compatibility between the thermal movement of the various constituents is important, particularly in conditions where thermal cycling occurs. Thermal expansion coefficients for various rock types range from approximately 1 to $14 \times 10^{-6}/^{\circ}C$ (Table 3.2). The values for cement paste are typically about $11–16 \times 10^{-6}/^{\circ}C$, but this value may be as high as $22 \times 10^{-6}/^{\circ}C$ depending on the age and degree of saturation of the paste. These disparate values indicate that an incompatibility may exist in concrete, particularly if the aggregates are low thermal coefficient materials such as certain granites, limestones, and dolomites. These aggregates should not be used in situations with large thermal cycles, where fire resistance is required, or where thermal curing is carried out (Zoldners, 1971; Lane, 1994). However, in most practical cases, the thermal incompatibility between paste and aggregate is not a problem (Smith, 1956).

Concrete thermal movements

The major factor influencing thermal movement is the type of aggregate. The volume concentration of aggregate, the moisture content of the concrete, and the mix proportions are of lesser importance. Other factors such as type of cement, strength and age of concrete, and curing methods are far less significant. The thermal coefficient of expansion for siliceous minerals such as quartz (about $12 \times 10^{-6}/^{\circ}C$) is higher than that for most other minerals, while calcareous aggregates and certain granites have the lowest (about $6 \times 10^{-6}/^{\circ}C$). As would be expected, quartzite concretes generally exhibit the highest coefficients of expansion. Table 5.7 gives typical coefficients of thermal expansion of concrete depending upon aggregate type, gleaned from US, UK, and SA sources, and reported in *Fulton* (2001). The lowest value of thermal coefficient is about $4 \times 10^{-6}/^{\circ}C$ for a marble aggregate concrete, while the highest is $12 \times 10^{-6}/^{\circ}C$ for a quartzite concrete. The influence of the volume concentration of aggregate, albeit minor, is shown in Figure 5.44.

The coefficient of linear thermal expansion of concrete is frequently taken as an average value of about $10 \times 10^{-6}/^{\circ}C$, but this value may vary from about 6 to $14 \times 10^{-6}/^{\circ}C$ depending on the factors given above. Aggregates with high coefficients of thermal expansion will produce concretes with similarly high coefficients, which may be viewed as concretes of low thermal volume stability. The importance of aggregate thermal coefficient to the performance of concrete is twofold: first it influences the thermal coefficient of the concrete and hence thermal movements in structures, and second it may contribute to the development of internal stresses if there are large differences between the thermal coefficients of the various constituents. This is particularly true when the concrete is subjected to temperature extremes

Table 5.7 Effect of aggregate type on coefficient of thermal expansion of concrete

Aggregate	Coefficient of thermal expansion of concrete($\times 10^{-6}/°C$)
Basalt	9.3
Blastfurnace slag	10.6
Chert	13.2
Dolerite	9.5
Dolerite (SA)	7.5–8.0
Dolomite (SA)	8.5–9.0
Felsite (SA)	9.0
Granite (SA)	8.0–9.5
Granite and rhyolite	6.8–9.5
Limestone	6.1–9.9
Malmesbury hornfels (SA)	10.0–11.0
Marble	4.1
Quartz	10.4
Quartzite	12.8
Quartzite (SA)	9.5–12.0
Sandstone	11.7

Source: After *Fulton's Concrete Technology*, 2001.

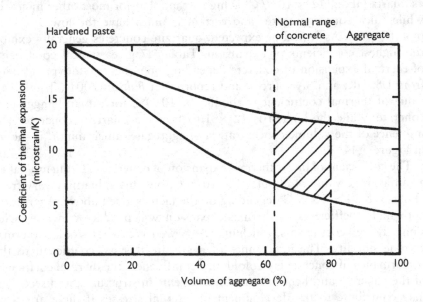

Figure 5.44 Thermal expansion of concrete having different aggregate types and contents (from Browne, 1972).

or thermal cycling, and will have an impact on internal cracking and thus on concrete durability (Pearson, 1942; Lane, 1994).

Debate continues about whether large differences between the thermal coefficients of the concrete constituents are important to concrete durability, but the weight of experimental evidence suggests this to be true. Lane (1994) reported on a number of experimental studies which appear to show that aggregates with particularly low thermal expansion coefficients (<5 or $6 \times 10^{-6}/°C$) give rise to cracking in concretes under severe thermal conditions such as frost action. Such cracking was not due to poor freeze–thaw resistance of the aggregate itself. The cracking developed between aggregate and matrix, and created pathways into the concrete for water to penetrate and cause subsequent freeze–thaw damage. As a broad rule, differences between thermal coefficients of coarse aggregate and mortar in which they are embedded should not exceed about $5 \times 10^{-6}/°C$ (Callan, 1952).

Other thermal properties

Other thermal properties of aggregates and concrete are conductivity, diffusivity, and specific heat. Conditions where these may be important relate to heat flow or heat dissipation in mass concrete and large concrete pours. Thermal properties of a concrete are strongly influenced by the corresponding thermal properties of the aggregates and by the aggregate volume concentration. Concretes with desired thermal conductivity properties will therefore need to be designed with the aggregate properties and proportions in mind.

THERMAL CONDUCTIVITY

Thermal conductivity is defined as the rate of heat flow through a body of unit thickness and unit area, with a unit temperature difference between two surfaces (units typically of W/mK). The thermal conductivity of concrete depends largely on the thermal conductivity of the aggregate, which in turn depends on rock and mineral type (Rhodes, 1978). Quartz has a relatively high conductivity while feldspars exhibit the opposite trend. Thus, quartzitic aggregates have high values of thermal conductivity (>3.0 W/mK), while basic igneous rocks and some limestones exhibit lower values (<1.5 W/mK). Porosity of an aggregate is also important in two respects: if the pores are air-filled, thermal conductivity is reduced since air is a good insulator; if the pores are water-filled, the opposite occurs due to the high thermal conductivity of water. Therefore, the moisture content of an aggregate has a large influence on thermal conductivity. Conductivity of *concrete* is influenced by its porosity and moisture content, which includes the corresponding aggregate properties. Consequently, if concrete with a high degree of insulation is desired, aggregates of high porosity yet relatively low absorption should be used. These requirements are most easily met with closed-cell artificial aggregates.

THERMAL DIFFUSIVITY

This is defined as the thermal conductivity divided by the specific heat and density (units of m^2/s); it is a physical material property that determines the time rate of change of temperature at any point within a body. As with thermal conductivity, the thermal diffusivity of a normal weight aggregate depends mainly on its quartz content. Likewise, specific heat is governed by mineral content (Lane, 1994; Scanlon and McDonald, 1994).

SPECIFIC HEAT

Specific heat is the amount of heat required to raise the temperature of a unit mass of material by one unit of temperature (units of J/kg K). The specific heat of concrete is largely governed by the specific heat of the aggregates. Methods of testing for thermal conductivity, thermal diffusivity, and specific heat are given in the literature (Lane, 1994).

The thermal properties of a material are related through the following formula

$$k = hc\rho \qquad (5.13)$$

where
k – thermal conductivity in W/mK
h – thermal diffusivity in m^2/s
c – specific heat in J/kg K
ρ – density in kg/m^3.

When determining these properties, it is normal to measure specific heat and then either diffusivity or conductivity and calculate the other property using the formula.

Fire resistance of aggregates and concrete

The thermal properties of aggregates have an important influence on the fire resistance of concrete. In general, low density and synthetic aggregates are more fire resistant than normal density aggregates due to their insulating properties and stability at high temperatures. As far as natural aggregates are concerned, carbonates fare better than siliceous aggregates in fire conditions. This is due to their ability to calcine at temperatures approaching 700–900 °C, depending on the mineralogy of the aggregate. As the calcined layer is formed, it insulates the concrete and reduces the heat transmission rate into the heart of the member. Quartz has a higher coefficient of thermal expansion than calcareous materials. More particularly quartz expands by 0.85 per cent at 573 °C, causing disruptive expansion and severe spalling of concrete with siliceous aggregates.

Closure

For many years, engineers and concrete technologists have regarded aggregates merely as inert fillers in concrete. This view needs to change. There is abundant evidence that aggregates exercise an important influence on the hardened properties of concrete, including strength. An understanding of the role of aggregates in hardened concrete can lead to a far more intelligent use of this important construction material.

References

Abrams, D.A. (1918) 'Design of concrete mixtures', *Struc. Materials Res. Lab., Lewis Institute Bulletin No. 1.*, Chicago.

Addis, B.J. (1992) 'Properties of high-strength concrete made with South African materials', PhD Thesis, Johannesburg, University of the Witwatersrand.

Aïtcin, P.-C. (1989) 'From gigapascals to nanometres: Advances in cement manufacture and use', *Proceedings of the Engineering Foundation Conference, Missouri, 1988*, New York: Engineering Foundation, 105–130.

Aïtcin, P.-C. and Mehta, P.K. (1990) 'Effect of coarse-aggregate characteristics on mechanical properties of high-strength concrete', *ACI Mater. J.*, 87(2): 103–107.

Alexander, K.M. and Taplin, J.H. (1962) 'Concrete strength, paste strength, cement hydration and the maturity rule', *Aust. J. Appl. Sci.*, 13(4): 277–284.

Alexander, K.M. and Taplin, J.H. (1964) 'Analysis of the strength and fracture of concrete based on unusual insensitivity of cement–aggregate bond to curing temperature', *Aust. J. Appl. Sci.*, 15(3): 60–70.

Alexander, M.G. and Ballim, Y. (1986) 'The concrete-making properties of andesite aggregates', *The Civil Engineer in South Africa*, 28(2): 49–57.

Alexander, M.G. and Ballim, Y. (1987) 'Use of andesite aggregate in concrete: Engineering properties and mechanism of strength enhancement', *Materials Science to Construction Materials Engineering, Proceedings of 1st International Congress, Paris, Vol. II, RILEM*, Cachan, France, 492–499.

Alexander, M.G. and Davis, D.E. (1991) 'Aggregates in concrete – a new assessment of their role', *Concrete Beton*, 59: 10–20.

Alexander, M.G. and Davis, D.E. (1992) 'The influence of aggregates on the compressive strength and elastic modulus of concrete', *The Civil Engineer in South Africa*, 34(5): 161–170.

Alexander, M.G. and Milne, T.I. (1995) 'Influence of cement blend and aggregate type on the stress–strain behaviour and elastic modulus of concrete', *ACI Mater. J.*, 92(3): 227–235.

Alexander, M.G., Mindess, S., Diamond, S. and Qu, L. (1995) 'Properties of paste/rock interfaces and their influence on composite behaviour', *Materials and Structures*, 28(183): 497–506.

Alexander, M.G. (1998) 'Role of aggregates in hardened concrete', in J.P. Skalny (ed.) *Materials Science of Concrete V*, Westerville, OH: American Ceramic Society, 119–147.

Alexander, M.G., Arliguie, G., Ballivy, G., Bentur, A. and Marchand, J. (1999) *Engineering of the Interfacial Transition Zone in Cementitious Composites*, State-of-the-Art-Report, Report No. 20, RILEM Publications.

American Concrete Institute Committee 318 (2002) 'Building code requirements for structural concrete', ACI 318M-02 (Metric Version), *American Concrete Institute Manual of Concrete Practice*, Farmington Hills, MI: American Concrete Institute.

Baalbaki, W., Benmokrane, B., Chaallal, O. and Aïtcin, P.-C. (1991) 'Influence of coarse aggregate on elastic properties of high-performance concrete', *ACI Mater. J.*, 88(5): 499–503.

Baalbaki, W., Aïtcin, P.-C. and Ballivy, G. (1992) 'On predicting modulus of elasticity in high-strength concrete', *ACI Mater. J.*, 89(5): 517–520.

Baker, S. (1931) *Cements and Aggregates*, Int. Correspondence Schools, Scranton, PA, 50.

Ballim, Y. and Alexander, M.G. (1988) 'Strength of concrete made with andesite aggregates', *Concrete Beton*, 50: 10–17.

Bennett, E.W. and Khilji, Z.M. (1964) 'The effect of some properties of the coarse aggregate in hardened concrete', *J. Br. Granite Whinstone*, 3(2): 1964.

Bentur, A. and Alexander, M.G. (2000) 'A review of work of the RILEM TC 159-ETC: Engineering of the interfacial transition zone in cementitious composites', *Materials and Structures*, 33: 82–87.

Bentz, D.P. and Garboczi, E.J. (1993) 'Digital-image-based computer modelling of cement-based materials', in J.D. Frost and J.R. Wright (eds) *Digital Image Processing: Techniques and Applications in Civil Engineering*, New York: American Society of Civil Engineers, 44–51.

Bentz, D.P. and Stutzman, P.E. (1994) 'SEM analysis and computer modelling of hydration of portland cement particles', in S.M. Dehayes and D. Stark (eds) *Petrography of Cementitious Materials*, ASTM STP 1215, West Conshohocken, PA: American Society for Testing and Materials, 60–73.

Bentz, D.P., Schlangen, E. and Garboczi, E.J. (1995) 'Computer simulation of interfacial zone microstructure and its effect on the properties of cement-based composites', in S. Mindess and J.P. Skalny (eds) *Materials Science of Concrete IV*, Westerville, OH: American Ceramic Society, 155–200.

Bloem, D.L. (1961) 'The problem of concrete strength relationship to maximum size of aggregate', *Joint Research Laboratory Publication No. 9*, National Sand & Gravel Association (or National Ready Mixed Concrete Association).

Bloem, D.L. and Gaynor, R.D. (1963) 'Effects of aggregate properties on strength of concrete', *J. Amer. Conc. Inst.*, 60(10): 1429–1454.

Brown, B.V. (1996) 'Alternative and marginal aggregate sources', in R.K. Dhir and T.D. Dyer (eds) *Proceedings of Int. Conf., Concrete in the Service of Mankind – Concrete for Environment Enhancement and Protection*, London: E&F N Spon, 471–484.

Browne, R.D. (1972) *Thermal Movement of Concrete*, Concrete Society Current Practice Sheet 3PC/06/1, London: Concrete Society.

Building Research Establishment (1968) *Shrinkage of Natural Aggregates in Concrete*, BRE Digest 35 (Withdrawn), Watford, UK: Building Research Establishment.

Building Research Establishment (1991) *Shrinkage of Natural Aggregates in Concrete*, BRE Digest 357, Watford, UK: Building Research Establishment.

Bush, T.D., Russell, B.W. and Freyne, S.F. (1997) 'High-performance concretes using Oklahoma aggregates', in L.S. Joha (ed.) *PCI/FHWA Int Symp. on HPC*, Chicago: Prestressed Conc. Institute.

Buyukozturk, O. and Lee, K.M. (1993) 'Assessment of interfacial fracture toughness in concrete composites', *Cem. Conc. Comp.*, 15(3): 143–152.

Callan, E.J. (1952) 'Thermal expansion of aggregates and concrete durability', *J. Amer. Conc. Inst. Proceedings*, 48: 504–511.

Chen, B. and Liu, J. (2004) 'Effect of aggregate on the fracture behaviour of high performance concrete', *Construction and Building Materials*, 18(8): 585–590.

Collepardi, M. (1989) *Use of Fly Ash, Silica Fume, Slag and Natural Pozzolans in Europe: Future Directions*, Italy: University of Ancona.

Collins, R.J. (1983) 'Concrete from crushed jurassic limestone', *Quarry Management and Products*, March: 127–138.

Counto, U.J. (1964) 'The effect of elastic modulus of the aggregate on the elastic modulus, creep and creep recovery of concrete', *Magazine of Concrete Research*, 16(48): 129–138.

Davis, D.E. and Alexander, M.G. (1992) *Properties of Aggregates in Concrete, Part 2*, Hippo Quarries: Hippo Quarries Technical Publication, 48.

de Larrard, F. and Belloc, A. (1997) 'The influence of aggregate on the compressive strength of normal and high-strength concrete', *ACI Mater. J.*, 94(5): 417–426.

Diamond, S. and Mindess, S. (1992) 'SEM investigations of fracture surfaces using stereo pairs: I, Fracture surfaces of rock and of cement paste', *Cem. Conc. Res.*, 22(1): 67–78.

Diamond, S., Mindess, S., Qu, L. and Alexander, M.G. (1992) 'SEM investigations of the contact zones between rock surfaces and cement paste', in J.C. Maso (ed.) *Proceedings RILEM Int. Conference on Interfaces in Cementitious Composites*, London: E&FN Spon, 13–22.

Diamond, S. (2003) 'Percolation due to overlapping ITZs in laboratory mortars? A microstructural evaluation', *Cem. Conc. Res.*, 33: 949–955.

Edwards, A.G. (1966) 'Shrinkage and other properties of concrete made with crushed rock aggregates from Scottish sources', *BGWF Journal*, Autumn, 23–41.

Elsharief, A., Cohen, M.D. and Olek, J. (2003) 'Influence of aggregate size, water cement ratios and age on the microstructure of the interfacial transitional zone', *Cem. Conc. Res.*, 33: 1837–1849.

Evans, R.H. and Kong, F.J. (1966) 'Estimation of creep in reinforced and prestressed concrete design', *Civil Engineering and Public Works Review*, 61(718): 593–596.

Fulton's Concrete Technology (1994) Ed. B.J. Addis, 7th edn, Midrand: Cement and Concrete Institute.

Fulton's Concrete Technology (2001) Eds B.J. Addis and G. Owens, 8th edn, Midrand: Cement and Concrete Institute.

Garboczi, E.L. and Bentz, D.P. (1991) 'Fundamental computer simulation models for cement-based materials', in J.P. Skalny and S. Mindess (eds) *Materials Science of Concrete II*, Westerville, OH: American Ceramic Society, 249–277.

Garboczi, E.J. (1993) 'Computational materials science of cement-based materials', *Materials and Structures*, 26: 191–195.

Garboczi, E.J. (2002) 'Three-dimensional mathematical analysis of particle shape using X-ray tomography and spherical harmonics: Application to aggregates used in concrete', *Cem. Conc. Res.*, 32(10): 1621–1638.

Giaccio, G., Rocco, C. and Zerbino, R. (1993) 'The fracture energy (G_F) of high-strength concretes', *Materials and Structures*, 26: 381–386.

Gilkey, H.J. (1961) 'Water-cement ratio versus strength – another look', *Journal ACI*, 57(4): 1287–1312.

Goldman, A. and Bentur, A. (1993) 'The influence of microfillers on enhancement of concrete strength', *Cem. Conc. Res.*, 23(4): 962–972.

Grieve, G.R.H. (2001) 'Aggregates for concrete', in B.J. Addis and G. Owens (eds) *Fulton's Concrete Technology*, 8th edn, Midrand: Cement and Concrete Institute.

Grills, F. and Alexander, M.G. (1989) 'Stiffness of concrete made with granite and andesite aggregates', *Civil Eng. South Africa*, 31(9): 273–285.

Hansen, T.C. (1965) 'Theories of multi-phase materials applied to concrete, cement mortar and cement paste', *The Structure of Concrete: Proceedings of an Int. Conf. London, September 1965*, London: Cement and Concrete Association, 16–23.

Hansen, T.C. and Nielson, K.E.C. (1965) 'Influence of aggregate properties on concrete shrinkage', *Journal ACI*, 63(7): 783–794.

Hashin, Z. (1962) 'The elastic moduli of heterogeneous materials', *Journal of Applied Mechanics*, 29(1): 143–150.

Higginson, E.C., Wallace, G.B. and Ore, E.L. (1963) 'Effect of maximum size of aggregate compressive strength of mass concrete', *Symp. Mass Concrete*, ACI SP-6, Farmington Hills, MI: American Concrete Institute, 219–256.

Hirsch, T.J. (1962) 'Modulus of elasticity of concrete affected by elastic moduli of cement paste matrix and aggregate', *Proceedings ACI*, 59(3): 427–451.

Hobbs, D.W. (1971) 'The dependence of bulk modulus, Young's modulus, creep, shrinkage and thermal expansion of concrete upon aggregate volume concentration', *Materials and Structures*, 4(20): 107–114.

Hobbs, D.W. and Parrott, L.J. (1979) 'Prediction of drying shrinkage', *Concrete*, 13(2): 19–24.

ISRM (1988) 'Suggested methods for determining the fracture toughness of rock', *International Journal of Rock Mechanics and Mining Science*, 25(2): 71–96.

Jones, R. and Kaplan, M.F. (1957) 'The effect of coarse aggregate on the mode of failure of concrete in compression and flexure', *Magazine of Concrete Research*, 9(26): 89–94.

Kaplan, M.F. (1959a) 'Flexural and compressive strength of concrete as affected by the properties of coarse aggregates', *Proceedings ACI*, 55: 1193–1207.

Kaplan, M.F. (1959b) *Ultrasonic Pulse Velocity, Dynamic Modulus of Elasticity, Poisson's Ratio, and the Strength of Concrete Made with Thirteen Difference Coarse Aggregates*, RILEM Bulletin No. 1 (New Series), Cachan: RILEM.

Katz, A., Bentur, A., Alexander, M.G. and Arliguie, G. (eds) (1998) *Proceedings RILEM Int. Conference on the Interfacial Transition Zone in Cementitious Composites*, London: E&FN Spon.

Kronlof, A. (1994) 'Effect of very fine aggregate on concrete strength', *Materials and Structures*, 27(165): 15–25.

Lane, D.S. (1994) 'Thermal properties of aggregates', *Significance of Tests and Properties of Concrete and Concrete Making Materials*, ASTM STP 169C, West Conshohocken, PA: American Society for Testing and Materials, 438–445.

La Rue, H. (1946) 'Modulus of elasticity of aggregates and their effect on concrete', *Proceedings Am. Soc. Test. Mater.*, 46: 1298–1310.

Leming, M.I. (1990) *Comparison of Mechanical Properties of High-Strength Concrete Made with Different Raw Materials*, Cement, Admixtures and Concrete. Transportation Research Record 1284, Washington, DC: Transportation Research Board, 230.

Lutz, M. and Monteiro, P.J.M. (1995) 'Effect of the transition zone on the bulk modulus of concrete', *Proceedings Mater. Res. Soc. Symp.*, 370: 413–418.

Mackechnie, J.R. (2003) *Properties of New Zealand Concrete Aggregates*, TR11, Wellington: Cement and Concrete Association, 57pp.

Maso, J.C. (1992) *Proceedings RILEM Int. Conf.*, Toulouse, London: E&FN Spon.

Mehta, P.K. and Monteiro, P.J.M. (1993) *Concrete: Structure, Properties, and Materials*, Englewood Cliffs, NJ: Prentice Hall.

Meininger, R.C. (1994) 'Degradation resistance, strength, and related properties', in *Significance of Tests and Properties of Concrete and Concrete Making Materials*, ASTM STP 169C, West Conshohocken, PA: American Society for Testing and Materials, 388–400.

Mindess, S. (1996) 'Tests to determine the mechanical properties of the interfacial zone', in J.C. Maso (ed.) *RILEM Report VI, Interfacial Transition Zone in Concrete*, London: E&FN Spon, 47–63.

Mindess, S. and Alexander, M.G. (1995) 'Mechanical phenomena at cement/ aggregate interfaces', in J.P. Skalny and S. Mindess (eds) *Materials Science of Concrete IV*, Westerville, OH: American Ceramic Society, 263–282.

Mindess, S. and Diamond, S. (1992) 'SEM investigations of fracture surfaces using stereo pairs: II, Fracture surfaces of rock-cement paste composite specimens', *Cem. Conc. Res.*, 22(4): 678–688.

Mindess, S. and Diamond, S. (1994) 'SEM investigations of fracture surfaces using stereo pairs: III, Fracture surfaces of mortars', *Cem. Conc. Res.*, 24(6): 1140–1152.

Mindess, S., Young, J.F. and Darwin, D. (2003) *Concrete*, Englewood Cliffs, NJ: Prentice Hall.

Mitsui, K., Li, Z., Lange, D.A. and Shah, S.P. (1994) 'Relationship between microstructure and mechanical properties of the paste–aggregate interfaces', *ACI Mater. J.*, 91(1): 30–39.

Neville, A.M. (1981) *Properties of Concrete*, 3rd edn, London: Pitman Publishing Limited.

Nilson, A.U. and Monteiro, P.J.M. (1993) 'Concrete: A three-phase material', *Cem. Conc. Res*, 23(1): 147–151.

Nimityongskul, P. and Smith, R.B.L. (1970) 'Effect of aggregate properties on the strength and deformation of concrete', *General Report, Proceedings of Symposium on Design of Concrete Structures for Creep, Shrinkage and Temperature Changes*, Madrid, Spain, International Association for Bridge and Structural Engineering, 193–200.

Odler, I. and Zurz, A. (1988) 'Structure and bond strength of cement–aggregate interfaces', in S. Mindess and S.P. Shah (eds) *Bonding in Cementitious Composites*, Materials Research Society, Symposium Proceedings, 114: 21–28.

Özturan, T. and Çeçen, C. (1997) 'Effect of coarse aggregate type on mechanical properties of concretes with different strengths', *Cem. Conc. Res.*, 27(2): 165–170.

Pearson, J.C. (1942) 'Concrete failure attributed to aggregate of low thermal coefficient', *Proceedings Journal ACI*, 38: 36–41.

Perry, C. and Gillott, J.E. (1977) 'The influence of mortar–aggregate bond strength on the behaviour of concrete in uniaxial compression', *Cem. Conc. Res.*, 7(5): 553–564.

Pickett, G. (1956) 'Effect of aggregate on shrinkage of concrete and a hypothesis concerning shrinkage', *Proceedings ACI*, 52: 581–590.

Pike, D.C. (ed.) (1990) *Standards for Aggregates*, Chichester: Ellis Horwood.

Ping, X. and Beaudoin, J.J. (1992) 'Modification of transition zone microstructure – silica fume coating of aggregate surfaces', *Cem. Conc. Res.*, 22(4): 597–604.

Powers, T.C. (1971) 'Fundamental aspects of concrete shrinkage', *Materiaux et Constructions*, 545: 79–85.

Reichard, T.W. (1964) *Creep and Drying Shrinkage of Lightweight and Normal Weight Concrete*, Monograph no. 74, Washington, DC: National Bureau of Standards.

Rhodes, J.A. (1978) 'Thermal properties', *Significance of Tests and Properties of Concrete and Concrete Making Materials*, ASTM STP 169B, West Conshohocken, PA: American Society for Testing and Materials, 242–262.

Road Research Laboratory (1955) *Concrete Roads – Design and Construction*, London: DSIR, HMSO, 19–34.

Rüsch, H., Kordina, K. and Hilsdorf, H. (1962) 'Der Einfluss des mineralogischen Charakters der Zuschläge auf das Kriechen von Beton', *Deutscher Ausschuss für Stahlbeton, Heft* 146, Berlin, 19–133.

Scanlon, J.M. and McDonald, J.E. (1994) 'Thermal properties', *Significance of Tests and Properties of Concrete and Concrete Making Materials*, ASTM STP 169C, West Conshohocken, PA: American Society for Testing and Materials, 229–239.

Schlangen, E. and Van Mier, J.G.M. (1992a) 'Simple lattice model for numerical simulations of fracture of concrete materials and structures', *Materials and Structures*, 25: 534–542.

Schlangen, E. and Van Mier, J.G.M. (1992b) 'Experimental and numerical analyses of micromechanisms of fracture of cement-based composites', *Cem. Conc. Comp.*, 14: 105–118.

Scrivener, K. (1989) 'The microstructure of concrete', in J.P. Skalny (ed.), *Materials Science of Concrete I*, Westerville, OH: American Ceramic Society, 127–162.

Scrivener, K. (1999) 'Engineering and transport properties of the interfacial transition zone in cementitious composites', in M.G. Alexander *et al.* (eds) *Characterisation of the ITZ and Its Quantification by Test Methods*, France: RILEM, 3–18.

Sengul, O., Tasdemir, C. and Tasdemir, M.A. (2002) 'Influence of aggregate type on mechanical behavior of normal- and high-strength concretes', *ACI Mater. J.*, 99(6): 528–533.

Sims, I. and Brown, B. (1998) 'Concrete aggregates', in P.C. Hewlett (ed.) *Lea's Chemistry of Cement and Concrete*, 4th edn, London: Arnold, 903–1012.

Smith, G.M. (1956) 'Physical incompatibility of matrix and aggregate in concrete', *Journal ACI*, Proceedings, 52: 791.

Stock, A.F., Hannant, D.J. and Williams, R.I.T. (1979) 'The effect of aggregate concentration upon the strength and modulus of elasticity of concrete', *Magazine of Concrete Research*, 31(109): 225–234.

Stutterheim, N. (1954) 'Excessive shrinkage of aggregate as a cause of deterioration of concrete structures in South Africa', *Transactions South African Institution of Civil Engineers*, 4(12): 351–367.

Swamy, N. (1970) 'Inelastic deformation of concrete', Highway Research Record No. 324, *Symposium on Concrete Deformation*, Highway Research Board, National Research Council, 89–97.

Teychenné, D.C. (1978) *The Use of Crushed Rock Aggregates in Concrete*, Garston, UK: Building Research Establishment.

Troxell, G.E., Raphael, J.M. and Davis, R.E. (1958) 'Long-time creep and shrinkage tests of plain and reinforced concrete', *Proceedings Am. Soc. Test. Mater.*, 58: 1101–1120.

Walker, S. and Bloem, D.L. (1960) 'Effects of aggregate size on properties of concrete', *J. ACI*, 57(3): 283–298.

Weinert, H.H. (1968) 'Engineering petrology for roads in South Africa', *Engineering Geology*, 2(5): 359–362.

Wright, P.J.F. (1955) 'A method of measuring the surface texture of aggregate', *Magazine of Concrete Research*, 7(21): 151–160.

Zoldners, N.G. (1971) 'Thermal properties of concrete under sustained elevated temperatures', *Temperature and Concrete*, SP 25, Farmington Hills, MI: American Concrete Institute.

Further reading

Barnes, B.D., Diamond, S. and Dolch, W.L. (1978) 'The contact zone between portland cement paste and glass aggregate surfaces', *Cem. Conc. Res.*, 8: 233–244.

Mindess, S., Qu, L. and Alexander, M.G. (1994) 'The influence of silica fume on the fracture properties of paste and microconcrete', *Adv. Cem. Res.*, 6(23): 103–107.

Neville, A.M. (1964) 'Creep of concrete as a function of its cement paste content', *Magazine of Concrete Research*, 16(46): 21–30.

Chapter 6

Aggregates in hardened concrete
Durability and transport properties

Durability is defined as the ability of a material, component, or structure to remain serviceable for a desired period within its design environment. Serviceability refers to the ability of a material, component, or structure to fulfil its design function, for example to retain water. A required property of concrete is long-term durability, and here aggregates also have a role. This chapter focuses on that role. Concrete is regarded as a 'durable' material, and this is generally true provided care is taken in the design of the material and precautions are exercised when it is required to function in particularly harsh or aggressive environments. Aggregates make a substantial contribution to concrete durability by normally being durable, hard, and longwearing. Aggregates in a concrete matrix reduce the cementitious paste component which is susceptible to deterioration, being more thermodynamically unstable. Aggregates are more resistant to external aggressive attack and are usually less permeable than the matrix. However, certain aggregates may exhibit properties that render them susceptible to deterioration and thus compromise concrete durability. In other cases, aggregates may be susceptible to attack from external agents.

Concrete durability is associated with 'transport properties' which refer to the movement of various substances or species (gases, liquids, ions) in or through the concrete. These are dealt with briefly to provide suitable background. Durability is also related to deterioration mechanisms, and this chapter considers these in the context of the influence that aggregates may have. Chapter 3 dealt with aggregate properties and included discussion on the durability of aggregates in their own right.

Concrete transport properties

Penetrability, percolation, and the ITZ

Penetrability is broadly defined as the degree to which a material permits the transport through it of gases, liquids, or ionic species. Penetrability embraces the concepts of permeability, sorption, diffusion, and conduction, and is quantifiable in terms of one of the transport parameters. Percolation refers to whether a material is penetrable or not; thus, percolation refers to a 'pass/fail' criterion.

The matrix component of concrete is inherently porous and penetrable. Even very dense, hydrated matrices display some penetrability related to the internal pore system and microcracking. However, well-hydrated cement paste has surprisingly low permeability, of the order of 10^{-12}–10^{-14} m/s, similar to many dense sound rocks used as concrete aggregates. Concrete is a composite material however, and the combination of the paste (or matrix) phase with an aggregate phase results in the introduction of an additional phase, that of the interfacial transition zone (ITZ). The ITZ, typically 20–50-μm thick, is characterised by a higher porosity and penetrability compared with the bulk paste. In some cases, the ITZ virtually 'disappears' as a distinct phase such as in dense heavily compacted dry mixes, in silica fume concretes, or in some very mature concretes. Chapter 5 dealt with this. Nevertheless, in many concretes the ITZ is an important phase, also in relation to concrete transport properties. Furthermore with typical ITZ thicknesses, about 30–60 per cent of the matrix may be comprised of the ITZ.

Regarding concrete penetrability, the presence of aggregates has several effects: first they produce ITZs in the composite, thereby increasing the penetrability; second, at sufficient aggregate volume concentration, the ITZ phases can become percolated, leading to an increase in penetrability; third, they tend to reduce overall penetrability due to their own relatively low permeabilities, sometimes called the dilution effect; and lastly, they influence the geometrical arrangement of the phases, introducing tortuosity into the flow paths, thus reducing penetrability.

The schematic diagram in Figure 6.1 illustrates the concept of percolation of the ITZ. The ITZ represents zones of higher penetrability that are preferential paths for transport. Provided aggregate particles do not approach each other too closely, these zones will not overlap or coalesce and unimpeded paths through the material are not available (Figure 6.1a). As aggregate concentration increases and aggregate particles approach each other more closely, the ITZs begin to overlap and create possible paths for transport of substances (Figure 6.1b). At a critical aggregate concentration, a point is reached when a path (or paths) through the material is created, which should be represented by a sudden increase in penetrability (measured for instance by permeability). At this point, the ITZ is said to be percolated (Figure 6.1c). Further increase of aggregate concentration beyond this point should result in additional penetrability. The percolation effect induced by aggregates is converse to the capillary discontinuity effect in pastes, where at a certain point governed by the water/binder (w/b) ratio and degree of hydration, capillaries become discontinuous, giving a lower degree of penetrability.

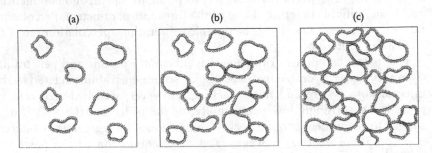

Figure 6.1 Schematic of penetrability and percolation related to ITZ: (a) Unpercolated, discrete isolated particles; (b) Unpercolated, some continuous ITZ
regions; and (c) Percolated, continuous ITZ regions.

In practice, this idealised concept is modified by several factors. The
matrix itself is penetrable, and as aggregate concentration increases and
ITZ boundaries move closer together, flaws and other discontinuities in the
matrix create 'bridges' so that penetrability rises more gradually, rather than
a sudden jump. The role of the coarse versus fine aggregate is also important.
Very fine aggregates (100 μm or less) will have less effect as the size of
these particles approaches that of the ITZ itself. With extreme fineness
such particles actually participate in the 'fine filler' effect, which generally
improves ITZ properties. However, if the coarse aggregate alone contributes
to a percolation effect, the required packing density for percolation will be
very high.

Computer modelling of percolation

It is possible to computer-model cementitious microstructure, including the
influence of ITZ and the phenomenon of percolation. The so-called 'NIST
model' (Bentz *et al.*, 1992, 1995) mimics concrete properties in simulated
microstructures incorporating aggregate particles at micrometre to millimetre scale. For real systems of multiple aggregate particles in a cementitious matrix, modelling is best achieved by a continuum hard-core/soft-shell
model. The 'soft shells' are permitted to overlap during initial packing of
the model, while the 'hard core' simulated aggregate particles are not. This
can be used to study the percolation problem relating to the connectivity of
ITZs in concrete (Winslow *et al.*, 1994). The soft shells may also be considered as enveloping ITZs which are penetrable and can therefore contribute
to transport of substances through the bulk material. However, the degree
of transport will depend on the overlapping of the soft-shell ITZs, with a
point being reached at which a permeable path is opened up through the
material, in effect a 'short-circuit' – the condition of percolation.

EXPERIMENTAL EVIDENCE

Experimental evidence of percolation is given in Figure 6.2 which shows the pore-size distributions for a series of sand–cement mortars all at a *w/c* ratio of 0.4, but with increasing sand volume fractions (Winslow *et al.*, 1994). A noticeable increase in the coarse (>0.1μm) pore volume occurs when the sand volume increases from 44.8 to 48.6 per cent, suggesting that percolation of the ITZs is achieved at this point. The experimental mortars in Figure 6.2 were modelled using a hard-core/soft-shell approach in which sand content and ITZ thickness were systematically varied. The results are given in Figure 6.3. Assuming that percolation is represented by a connected fraction approaching 100 per cent, an ITZ thickness of 30–40 μm is required for the critical sand fraction of between 44.8 and 48.6 per cent at which percolation occurs experimentally. This is very similar to experimentally measured ITZ thicknesses using SEM techniques (Scrivener and Gartner, 1988). Figure 6.3 suggests that at total aggregate volume fractions of 55 per cent or higher in typical real mortars and concretes, percolated pathways will be present due to overlapping ITZ regions. The results also suggest that pore refinement, and in particular ITZ refinement achieved by use of ultra-fine mineral additives such as CSF, will control or even eliminate percolation phenomena from real concretes or mortars.

Further experimental evidence of the influence of ITZ on transport properties comes from several investigators (Costa *et al.*, 1990; Ping *et al.*, 1991; Bretton *et al.*, 1992; Houst *et al.*, 1992), who show that increases in properties such as chloride ion diffusivity and gas permeability are observed at

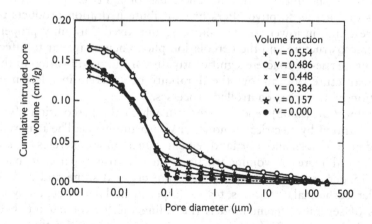

Figure 6.2 Mercury intrusion porosimetry for mortars with various sand contents. (Curves normalised based on intruded pore volume per gram of cement paste) (from Winslow *et al.*, 1994; ©1994, reprinted with permission from Elsevier).

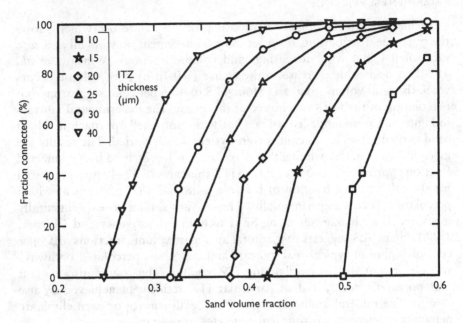

Figure 6.3 Outcome of hard-core/soft-shell model to determine ITZ percolation in mortars with varying sand contents and ITZ thicknesses (from Winslow *et al.*, 1994; ©1994, reprinted with permission from Elsevier).

critical aggregate volume concentrations. Other studies have shown that the ITZ can be modified by, for instance, use of lightweight or clinker aggregates that act as sorptive filters or contribute hydration products to the ITZ, resulting in lower penetrability and improved durability properties. ITZ microstructure and the percolation phenomenon appear to affect transport properties of concrete significantly. Intelligent modification of the ITZ microstructure can improve the durability of concrete in relation to deterioration by transport-controlled processes.

Princigallo *et al.* (2003) report a study in which the percolation effect was demonstrated by electrical conductivity measurements. The materials comprised high performance portland cement – silica fume concretes ($w/c = 0.37$) with total aggregate volume fractions varying from 0 per cent (plain paste) to 75 per cent, and comprising 16-mm crushed stone and natural sand, under isothermal conditions, tested. Experimental data were analysed by means of several different models (including the 'hard-core/soft-shell' model) and showed an ITZ thickness of about 9 μm and a ratio of the conductivity of the ITZ to the bulk paste of approximately 2.5. Percolation appeared to occur when the total aggregate volume fraction exceeded 60 per cent, from both experimental measurements and modelling. These results suggest that most normal strength concretes at typical aggregate volume fractions of 65–75 per cent would be percolated.

The inferences drawn from Figures 6.2 and 6.3 have been challenged recently by Diamond (2003), who conducted experiments on 28-day-old mortars that were prepared in the same way as the samples quoted in Winslow et al. (1994). Results of backscatter mode SEM were not consistent with the ITZ-percolation interpretation: as discussed in Chapter 5, the hardened cement paste (hcp) was found to consist of patches of dense, non-porous regions and highly porous patches, such patches indifferently occupying both classical 'ITZ' and 'bulk' locations. Dense, non-porous hcp was observed to surround many sand grains. The porous regions appeared to be capable of supporting fluid transmission. The observations were interpreted to mean that if percolation does in fact occur in high sand content mortars, it likely results from the interconnection or overlap of the porous patches rather than the ITZs.

The computer modelling approach of Bentz and co-workers can be exploited to study concrete properties such as conductivity, diffusivity, and fluid permeability. The pixel-based microstructure can be mapped into finite element or finite difference schemes and appropriately solved, assuming that the relevant properties of the individual phases in the microstructure are known. As further experimental evidence becomes available, the computer modelling approach can be better calibrated and applied.

Transport mechanisms and parameters

Concrete transport parameters relate to the mechanisms of permeation, sorption, diffusion, and conduction. Aggregates play a role in each of these mechanisms, and further fruitful work on this remains to be carried out.

Permeation

Permeation is the transport of substances through saturated concrete under a pressure gradient, measured by a permeability coefficient. The simplest permeability expression is that of D'Arcy:

$$\frac{\partial m}{\partial t} = \frac{k}{g}\frac{\partial P}{\partial z} \tag{6.1}$$

where
$\partial m/\partial t$ is the rate of mass flow per unit cross-sectional area
$\partial P/\partial z$ is the pressure gradient in the direction of flow
k is the coefficient of permeability
g is the acceleration due to gravity.

For concrete, permeation is important in situations such as water-retaining structures where water transport through the structure is undesirable.

Normal density aggregates possess a permeability which is usually lower than that of concrete, but similar to the paste. The permeability of concrete will depend on numerous factors, many related to the matrix. Aggregates affect the permeability of the composite (a) by the introduction of ITZs which have a higher permeability than the bulk paste, increasing the overall permeability, (b) by the interruption of the capillary pore system in the matrix, and (c) by reducing the effective area for flow, thus lowering the permeability, particularly of high w/c matrixes. Increasing aggregate concentration also causes the permeability of the composite material to increase as overall porosity reduces, in contrast to the response of pastes, since the porosity of normal aggregates is less than that of paste. Figures 6.4a and b show typical experimental results. These phenomena can best be explained by the presence of the ITZ, and the fact that its influences become more severe as the maximum aggregate size increases.

In the practical range of aggregate concentrations, that is 65–75 per cent, these effects are unlikely to vary much with aggregate content, since 'normal' concrete mixes are generally percolated at conventional aggregate contents. However, certain aggregates, notably low density porous aggregates, may have coefficients of permeability similar to or even higher than those of their embedding matrix, resulting in a more marked increase in permeability with aggregate concentration, also illustrated in Figure 6.4a.

ITZs induce percolation when a critical aggregate content is reached. However, this is neither universal nor inevitable since this zone can be altered by use of fine fillers or by continuing densification in maturing matrixes. Further, macro-defects such as compaction voids and cracking will usually have a far greater influence on concrete permeability than aggregate concentration or the ITZ effect.

Sorption

Sorption refers to the uptake of liquids into a solid by capillary suction. It is important in applications such as water-retaining structures, piers and quay walls partially in water, and even building facades that can be impinged by rain. Sorption is measured by parameters such as bulk absorption, or alternatively sorptivity S, defined as the rate of advance of a wetting front in a dry or partially saturated porous medium:

$$S = \frac{\Delta M_t}{t^{1/2}} \left[\frac{d}{M_{sat} - M_0} \right] \qquad (6.2)$$

where
$\Delta M_t/t^{1/2}$ is the slope of the straight line produced when the mass of water absorbed is plotted against the square root of time
d is the sample thickness
M_{sat} and M_0 are saturated mass and dry mass of concrete respectively.

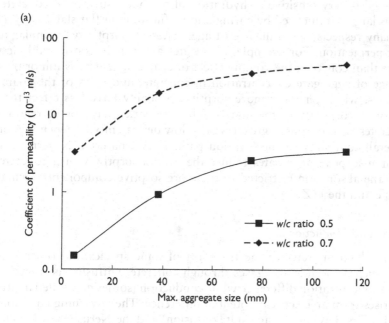

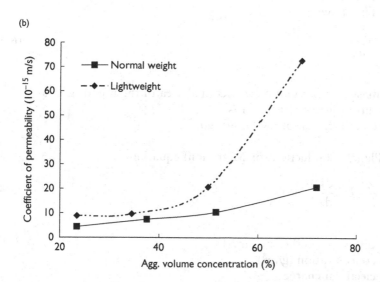

Figure 6.4 Water permeability of concrete and mortar: effect of (a) *w/c* ratio and maximum aggregate size in concrete (adapted from USBR, 1975), and (b) aggregate volume concentration in mortar (adapted from Nyame, 1985).

Sorptivity is influenced by the larger capillaries and their degree of continuity and is very sensitive to hydration of the outer surface of concrete, which is largely influenced by curing and by finishing in flat slabs.

In many respects, the influence of aggregates on sorptivity is similar to that of permeation. For example, aggregates are usually considerably less sorptive than concrete; the volume effect of coarse aggregate is minor over the range of aggregate concentration in concrete; and ITZs of the coarse aggregates, while generally more sorptive, probably have less effect than the intrinsic sorptivity of the matrix. These remarks may not be relevant in concretes with porous aggregates or in low density mixes, depending on the overall continuity of the sorption paths in the material. Nevertheless, dense non-sorptive aggregates reduce the overall sorptivity of a concrete, leaving the absorption restricted to the more sorptive components such as the paste and the ITZ.

Diffusion and conduction

These mechanisms refer to the transport of ionic species and other substances such as liquids and gases through concrete. Diffusion is transport due to a concentration difference while conduction (sometimes called migration) arises from an electrical potential difference. The governing equations are Fick's First Law (for steady state diffusion) and the Nernst-Planck equation respectively:

Fick's First Law

$$J = D\frac{\partial C}{\partial x} \tag{6.3}$$

where
J is the mass flux per unit cross-sectional area of the liquid (g/m^2s)
D is the diffusion coefficient (m^2/s)
C is the concentration of the liquid (g/m^3).

Nernst-Planck Conduction (or Migration) equation

$$J = \frac{ZF}{RT}DC\frac{dE}{dx} \tag{6.4}$$

where
D is the diffusion coefficient (m^2/s)
C is the concentration (g/m^3)
Z is the electrical charge
F is the Faraday constant ($J/V \cdot mol$)
R is the gas constant ($J/mol \cdot K$)
T is the absolute temperature (K)
dE/dx is the gradient of the electrical field (V/m).

These relationships have been extensively studied in concrete mainly in relation to binder type, proportions, and hydration, but not much in relation to
aggregates. The effects of the aggregates will generally be less than those of
the paste component. For transport of highly mobile ions such as chlorides,
the binder chemistry and pore structure will have a far larger influence
than the aggregate volume concentration or the ITZ. If the aggregates in
a well-proportioned and compacted matrix are non-penetrable and non-
conductive, they are likely to reduce the overall diffusivity and conductivity
of the concrete, particularly if the ITZ zone densifies with time.

Experimental evidence for the influence of aggregates on chloride migration through mortars is given in Figure 6.5. The effects shown are due
only to dilution, tortuosity, and the ITZ, and not percolation due to the
relatively low aggregate volume concentrations. Aggregates over this range
of values do indeed have an effect on chloride migration. Depending on
the assumed thickness of the ITZ, the effective migration coefficient of
the ITZ was computed as being as much as 2.83 times that of the plain
matrix.

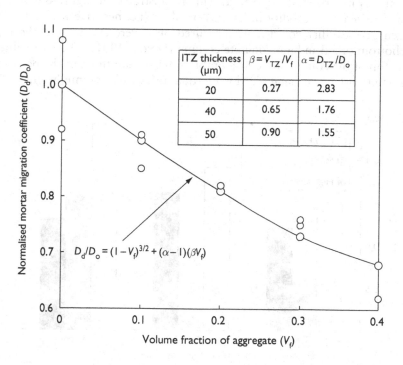

Figure 6.5 Experimental chloride migration coefficients of mortars as a function
of aggregate volume concentration. $\beta =$ ITZ volume fraction; $\alpha =$ ITZ
migration coefficient ratio (adapted from Yang and Su, 2002).

Aggregates and 'covercrete' transport

'Covercrete' is defined as the cover zone of concrete, the outer layer covering the reinforcing steel. The covercrete may be regarded as a large 'transition zone' with properties generally being inferior to those of the bulk concrete, particularly as the exposed surface is approached. Due to the effects of less efficient particle packing against formwork faces there is less aggregate and more matrix in the covercrete, and bleeding effects are often exacerbated in this zone. Curing will affect the covercrete far more than the bulk concrete. Transport mechanisms described above apply equally to the covercrete. However, due to the critical nature of this zone in providing corrosion protection to the steel, transport of aggressive gases, liquids, and ions through this zone becomes very important (Alexander *et al.*, 2001).

The maximum size of the aggregates may be similar to the thickness of the covercrete and so ready paths may be opened up to the steel along aggregate interfaces or other permeable flaws. The key factors are maximum aggregate size in relation to covercrete thickness and effects such as bleeding or plastic settlement that produce flaws and voids under or around aggregate particles. Such flaws may not be visible to the naked eye, but their influence is nevertheless important. This is illustrated in Figure 6.6 which shows oxygen permeability index (OPI) values (i.e. negative log of Darcy oxygen permeability coefficient) measured on covercrete discs at the top and bottom of a 3-m high concrete column (Dixon, 2001). The OPI values at the top of the column are considerably smaller than those at the bottom, attributed to the increasing bleed water quantities with column height (OPI

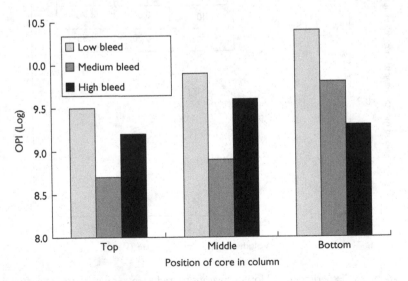

Figure 6.6 Effects of bleeding in 3-m high concrete column, in terms of OPI (adapted from Dixon, 2001).

values are on a log scale – the lower the value, the more permeable the concrete). Bleed water becomes trapped under aggregate particles and creates bleed channels through the concrete, providing paths for gas transport (a photograph showing this is in Figure 4.10).

Concrete deterioration mechanisms relating to aggregates

The deterioration of concrete in natural environments is most commonly a consequence of the breakdown of the binder or matrix phase, the aggregates being harder and more durable. However, there are several important instances when the aggregates themselves contribute directly to concrete deterioration or are susceptible to external aggressive environments. These can be divided into *intrinsic* deterioration mechanisms, referring to processes that occur internally as a consequence of the nature of the constituent materials (e.g. chemical or physical incompatibility), and *extrinsic* deterioration mechanisms, whereby concrete deteriorates from the actions of external agents such as weathering. Deterioration mechanisms can also be divided into *physical* or *chemical* mechanisms and aggregates may play a role in these. Physical deterioration involves surface wear or mass loss due to abrasion, erosion and cavitation, and cracking due to volume changes, structural loading, or exposure to temperature extremes. Typical chemical deterioration mechanisms are:

a Dissolution or leaching of the products of cement hydration with an accompanying loss of strength.
b Exchange reactions between acids and hcp, in which deterioration starts from the surface and continues into the concrete.
c Conversion of the products of hydration by external agents with associated expansive forces causing deterioration – the effect of sulphates on the concrete matrix being an example of this form of attack.
d Internal processes of deterioration such as incompatibilities between the constituents of the concrete that produce secondary products internally and cause the concrete to deteriorate. These processes usually require external water to be available. Alkali–silica reactions are typical of this form of deterioration.
e Steel corrosion in reinforced concrete caused by depassivation of the steel.

Aggregates may not necessarily influence all of the above mechanisms, particularly a, b, and c, but since they interact with the binder phase during deterioration, the overall material response to degradation must be considered.

Table 6.1 Matrix of deterioration mechanisms relating to aggregates in concrete

	Physical mechanisms	*Chemical mechanisms*
Intrinsic mechanisms	Dimensional incompatibility: • Thermal effects • Moisture effects	Alkali-aggregate reaction Sulphides in aggregates Thaumasite sulphate attack
Extrinsic mechanisms	Freeze–thaw Surface wetting and drying (moisture cycles) Surface abrasion and erosion	Acid attack Alkali attack Other aggressive chemicals, e.g. sulphates

Table 6.1 provides a matrix in which the primary deterioration mechanisms are summarised as they relate to the influence or presence of aggregates. A number of the more important mechanisms will be discussed in this chapter.

Table 6.1 overlaps with other sections dealing with durability and deterioration. Chapter 3 not only dealt with the resistance of aggregates to various aggressive substances (Table 3.13), but also stressed that aggregates themselves may be a potential source of aggressive or deleterious substances in concrete, for instance when they contain sulphates. Table 3.15 covered this aspect. Chapter 3 also dealt with freeze–thaw resistance of aggregates, and this topic will be covered here again briefly. Chapter 5 discussed concrete physical durability from an aggregate perspective, in regard to dimensional incompatibility and surface abrasion and erosion. Consequently, this chapter will cover the relevant topics only insofar as they have not been dealt with in previous chapters, and from a composite-material perspective.

Intrinsic mechanisms

Alkali–aggregate reaction

Alkali–aggregate reaction (AAR) is a chemical reaction between alkalis in the matrix and reactive components of the aggregates. It was first observed in the USA in the 1920s, and pioneering research was done on the mechanism by Stanton in the USA in 1940. Deterioration of concrete structures due to AAR has been observed in many countries including Australia, Canada, China, Denmark, Germany, Iceland, Japan, New Zealand, South Africa, UK, and USA. AAR reactions require the presence of sufficient moisture before they induce substantial internal damage which is usually in the form of micro- and macro-cracking within the matrix and the aggregates.

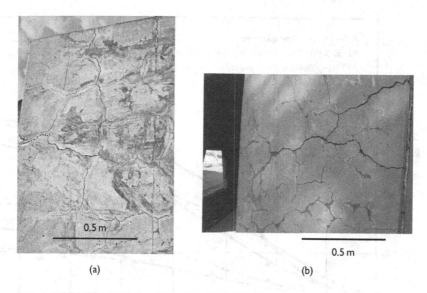

Figure 6.7 Surface cracking of (a) haunch of portal structure and (b) column of bridge structure due to AAR (photograph 6.7b courtesy of Y. Ballim).

Externally, these effects appear as unsightly and sometimes alarming surface cracking with crack widths of the order of mm or even cm. Figure 6.7 shows typical surface manifestations of AAR cracking on a large portal frame motorway structure and on a bridge column in South Africa. Cracking usually becomes evident several years after construction and the reaction may then continue for many years.

Alkali–aggregate reaction usually results in the formation of a gel which, in the presence of sufficient moisture, swells and causes internal expansion, with extensional strains that exceed the tensile strain capacity of the concrete, causing the cracking that is characteristic of AAR. (An exception to this rule is alkali–carbonate-rock reaction, mentioned later.) Figure 6.8 shows typical expansions measured on laboratory specimens containing different silica-bearing coarse aggregates.

Other distinguishing external features of AAR may be gel weeping from cracks, unsightly staining of cracks, and characteristic crack patterns that vary from random 'map cracking' (Figure 6.7) where structural stresses are low, to directional cracking in the presence of predominant stress systems (e.g. beams or columns). Figure 6.9 shows such a case, where the predominant cracks in the bridge pier are vertical and parallel to the principal compressive stress, although other transverse cracks do exist. The figure shows a prominent weep-hole at the top of the pier. A further manifestation of AAR occurring in certain Midwest areas of the US is small pop-outs on slab surfaces (Landgren and Hadley, 2002). This problem has occurred

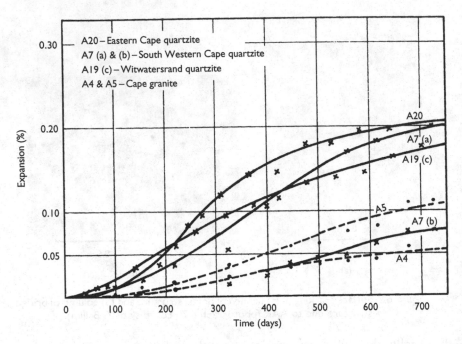

Figure 6.8 Laboratory expansions of different aggregates due to AAR. Cement content of concrete 350 kg/m³. Stored under ASTM C227 conditions above water in sealed containers at 38 °C (from Oberholster, 1986a).

with shale aggregates containing an opaline component and is aggravated by harsh drying conditions during and after the finishing operation, which tends to bring additional alkalis to the surface by evaporation. The problem can largely be avoided by adopting appropriate curing techniques such as water ponding or use of damp sand. In the Vancouver area of British Columbia, certain aggregates containing an iron compound undergo an AAR when used with high alkali cements. The reaction produces brown stains and pop-outs, but not excessive expansion, and the problem is mainly an aesthetic one (Mindess and Gilley, 1973).

There are three main types of AAR depending on the nature of the aggregates: alkali–silica reaction, alkali–silicate reaction, and alkali–carbonate-rock reaction.

ALKALI–SILICA REACTION (ASR)

This refers to reactions between alkaline pore solution and metastable forms of silica such as volcanic glasses, cristobalite, tridymite, and opal. ASR is also taken to refer to reactions between pore solution alkalis and aggre-

Figure 6.9 Directional AAR cracking – parallel mainly to major stress system (Photo Courtesy of Prof. Ballim University of the Witwatersrand).

gates containing or comprising chert, chalcedony, microcrystalline quartz, cryptocrystalline quartz, or strained quartz. Occurrences are found in rocks such as greywacke, quartzite, hornfels, phyllite, argillite, granite, granite-gneiss, granodiorite, and so on. The essential feature is a reactive form of silica of amorphous or altered forms that are not chemically stable. ASR is associated with the formation of expansive alkali–silica gel in concrete.

The distinctives of ASR in broken sections of concrete are usually whitish product and reaction rims around aggregate particles, cracks through aggregates sometimes filled with gel, matrix cracks often contiguous with the

(a)   (b)

Figure 6.10 Distinctive features of ASR in (a) broken chunk of concrete and (b) sliced
core. Note aggregate reaction rims, white staining, cracks through aggregate
and matrix (photograph 6.10b courtesy of Y. Ballim).

aggregate cracks, and voids filled with reaction products – Figure 6.10. A
loss of bond may also occur between aggregate and matrix. Cracks in aggregates that show evidence of ASR are often incipient or pre-existing microc-
racks that are planes of weakness into which alkaline pore solution penetrates
and reacts. Crushed aggregates may have abundant incipient fractures.

ALKALI–SILICATE REACTION

According to Oberholster (2001), this term was originally introduced to
differentiate ASR from reactions involving aggregates such as greywacke
and argillite found in Nova Scotia, Canada. This reaction is surmised to
involve expansion and exfoliation of certain clay minerals (phyllosilicates),
but there seems to be no clear evidence that the basic reaction is different
from conventional ASR.

ALKALI–CARBONATE-ROCK REACTION (ACR)

This does not produce a gel. Instead, the coarse aggregate particles expand
due to the alkali hydroxides reacting with small dolomite crystals in a clay
matrix, resulting in a dedolomitisation reaction. This type of AAR is limited to carbonate aggregate containing clay, such as certain argillaceous
dolomitic limestones or argillaceous calcitic dolostones, and causes expansion and extensive cracking (Swenson and Gillott, 1964; Dolar-Mantuani,
1983). Alkali–carbonate-rock reaction is not widespread and is encountered
mainly in Canada where alkali-susceptible carbonate rocks occur in the

Gull River Formation along the southern margin of the Canadian Shield from Midland to Kingston in Southern Ontario. The same reactive rock outcrops in the Ottawa-St Lawrence lowlands near Cornwall and in the Ottawa area. While ACR is locally very important, it will not be further covered here due to its limited occurrence, and this chapter will deal with the much more common phenomenon of ASR.

Table 6.2 summarises the major rock and mineral types that are alkali-reactive, indicating the reactive component in each case. All three major rock classifications contain alkali-reactive rocks. The table indicates that AAR is indeed a global phenomenon – there are very few areas that are completely devoid of this problem. However, concrete is successfully manufactured even where alkali-reactive rocks occur by applying principles dealt with later.

MECHANISM OF ASR

An explanation for the ASR mechanism has been provided by Helmuth and Stark (1992), reviewed by Mindess *et al.* (2003) as follows:

> Helmuth and Stark observed that the alkali–silica reaction results in the production of two-component gels – one component is a non-swelling calcium-alkali-silicate-hydrate (C-N(K)-S-H) and the other is a swelling alkali-silicate-hydrate (N(K)-S-H). When the alkali–silicate reaction occurs in concrete, some non-swelling C-N(K)-S-H is always formed. The reaction will be safe if this is the only reaction product, but unsafe if both gels form. The key factor appears to be the relative amounts of alkali and reactive silica. The overall process proceeds in a series of overlapping steps:
>
> 1. In the presence of a pore solution consisting of H_2O and Na^+, K^+, Ca^{2+}, OH^- and $H_3SiO_4^-$ ions (the latter from dissolved silica), the reactive silica undergoes depolymerization, dissolution, and swelling. The swelling can cause damage to the concrete, but the most significant volume change results from cracking caused by subsequent expansion of reaction products.
> 2. The alkali and calcium ions diffuse into the swollen aggregate resulting in the formation of a non-swelling C-N(K)-S-H gel, which can be considered as C-S-H containing some alkali. The calcium content depends on the alkali concentration, since the solubility of CH is inversely proportional to the alkali concentration.
> 3. The pore solution diffuses through the rather porous layer of this C-N(K)-S-H gel to the silica. Depending on the relative concentration of alkali and the rate of diffusion, the result can be safe or unsafe. If CaO constitutes 53 per cent or more of the C-N(K)-S-H

Table 6.2 Minerals, rocks and other substances that are potentially deleteriously reactive with alkalis in concrete

Minerals

Opal
Tridymite
Cristobalite
Chalcedony, cryptocrystalline, microcrystalline or glassy quartz
Coarse-grained quartz that is intensely fractured, granulated and strained internally or rich in secondary inclusions
Siliceous, intermediate and basic volcanic glasses
Vein quartz

Rocks

	Rock	Reactive component
Igneous	Granodiorite	
	Charnockite	Strained quartz; microcrystalline quartz
	Granite	
	Pumice	
	Rhyolite	
	Andesite	
	Dacite	Silicic to intermediate silica-rich volcanic
	Latite	glass; devitrified glass, tridymite
	Perlite	
	Obsidian	
	Volcanic tuff	
	Basalt	Chalcedony; cristobalite; palagonite; basic volcanic glass
Metamorphic	Gneiss	Strained quartz; microcrystalline quartz
	Schist	
	Quartzite	Strained and microcrystalline quartz; chert
	Hornfels	Strained quartz; microcrystalline to
	Phyllite	cryptocrystalline quartz
	Argillite	
Sedimentary	Sandstone	Strained and microcrystalline quartz; chert; opal
	Greywacke	Strained and microcrystalline to cryptocrystalline quartz
	Siltstone	Strained and microcrystalline to
	Shale	cryptocrystalline quartz; opal
	Tillite	Strained and microcrystalline to cryptocrystalline quartz
	Chert	Cryptocrystalline quartz; chalcedony;
	Flint	opal

Rock	Reactive component
Diatomite	Opal; cryptocrystalline quartz
Argillaceous dolomitic limestone and calcitic dolostone	
Quartz-bearing argillaceous calcitic dolostone	Dolomite; clay minerals exposed by dedolomitisation

Other substances
Synthetic glass; silica gel

Source: From Oberholster, 2001.

Notes
1 Reactive aggregates vary widely in their reactivity depending on geological origin, location within a given geological formation, and location within a given source such as a quarry. Thus, where an aggregate may be suspected of being alkali-reactive, it is necessary to test the specific source from which it is derived.
2 Only dense reactive aggregates are potentially damaging; porous rocks, even if reactive, generally contain sufficient pore volume to absorb the expansive gel.
3 Rocks listed above, although being siliceous in character, may be innocuous if their siliceous minerals are not alkali-reactive.
4 Alkali-susceptible rocks are found worldwide, and there are few countries or regions that have not had AAR problems in concrete. Specific detail on suspect rocks and aggregate sources must be sought in the region or country concerned. Sources such as CSA A23.1-00 – Appendix B, Pike (1990), Oberholster (2001), Hobbs (1987), and Swamy (1992) can be consulted.

on an anhydrous (without water) weight basis of the gel, only a non-swelling gel will form. For high-alkali concentrations, however, the solubility of CH is depressed, resulting in the formation of some swelling N(K)-S-H gel that contains little or no calcium. The N(K)-S-H gel by itself has a very low viscosity and could easily diffuse away from the aggregate. However, the presence of the C-N(K)-S-H results in the formation of a composite gel with greatly increased viscosity and decreased porosity.

4. The N(K)-S-H gel attracts water due to osmosis, which results in an increase in volume, local tensile stresses in the concrete, and eventual cracking. Over time, the cracks fill with reaction product, which gradually flows under pressure from the point of its initial formation.

(Mindess *et al.*, 2003: 148)

In their review, Mindess *et al*. point out that the higher the alkalinity of the cement (i.e. the OH⁻ concentration of the pore solution), the greater the solubility of amorphous silica and the rate at which it dissolves. The rate and extent of step 1 will depend on the initial porosity of the aggregate, which will also govern whether alkali attack takes place throughout the particle or initially only on the surface. Further, whether C-N(K)-S-H represents the sole component of reaction depends on the relative amounts of silica and alkali. For low S/N ratios, the pH of the pore solution remains high and the solubility of calcium remains low, resulting in formation of swelling N(K)-S-H. As the S/N ratio increases, the greater amount of reactive silica results in an increase in the total reaction product, until eventually a greater portion of the alkali is tied up and the pH of the pore solution decreases, thus increasing both the amount of calcium in solution and the non-swelling C-N(K)-S-H component of the gel. This explains the occurrence of the 'pessimum percentage'.

Expansion due to ASR is slowed as compressive stress is applied to concrete. Also, accelerated tests have shown that alkalis, in the form of salts, can accelerate the ASR. Conversely, it is well known that pozzolans (i.e. reactive silicas) reduce the severity of the reaction if used in adequate quantities; finely divided silica (<0.15 mm) can encourage a rapid reaction without deleterious effects, with the reaction moving rapidly through step 4 and the reaction products being well distributed, resulting in little effect of gel viscosity and a uniform distribution of reaction products. This decreases local concentration gradients and subsequent osmotic pressures. The action of silica fume in eliminating harmful effects of ASR is an example of this effect.

FACTORS GOVERNING ASR

There are three necessary requirements for ASR: a source of alkali, a source of reactive silica in the aggregate, and an environment contributing sufficient moisture to cause swelling of the gel.

Alkalis   The primary source of alkalis in the pore solution is the cement or binder, which contains metal alkalis – sodium and potassium hydroxide – as well as liberal amounts of calcium hydroxide. Cement alkalis are quantified by their oxide values, in particular by the *equivalent sodium oxide* expressed as a percentage by mass of the cement, namely

$$\%Na_2O_{eq} = \%Na_2O + 0.658\%K_2O \tag{6.5}$$

The constant 0.658 in Equation (6.5) derives from the ratio of the atomic mass of Na_2O to K_2O. The equation describes the *equivalent* effect of potassium in contributing alkalis. In measuring these constituents in cement the acid-soluble component is usually used. Other sources of alkalis in concrete can be the environment, such as marine or de-icing salt conditions, or chemical admixtures usually containing sodium. Alkalis from de-icing salts can have particularly serious consequences for bridge decks and other structures, even where the concrete contains only slowly reactive aggregates (Swamy, 1994). Alkalis sometimes derive from the aggregates themselves, for example alkali-containing minerals such as feldspars which react with the calcium hydroxide released by cement hydration, aggregates containing artificial glass, and volcanic glass present in some rhyolites, basalts, and andesites (Stark, 1978; Stark and Bhatty, 1985). The alkali contribution from aggregates, though difficult to assess, may be important enough to take into account.

Cements with Na_2O_{eq} greater than 0.6 per cent are regarded as high alkali cements. However, while cements may be classified as high, medium, or low alkali, it is the *total* alkali content in the concrete that is more important than the alkali level of the cement itself. The total alkali content is determined from the alkali content of the cement, the cement content, and the proportion of alkalis released during hydration and available for reaction (termed 'active alkalis'). This latter factor varies for different cements but may range between 70 and 100 per cent of total Na_2O_{eq}, with a typical value for South African portland cement clinkers of 85 per cent. Figure 6.11

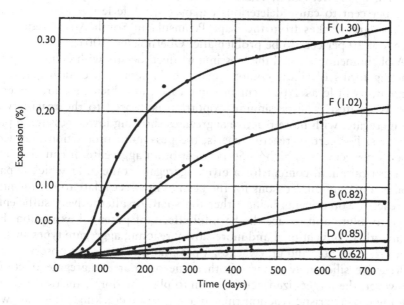

Figure 6.11 Effect of cement alkalinity on ASR. Malmesbury coarse aggregate (greywacke) in combination with different cements. Cement content of concrete = 350 kg/m³. Stored under ASTM C227 conditions above water in sealed containers at 38 °C. The figures in brackets refer to the total %Na_2O equivalent of the cement (from Oberholster, 1986a).

illustrates the effect of increasing alkali content on expansion of concrete specimens stored at 38 °C. Practically, one calculates the total alkali content of the concrete as the product of the cement alkali content and the cement content of the mix and limits this value depending on the type of reactive aggregate. This is covered later.

Reactive silica   The severity of ASR will be governed by the nature and quantity of reactive silica present. Table 6.2 indicates the rocks and minerals that are susceptible to alkali attack. In ASR, *reactivity* of the silica is important. Glassy, amorphous silica such as opal and chalcedony are highly reactive, while crystalline varieties such as stable quartz are not. Reactive silica may be present in the form of poorly crystallised minerals (cryptocrystalline or microcrystalline) or as strained quartz crystals. The latter occur in rocks subjected to internal shearing and distortion from tectonic forces often resulting in metamorphism, or where intrusive igneous rocks have induced re-crystallisation of existing sedimentary or metamorphic rocks. Such strained silica lattices are readily identifiable under polarised light in a petrographic microscope.

The *amount* of reactive silica present is also an important factor governing the severity of ASR, but depends critically on the nature of the reactive aggregate. Highly reactive forms such as opal may require as little as 2 per cent to cause deleterious expansion, while less reactive varieties such as greywackes from the Cape Peninsula in South Africa require in excess of 20 per cent to be problematic (Oberholster, 2001).

A phenomenon termed the 'pessimum effect' occurs with certain rapidly reacting forms of silica, notably opal. This refers to a critical aggregate content, as little as 1 per cent in some cases, at which measured expansion is greatest. The pessimum proportion is related to the reactivity of an aggregate, with more reactive aggregates showing lower pessimum proportions. For certain reactive cherts, the pessimum proportion may vary from 5 per cent to as high as 50 per cent of an aggregate. It can also refer to a critical alkali content for a given aggregate content, at which expansion is a maximum. Reasons for the pessimum effect relate to the quantity of reactive components being either too small or alternatively sufficiently large to dilute and reduce the harmful effects of potential expansion. For many other conventional and more slowly reacting aggregate types such as reactive quartzites and greywackes, a pessimum effect is not observed.

Reactive silica may occur in either fine or coarse aggregate fractions. However, the larger size fractions seem to play the dominant role, exhibiting the characteristic reaction rims and aggregate cracking. This may well be an example of the pessimum effect whereby the finer fractions do not contribute significantly to internal damage by virtue of their much higher surface area which distributes the expansive sites and dilutes the reactive effect. The same argument can be applied to highly reactive forms of finely divided silica such as silica fume or fly ash which ameliorate ASR by inducing multiple reaction sites that effectively immobilise the alkalis.

Environment and moisture   The environment plays a crucial role in ASR expansion and damage of structures by governing the availability of external moisture and ambient temperature. Without the presence of sufficient moisture the damaging effect of ASR will not occur. These effects may be greater in thinner elements where the interactions between external environment and internal conditions are greater, assuming a moist environment. In larger members even in drying environments, internal moisture will usually be retained, while the outer portions dry. This results in the expansive reaction occurring internally but not in the outer zones, giving surface cracking while the interior of the member may appear uncracked. Figure 6.12 shows the outer 8 cm of a concrete core with the surface zone (RHS) showing little or no evidence of ASR but major cracking due to the low moisture content in this zone (from a low relative humidity), while the interior has ASR present but no cracking. In general, structures in moist, humid climates are more susceptible to ASR than those in drier regions.

There is usually sufficient moisture 'locked into' a concrete structure from the mixing water to trigger ASR, other conditions being met. Figure 6.12 also shows this, where deeper parts of the member were able to retain sufficient moisture to drive the ASR. It has been suggested that a minimum internal RH of 75–85 per cent is required to permit ongoing harmful reactions and expansion (Jones and Poole, 1986). Research indicates that moisture must be available during gel formation as well as subsequently to increase expansion (Sims, 2000).

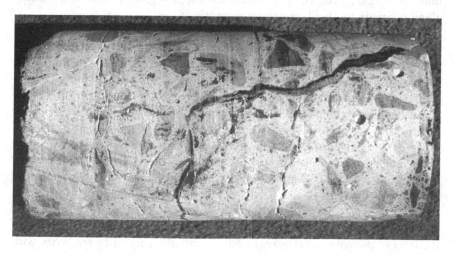

Figure 6.12 Cracking of core due to internal ASR and external drying.

ASR may also be caused or exacerbated by external sources of moisture such as precipitation, condensation, or ponding. Water may be locally directed into or onto a structure, leading to localised zones of ASR damage. Poor drainage (e.g. leaking bridge joints) or capillary uptake of moisture from the subgrade that may occur in concrete pavements are typical causes. A notable failure of a major concrete highway due partly to this cause occurred in Cape Town in the mid-1970s. Retaining walls and abutments containing poorly drained saturated fills can also suffer from ASR even in relatively dry natural conditions. Conditions promoting cyclic moisture fluctuations also favour more rapid ASR damage.

ASSESSING AGGREGATES FOR POTENTIAL ASR

Engineers need reliable and easy-to-use tests or procedures to evaluate whether an aggregate source is likely to be alkali-reactive. Sometimes this may simply be a 'pass/fail' criterion for initial screening. For other critical applications, detailed information will be required such as the threshold alkali content below which the aggregate is unlikely to show deleterious expansion. The nature of the assessment must be linked to the desired information and to the criticality of the assessment for the structure in question. The issue is complicated by differing national standards and test methods in which criteria and approaches may differ.

A summary of aggregate tests for susceptibility to ASR is given in Table 6.3, where aggregate assessment is considered in three main divisions: initial non-quantitative screening tests (used to make a provisional assessment); indicator tests to differentiate between potentially reactive and innocuous aggregates; and performance tests, giving information on limiting alkali contents to avoid damaging expansions.

The accelerated mortar bar test (e.g. ASTM C1260) is probably the most common test used worldwide at present. It was first reported in 1986 and has gained international acceptance. The method involves monitoring the expansion, typically for 14 days, of mortar prisms containing the test aggregate and immersed in a 1-M sodium hydroxide solution at 80 °C. The method derives from work carried out at the NBRI in South Africa by Oberholster and co-workers (Oberholster, 1983). It is the basis of the test accepted and developed by RILEM TC 106 (AAR-2) and has been included in various national test methods, for example ASTM C1260, BS DD 249, CSA A23.2-25A, and SANS 6245. It is useful for rapid screening but does not necessarily remove the need for conventional concrete prism testing (such as BS 812: Part 123, or RILEM AAR-3). ASTM approved (in 2004) a modification of the C1260 test, C1567, which can be used to determine the level of pozzolan or slag required to control alkali–silica reactivity with a particular aggregate.

Table 6.3 Aggregate assessment for AAR (excluding carbonate rock assessment)*

Test designation and purpose	Material tested	Procedure and duration	Assessment criteria and test outcomes	Test standards	Limitations and remarks
I Initial screening tests (non- or semi-quantitative)					
Petrographic Examination Petrographic examination procedures for aggregates, according to standard descriptive nomenclature, as an aid to determine their performance	Coarse and fine aggregates, or rock cores	Standard petrographic techniques, including optical microscopy, XRD analysis, differential thermal analysis. Presence and quantities of deleterious components such as: **Minerals:** opal, tridymite, cristobalite, chalcedony, chert **Rocks:** crypto- and microcrystalline quartz; evidence of deformation of quartz, such as undulatory extinction, intergrowth, or reaction with matrix. Also evidence of silica gel formation, cracking of aggregates and matrix, reaction rims, formation of crystalline silicates.	ASTM C294 & C295 BS 812: Part 104 BS 7943 RILEM TC 191-ARP (AAR-I)	These tests are for identification of potentially reactive constituents, and characterisation of minerals making up concrete aggregates. They are essential to confirm whether cracking in a structure is AAR-related or not – see ASTM C856 (Note below this table).
Gel-pat test Detection of reactive silica, e.g. opaline silica	Gravel-sized particles	Particles embedded in a cement pat, at 20°C or at 80°C; examined for 10 days for signs of reaction.	Evidence of AAR gel and reaction.	Appendix to BS 7943	Does not require sophisticated equipment, and can be performed on site.

Table 6.3 (Continued)

Test designation and purpose	Material tested	Procedure and duration	Assessment criteria and test outcomes	Test standards	Limitations and remarks
2 Rapid indicator tests (to determine whether aggregates are potentially reactive or innocuous)					
Potential alkali reactivity of aggregates (chemical method) _Determination of potential reactivity of siliceous aggregates_	Siliceous aggregates, crushed and sieved (150–300 μm)	Aggregates placed in 1-M NaOH solution at 80°C for 24 h. Analysed for dissolved silica and reduction in alkalinity.	Results checked against calibration curve; pass/fail criterion (deleterious or potentially deleterious).	ASTM C289	Not recommended for aggregates such as greywacke, hornfels, quartzite, granite, etc. Test shows some aggregates to be innocuous when they are known to have a poor service record.
Potential alkali-reactivity of cement–aggregate combinations (mortar-bar method) _Determination of susceptibility of cement–aggregate combinations to expansive reactions with alkalis_	Cement–aggregate combinations. Particular size fractions <4.75 mm required, obtained if necessary by crushing	Storage of 25 × 25 × 285-mm mortar bars at 38°C and 90% RH, 3 months to 1 year. Essential to control RH at sufficiently high level during test.	Criteria given in ASTM C33: Harmful reactivity if expansion >0.05% at 3 months or >0.10% at 6 months. Revised criteria for quartz-bearing rocks: Deleterious if expansion >0.05% in 52 weeks. See Brandt and Oberholster (1983).	ASTM C227	Results take 3 months to 1 year. Quartz-bearing rocks require the longer test period. Useful test to determine susceptibility of combinations of cements and aggregates to harmful expansion. Reactive dolomitic aggregates not revealed by this test.
Accelerated mortar-bar (mortar-prism) test (RILEM TC 106: 'Ultra-accelerated mortar-bar test')	Mortar prism comprising susceptible aggregates, with specified	Prisms 25 × 25 × 285 mm stored in 1-M NaOH at 80°C for 14 days.	Expansion after 14 days (12 days in South Africa): <0.10% – non-expansive 0.10–0.20% – slowly reactive or potentially reactive	ASTM C1260 BS DD 249 CSA: A23.2-25A RILEM TC	Based on work at NBRI in South Africa, Oberholster and Davies (1986); also used in USA and

Determination of potential for deleterious alkali-silica reaction of aggregate in mortar bars	grading from 4.75 mm to 150 μm, obtained by crushing as necessary	≥0.20% – deleteriously reactive CSA: >0.15% – potential deleterious expansion.	191-ARP (AAR-2) SANS 6245	Canada (Hooton and Rogers, 1992), and recommended by RILEM. BS draft version. Rapid test, useful for slowly reacting aggregates or those producing expansions late in the reaction; generally reliable and reproducible, but not reliable for aggregates containing more than 2% porous flint.

3 Performance tests (e.g. to provide information on limiting alkalis or structural performance)

Concrete prism method *Determination of the potential ASR expansion of cement–aggregate combinations*	Concrete aggregates proposed for actual construction. Na_2O_{eq} content in test = 5.25 kg/m³ — Prisms 75 × 75 × 300 mm stored over water (100% RH) at 38°C. Cement and alkali contents stipulated in ASTM C1293 and other standards. Duration 3 months to 1 year.	Any combination of cement and aggregate giving expansion after 52 weeks: <0.05% – non-expansive (CSA: <0.04%) 0.05–0.10% – potentially or moderately expansive (CSA: 0.04–0.12%) >0.10% – expansive (CSA: >0.12%).	ASTM C1293 BS 812: Part 123 CSA A23.2-14A RILEM TC 191-ARP (AAR-3)	Advantage is that actual mixes can be tested in proportions specified. Can be used to evaluate effect of supplementary cementitious materials, and to assess alkali–carbonate reactive aggregates. Long test duration required for meaningful results.

Table 6.3 (Continued)

Test designation and purpose	Material tested	Procedure and duration	Assessment criteria and test outcomes	Test standards	Limitations and remarks
Ultra-accelerated concrete prism test	Concrete prisms containing aggregates proposed for construction	Prisms 75 × 75 × 250 mm stored over water at 60 °C. Duration at least 20 weeks; longer for some slowly reacting aggregates.	Expansion after 3 months: <0.02% – non-expansive <0.02% after 6 months indicates minimal risk of ASR. Shape of expansion curve to be considered also.	RILEM TC 191-ARP (AAR-4)	Used to assess reactivity performance of particular concrete mixes.
Long-term structural monitoring	Actual structures and structural members	Procedures designed for particular structures; generally involve monitoring expansions, deflections, and cracking with time; full-scale load testing may also be carried out. Criteria depend on particular structure. Excessive expansions, deflections or cracking taken as appropriate criteria.		Inst. of Struc. Engineers (1989); RILEM TC 20-TBS (1984)	On occasions, structural monitoring is the only way to assess the performance of an AAR-affected structure and to assess its ongoing integrity.

Sources: ASTM, CSA, and RILEM sources quoted in table; Oberholster, 2001; Jones and Tarleton, 1958; Brandt and Oberholster, 1983; Oberholster and Davies, 1986; RILEM TC106-AAR, 2000; Sims and Nixon, 2003a; Sims and Nixon, 2003c.

Notes

* The shaded section represents most commonly used tests worldwide at present.
• Aggregates differ in their alkali-reactivity. Furthermore, the demarcation lines between reactive and innocuous aggregates are not sharp (see Figure 6.8); therefore judgement must be exercised when evaluating an aggregate.
• Other tests are: ASTM C342: Potential volume change of cement–aggregate combinations – *determines the potential ASR expansion of cement aggregate combinations* (primarily used for aggregates from Oklahoma, Kansas, Nebraska, and Iowa); ASTM C441: *Effectiveness of mineral admixtures or GGBS in preventing excessive expansion of concrete due to ASR – determines effectiveness of supplementary cementing materials in controlling ASR expansion* (also covered in CSA Standard A23.2-28A); ASTM C856: Petrographic examination of hardened concrete – *outlines petrographic examination procedures for hardened concrete, useful in determining condition or performance.*
• ASTM approved a further test in 2004: C1567, Test Method for Determining the Potential Alkali–Silica Reactivity of Combinations of Cementitious Materials and Aggregate, Accelerated Mortar Bar Method. This test evaluates pozzolans and slag for controlling ASR, and can be used to determine the level of extender required to control ASR with a particular aggregate. It is a modification of test method C1260, which is strictly an aggregate test. Test results are produced in 14 days.
• RILEM tests have been produced by Technical Committees (TC) of RILEM. The TC previously concerned with AAR was RILEM TC 106 (2000), and is currently (2004) RILEM TC 191-ARP.
• RILEM TC 106 (2000) recommends that chemical testing such as ASTM C289 be used only as a secondary method, due to difficulties of interpretation.
• RILEM TC 191-ARP envisages that the AAR-4 test (Ultra-accelerated concrete prism test) may be used in three modes: for testing potential reactivity of an aggregate combination; as an ultra-accelerated version of the AAR-3 test, i.e. as a performance test for assessing the alkali-reactivity of a particular mix; and as a test for establishing the critical alkali threshold of a particular aggregate combination.
• An accelerated concrete prism test exists as an AFNOR (French) standard; procedures are identical to the Concrete Prism Method in CSA A23.2-14A covered in the table above, except that storage is at 60°C and 100% R.H. for a period of 3 months (91 days). A number of task groups including ASTM are investigating the test; concerns relate to expansion limits in relation to measurement precision.
• Developments in BS standards on AAR depend on parallel developments in European Standards. For example, RILEM AAR-2 may in due course be included as a European Standard, replacing BS DD 249, for which no steps are presently being taken to convert it into a formal British Standard.
• Canadian Standards Association document CSA A23.1-00, Appendix B, contains useful information on AAR testing specific to Canadian conditions.
• The above table is based on an outline given in Oberholster (2001) (Chapter 10 of *Fulton*, 2001).

RILEM Technical Committee TC 191-ARP, the successor committee to TC 106, has developed a scheme for the integrated assessment of AAR, illustrated in Figure 6.13 (Nixon and Sims, 1996; Sims and Nixon, 2001, 2003a). Various elements are proposed: petrographical examination (AAR-1); rapid testing using the accelerated mortar bar expansion test (AAR-2); and a concrete prism expansion test (AAR-3) which is viewed as a long-term reference test. An ultra-accelerated performance test for concrete is also envisaged (AAR-4) in which concrete prisms are stored at 60 °C, and which might be appropriate for a project-specific performance test. The actual concrete mix for the job is tested and it is anticipated that useful results could be obtained in 6 months. It is undergoing international precision trials as a recommended method at present (2004). Initial results indicate that it has merit for being able to distinguish between non-reactive and reactive combinations, and is reproducible. (An additional procedure, AAR-5, is proposed to identify susceptible carbonate aggregates, but this is

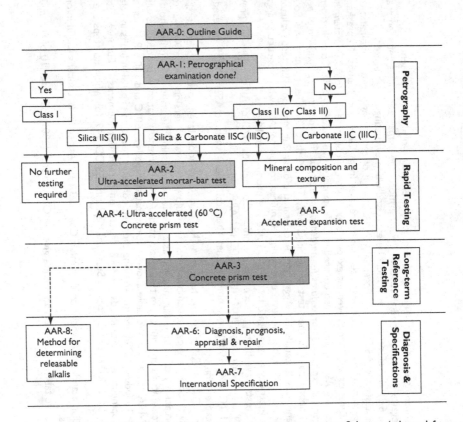

Figure 6.13 RILEM TC 106/TC 191 Integrated AAR Assessment Scheme (adapted from Sims and Nixon, 2003a,b). Shaded boxes represent those sections that have been published (2004).

not further dealt with here.) Also, an outline guide (AAR-0) is available for using the RILEM methods of assessment (Sims and Nixon, 2003b). It must be stressed that all these tests are useful only as tests for AAR susceptibility; they do not indicate if a structure has been affected by ASR.

Petrographic examination, which should be undertaken first, is intended to establish the types and concentrations of any potentially reactive constituents. This permits the aggregate combinations to be classified in terms of their alkali-reactivity as either: Class I – very unlikely; Class II – potentially, or reactivity uncertain; and Class III – very likely. For new aggregate sources, Class II is common and further testing will be required, while Class III would generally imply the presence of opal or opaline silica. For existing aggregate sources, Classes I and III are the common possibilities. The suite of tests indicated in Figure 6.13 can be utilised with the long-term concrete prism method (AAR-3) regarded as the standard reference test. Generally, it is unwise to rely solely on the accelerated mortar bar test and results should be confirmed by the longer-term concrete prism test. Further recommended methods or procedures are AAR-6, dealing with diagnosis, appraisal, and repair, and AAR-7 on international specifications. A method for determining releasable alkalis (AAR-8) is also envisaged.

Yet a further test, based on a Chinese procedure and following similar protocols as ASTM C1260, has recently been proposed (Grattan-Bellew et al., 2003). Termed 'The Concrete Microbar Test', it uses $40 \times 40 \times 160$-mm bars, aggregate graded between 12.5 and 4.75 mm, w/c ratio of 0.33, and immersion in 1-M NaOH solution for 30 days at 80 °C. The method is claimed to be applicable to both alkali–carbonate and alkali–silica reactive aggregates (differentiated in the test by replacing a portion of the portland cement by a supplementary cementing material). Moderate correlations were found between 30-day expansions in this test and 1-year expansions measured in the modified concrete prism test (CSA A23.2-14A).

ENGINEERING CONSEQUENCES OF ASR

Because ASR involves destructive breakdown of the internal microstructure of concrete, including cracking, it has inevitable structural consequences. Other consequences include aesthetics, effects on services, and long-term maintenance and repair. The primary effect of ASR is to cause concrete expansion, which can be as high as 1 per cent or more.

Despite the often-severe nature of material deterioration, the overall stability or strength of a structure may not be excessively compromised, although serviceability may be adversely affected. For example, extensive full-scale load testing with deflection measurements on a large portal frame of a double-deck motorway in Johannesburg showed that, despite the alarming appearance of ASR-related surface cracking, structural integrity was maintained (Blight et al., 1983). This favourable outcome was, however, not true for an ASR-affected car park at Charles Cross in Plymouth, England, where the factor of safety of the structure was judged to be compromised. Thus, the effects of ASR on a structure will be governed as much by the nature of the structure and structural system as by the severity of material deterioration.

It may be necessary to test an affected structure to determine the material and other properties. Full-scale load testing can yield valuable results, although it is usually expensive (RILEM TC 20-TBS, 1984; Institution of Structural Engineers, 1989). Instrumenting the structure in service and observing its behaviour can yield valuable information on its state. Drilled cores can also provide useful information not only on mechanical properties, but also on depth of cracking, internal state of the concrete, and steel corrosion. Mechanical properties obtained from individual cores can be highly variable, requiring more cores than normal to be tested. Cores should also be as long as possible to include as many cracks or defects as possible. When drilling cores, attention should be paid to the cracking and stress systems in the structure, in order to orientate the cores either to intersect or to avoid the cracks depending on the purpose for drilling.

When assessing the overall structural effects of ASR, the engineer must take into account both the *magnitude* of total expansion and the *rate* of expansion. The former can be measured on representative cores drilled from the structure, while the latter is best assessed by in-situ measurements of expansion over time. Both effects depend on aggregate and cement type, cement content, and the environment, so that it is not possible to generalise about the resulting engineering properties. The following sections, however, give broad indications of the likely effects of ASR on concrete properties.

It is very useful to have an estimate of the residual ultimate expansion of concrete in an ASR-affected structure. This can be done using tests on cores from the structure. Swamy (1997) suggests accelerated laboratory expansion tests on cores immersed in 4 per cent NaCl solution at 38 °C, and field expansion tests on cores exposed in the same environment as the structure. The first test provides information on the residual ultimate expansion, while the second test, although requiring a long time of several years, indicates the likely rate of expansion in the real structure. Stark (1991) on the other hand proposes the following procedure. Three 100-mm diameter companion cores are stored as follows: (i) one continuously over water in a sealed container at 38 °C, to give an indication of the potential for continued expansion, (ii) one continuously in 1-N NaOH solution at 38 °C, to indicate the presence of unreacted reactive particles, and (iii) one continuously in water at 38 °C, to allow corrections to be made for wetting expansion. South African experience favours monitoring the structure itself, since relatively small specimens such as prisms do not always expand under natural exposure conditions, while larger samples do. Thus, laboratory tests can and should be supplemented by expansion monitoring of the real structure.

Concrete strength The visually alarming appearance of ASR on a structure generally leads one to assume that the structural strength has been severely reduced. While this may be the case in advanced or very severe incidences of ASR, it sometimes comes as a surprise that a substantial amount of concrete compressive strength is retained after the damaging effects of ASR. The tensile strength of an ASR-affected member will be more severely affected. Concrete microstructure resembles a highly internally redundant lattice frame in the way it resists compressive stresses, and the presence of internal damage is akin to the removal of certain lattice members without rendering the entire structure unstable or excessively weakened (depending, of course, on the extent and severity of ASR-induced cracking). Each case must be judged on its own merits, but residual compressive strength may not be the critical factor governing structural performance.

The extent of strength reduction due to ASR depends on, inter alia, expansion of the concrete, size of member, and type of aggregate. Hobbs (1988) quotes the results of a number of investigations in which there were reductions in compressive and indirect tensile strength for various ranges of expansion as shown in Table 6.4.

Oberholster (2001), drawing on data from mainly UK sources including the Institution of Structural Engineers Guidance Document on ASR (ISE, 1992), gives lower-bound residual mechanical properties for ASR-affected concrete, reproduced in Table 6.5. The residual strengths and stiffnesses in actual structures will not necessarily be as in Table 6.5 due to two important reasons. First, restraint effects induced by reinforcement in real structures will generally have a beneficial effect, and second strength reduction is highly variable and a range of possible values should be considered in critical cases. Figure 6.14 illustrates the substantially greater strength reduction that can be expected for tensile properties.

Data above were derived mainly from laboratory tests and there is evidence that strength in real structures affected by ASR may be somewhat less reduced. However, this may only be true for large, massive sections in which the major cracking damage occurs in the outer 'skin' of the member, the inner core remaining intact despite ASR damage. For example, core

Table 6.4 Reduction in compressive and indirect tensile strength due to ASR expansion

Expansion (%)	Reduction in	
	Compressive strength (%)	Indirect tensile strength (%)
0.1–0.2	0–25	20–25
0.3–1.0	20–65	20–30
>1.5	>60	–

Source: Adapted from Hobbs, 1998.

Table 6.5 Lower-bound residual mechanical properties of concrete affected by ASR as percentage of values for unaffected concrete at 28 days

Property	Percentage of property as compared with unaffected concrete, for free expansion (%)				
	0.05	0.1	0.25	0.5	1.0
Compressive strength*	100	85	80	75	70
Uniaxial compressive strength[†]	95	80	60	60	–
Splitting tensile strength	85	75	55	50	–
Elastic modulus	100	70	50	35	30

Source: From Oberholster, 2001, based on data in ISE, 1992.

Notes
* Cube.
[†] Core, length:diameter ratio = 2.5 or greater.

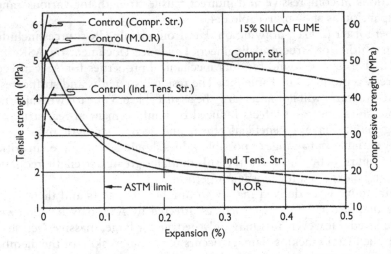

Figure 6.14 Loss in compressive and tensile strength due to ASR (from Swamy, 1994; © Minister of Natural Resources Canada 1994, reproduced with permission).

tests from a haunch and beam of the extensively ASR-damaged, elevated motorway structure in Johannesburg referred to earlier showed that mean compressive strength was still in excess of the design strength of 30 MPa, although the splitting tensile strength had been reduced (Blight *et al.*, 1984).

A number of cautionary comments are needed regarding the strength of ASR-affected structures. First, if it is intended to retain a damaged structure in service, core testing should be carried out to establish the range and variability of in-situ strength. This may reveal that certain sections of a structure are more badly damaged than others and may require different treatment. Those sections that are critical to its overall stability and load-carrying capacity will need to be more carefully examined. Second, the reinforcing steel details of the structure must be established, preferably from as-built drawings. If necessary, probing in the form of small cores drilled at critical sections may be required particularly if covermeter or other NDT surveys are not accurate or conclusive. The amount and disposition of reinforcing steel will be critical in resisting tensile stresses and more importantly in coping with shears. Inadequate steel in this regard could render the structure unsafe for continued service. Third, reinforcing in the form of confining links or hoops is essential to confine the concrete in compressive regions to enable it to carry the compressive stresses. Such steel can impose a biaxial or triaxial restraint on the cracked concrete thus improving its compressive capacity. Confining steel is also very important in ASR-affected members in restraining expansions and reducing cracking.

Concrete stiffness   Stiffness refers to effects of elastic and creep strains (with shrinkage being considered separately). Damage and cracking from ASR generally has a more detrimental effect on stiffness than on compressive strength. This is consistent with the presence of multiple cracks resulting from expansion and which can move or close under stress. To illustrate this point, consider Figure 6.15 which shows loss in elastic modulus and pulse velocity due to ASR (Swamy, 1994). Similar data are reported by

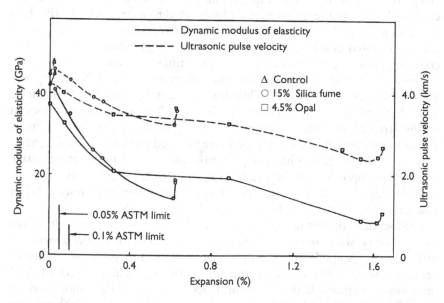

Figure 6.15 Loss in elastic modulus and pulse velocity due to ASR (after Swamy, 1994; © Minister of Natural Resources Canada 1994, reproduced with permission).

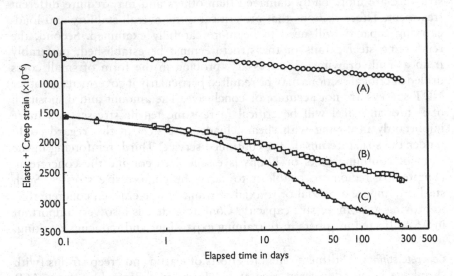

Figure 6.16 Elastic and creep strains on (A) undeteriorated, and (B) and (C) ASR-deteriorated specimens (concrete stressed to 15 MPa, half of design compressive strength) (after Blight *et al.*, 1984).

Hobbs (1988). Figure 6.16 (Blight *et al.*, 1984) shows instantaneous and creep strains measured on sound (A) and ASR-affected (B and C) concrete cores drilled from a real structure and subjected to a stress of 15 MPa, half the design strength. The results indicate that the elastic deformations of the deteriorated concrete were about 3 times larger than those of sound concrete, while the creep strains were 2.5–4 times as large.

These results indicate significant losses in strength and stiffness depending on the type of aggregate. While losses in compressive strength can be as high as 50 per cent or greater under very high expansions, losses in tensile strength and elastic modulus can approach 75 per cent or more. Flexural members therefore will be more affected than compression members. Also, losses in mechanical properties do not all occur at the same rate or in proportion to the expansion, and can be highly variable. This makes it imperative to determine actual concrete properties by testing, when assessing critical structures for the overall effects of ASR.

The fact that stiffness parameters such as dynamic modulus and ultrasonic pulse velocity are sensitive to ASR expansion makes these non-destructive methods very suitable to monitor the initiation and progress of deterioration due to ASR. The consequence of the substantial effect of ASR on concrete stiffness is that structural deflections may be a more serious issue than strength, and remedial measures may have to address this aspect more directly.

Corrosion of reinforcing steel Surface cracking resulting from ASR may increase the possibility of reinforcement corrosion, particularly in environments promoting corrosion (e.g. de-icing salts or marine conditions) where such corrosion will be virtually inevitable. However, in milder or more benign environments corrosion will not necessarily follow, particularly if there is insufficient moisture at the steel. For instance, when the concrete haunch shown in Figure 6.7 was demolished, the steel reinforcement was found to be uncorroded despite the large surface cracks present. Similar experience is cited on a large sports stadium in Stellenbosch, South Africa, which suffered major ASR cracking with crack widths up to 5 mm or more in places. Despite the structure experiencing cool wet winters, there was little or no evidence of active corrosion after many years of service.

Secondary effects The expansive reaction associated with ASR may cause disruption to adjoining parts of a structure or other services such as pipes or conduits, or misalignment of turbines and generators in dam powerhouses. It may also cause jammed expansion joints resulting in stress-related cracking. Figure 6.17 shows the buckled web of an I-beam between expanding and static sections of a hydro power structure in Quebec, indicating the distortions that can occur.

Long-term maintenance and repair Other texts (Hobbs, 1988; Swamy, 1994; Oberholster, 2001) deal more comprehensively with this subject. The presence of moisture is usually the key to arresting or slowing the expansion due to ASR. Thus, attention should be paid to improving drainage, cutting off any external sources of moisture if possible, and preventing environmental ingress of moisture by the use of hydrophobic coatings or other moisture barriers (Hoppe, 1990). Attempts to seal or 'stitch' active cracks will be

Figure 6.17 Distorted web of I-beam due to AAR in a hydroelectric plant in Quebec, Canada. Note directional arrows, showing how structure has moved due to AAR.

to no avail, and monitoring of crack movement should be undertaken to indicate when cracks become dormant, before crack sealing is done.

Relating results of AAR tests to performance of real structures   Blight (2004, pers. comm.) states that one of the major problems with AAR testing and research is the difficulty of relating the results to the performance, rates of deterioration, remaining service life and so on of real structures under natural climatic conditions. Correspondence between laboratory tests, even tests on model structures in which AAR has been induced artificially, and real structures exists not at all or only tenuously. Thus, it is very difficult to relate results such as those in Figures 6.14 and 6.15 to real structural behaviour. These comments lend weight to the necessity of full-scale monitoring of affected structures over several years together with inspections and core tests, in-situ load tests, and measurements of in-situ strains and deformations under working loads. Without such work, little useful information is available for an affected structure on the stage reached by AAR, the rate at which it is progressing, or the structural effect it has had.

PRACTICAL MEASURES TO AVOID OR MINIMISE ASR

Alkali–silica reaction can occur only if three necessary conditions are met: an alkali source, reactive aggregate, and sufficient moisture to sustain the reaction and induce expansion. Thus, preventative measures must address one or more of these factors. Preventative measures have been developed in most countries where ASR is a problem and these seem to have been effective (see for instance Concrete Society, 1999; The Institution of Structural Engineers, 1992; and the Canadian Standards Association, CSA A23.1-00, Appendix B, 2000). Several important factors must be considered for measures to minimise or mitigate ASR damage: the environment in which the structure is located, the importance of each element or the structure as a whole, the costs of the measures, and the materials likely to be used. The Concrete Society document (1999) gives comprehensive recommendations appropriate to the UK, dealing with classifications of different binders including extenders, environmental exposure conditions, susceptible aggregates, and the alkalis that can be contributed by the binders and other sources such as mixing water, aggregates, admixtures, and the environment. Model clauses for a specification are included and worked examples are useful in illustrating the approach. The Canadian document covers preventative measures to mitigate AAR reactions, including ACR and ASR. The recommended measures relate to selective extraction or beneficiation of the aggregate to reduce or eliminate the reactive material, reduction of the alkali content of the concrete by reducing the cement content or using lower alkali cement, and use of supplementary cementitious materials such as fly ash or slag in suitable proportions.

The following sections give some detail on measures to avoid or minimise ASR. These relate to the three necessary conditions for ASR, and how the problem is mitigated by addressing one or more of them.

Reducing the effect of alkalis, including use of cement extenders   There is a critical total alkali content with a given aggregate above which ASR is virtually inevitable, and a lower alkali content below which ASR is highly unlikely. Between these two limits, the incidence and severity of the reaction will vary depending on other factors such as environment, size of member, and so on. This is shown in Figure 6.18 (Oberholster, 2001) for a greywacke aggregate from the Malmesbury group in South Africa. To avoid ASR completely, it is therefore necessary to limit the total alkalis in the mix to less than the lower limit, appropriate to the environment of the structure. This limit varies for different aggregate types and binder alkali classification. Various countries have evolved different schemes for limiting the alkali content in a mix to avoid or minimise the risk of ASR. Typical values for South Africa, Canada, and the UK for a variety of aggregates are given in Table 6.6, with interpretive notes. The UK approach is based on recent research which shows that different aggregate types required markedly different amounts of alkali in the concrete to produce damaging expansion, and that the effectiveness of extenders such as slag or fly ash depends on the reactivity of the aggregate used. The table shows that permissible alkali contents vary from about $2\,kg/m^3$ (Na_2O_{eq}) to a maximum of $5\,kg/m^3$ over the range of aggregates considered. Assuming a concrete with a portland cement component having Na_2O_{eq} of 0.8 per cent, the corresponding range of cement contents is from about 250 to $625\,kg/m^3$. The upper limit is far in excess of normal cement contents, while the lower limit implies that it will be possible to avoid ASR only in lower grades of structural concrete if based purely on limiting the alkali content of the mix. Consequently, the use of supplementary cementitious materials is favoured in many cases and for good reasons, as discussed later.

Hobbs (1987) gives data showing that no severe cracking due to ASR was observed in controlled tests with specimens at total alkali contents less than $5.0\,kg/m^3$. The aggregates were a Thames Valley sand and a Mendip limestone coarse aggregate. CSA A23.1-00, Appendix B, states that in general when alkali–silica reactive aggregates are used in concrete containing less than $3.0\,kg/m^3$ of total alkali (Na_2O_{eq}), deleterious expansion will not take place. However, it also cautions that a limit of $3.0\,kg/m^3$ may not be effective with massive concrete structures such as dams, where problems with slight expansion have occurred when the alkali content was as low as $2.0\,kg/m^3$. The limit may also need to be lower when the concrete is exposed to external sources of alkali (e.g. de-icing salts) or when the aggregate itself may contribute alkalis. Table 6.6 indicates that most South African aggregates will not cause ASR problems as long as the alkali content is less than about $3.0\,kg/m^3$, although

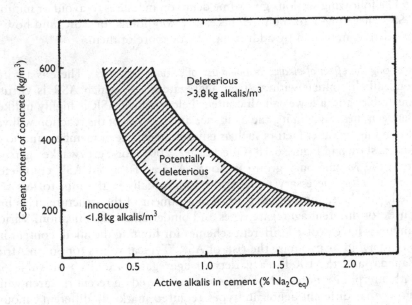

Figure 6.18 Potential of concrete for ASR as function of active alkali content of concrete (after Oberholster, 2001).

Table 6.6 Limits for alkalis to avoid ASR

	Alkalis (kg/m^3 Na$_2$O$_{eq}$) *not to exceed*
South African aggregates: Limits to total alkalis*, depending on aggregate type	
Very slowly reactive aggregates:	
Cape granite	4.0
Slowly reactive aggregates:	
Orthoquartzite (Cape, KwaZulu Natal)	2.8
Tillite (KwaZulu Natal)	
Highly reactive aggregates:	
Malmesbury meta-sediments ('Malmesbury	2.1
shale')	
Witwatersrand meta-quartzite	
Other non-reactive aggregates:	
Dolerite, siltstone, andesite, dolomite, felsite, granite (Gauteng), quartzite (Pretoria)	
Canadian aggregates: Type of preventive action required[†]	
Mild – e.g. required service life <50 years,	3.0
moderately reactive aggregate in large members	
in relatively dry conditions	
Moderate – e.g. required service life <50 years,	2.4
moderately reactive aggregate in concrete	
exposed to humid air, buried or immersed	

Strong – e.g. required service life <50 years, highly reactive aggregate in concrete exposed to humid air, buried or immersed	1.8		
Exceptional	1.8, plus use of a sufficient amount of cement extender or combination of effective extenders		

UK aggregates: Initial limits to alkalis to minimise risk of damage due to AAR (kg/m^3 Na$_2$O$_{eq}$)

Aggregate classification	Binder alkali classification[‡]		
	Low	Moderate	High
Low reactivity: These aggregates include andesite, dolerite, granite, and limestone among others. (Table 4 of Concrete Society document [1999] gives details)	Self-limiting	Self-limiting	≤5.0
Normal reactivity: Refers to all aggregates other than those classified as low or high reactivity; applies to most UK aggregates	Self-limiting	≤3.5	≤3.0
High reactivity: Aggregates containing more than 10% crushed greywacke, and recycled demolition waste (crushed concrete and masonry[§]). (Uncrushed greywacke-type material is classified as normally reactive)	≤2.5	≤2.5	≤2.5

Sources: Based on data in CSA Standard practice A23.2-27A; Concrete Society, 1999; Alpha, 2000; Oberholster, 2001.

Notes
* The limits in Na$_2$O-equivalent apply when using CEM I.
[†] See Tables 3, 4, and 5, CSA Standard Practice A23.2-27A.
[‡] Refers to the alkali content of the portland cement component:
  Low alkali: ≤0.60% Na$_2$O$_{eq}$; Moderate alkali: ≤0.75% Na$_2$O$_{eq}$; High alkali: >0.75% Na$_2$O$_{eq}$.
  Table 6.7 refers to alkalis in cementitious materials, and for the UK, these alkalis are taken into account under certain conditions mentioned in Table 6.7.
[§] The classification as high reactivity of recycled aggregates from crushed concrete and other demolition wastes is a precautionary measure pending further knowledge about their long-term behaviour.

Notes to section on UK aggregates
1 The UK approach is based on a matrix incorporating the reactivity classification of the aggregates and the classification of the alkali level in the binder. Aggregates are classified as low, normal, or high reactivity according to their petrographical types or by use of the concrete prism test (BS 812: Part 123). Depending on the alkali content of the binder and the reactivity of the aggregate, initial limits are set on the alkali content of the concrete. These limits are amended depending on the contribution of alkali from other constituents, and where slag or fly ash are used below the recommended minimum proportions (see also Table 6.7).
2 If alkali from other sources (cement extenders, aggregates, etc.) exceeds 0.6 kg/m^3 Na$_2$O$_{eq}$, the above limits do not apply. Table 1 of the Concrete Society document (1999) gives comprehensive detail to limit alkali contents when significant alkalis are contributed from other sources.

two of the most alkali-reactive aggregates, the Western Cape greywacke ('Malmesbury Shale') and the Witwatersrand quartzite, require alkali contents less than about 2.0 kg/m^3 in order to avoid long-term problems.

It may not always be possible to limit the total alkalis, particularly in mixes with high cement contents. In this case, the influence of the aggressive alkalis can be mitigated by use of a cement extender (supplementary cementitious material) such as ground-granulated blastfurnace slag (GGBS), fly ash, or silica fume. These materials reduce and immobilise the alkalis sufficiently to prevent or control ASR. This practice has been accepted worldwide and contributes to environmental improvement. However, cement extenders can themselves contribute some alkalis to the mix. The advantage of extenders is that they have considerably lower active alkali contents than portland cement. This is because the extender alkalis are generally tied up in their glassy phases and are released at a much slower rate than for portland cement. Typical values are given in Table 6.7.

The mechanisms by which cement extenders suppress ASR expansion involve complexing alkalis to reduce hydroxyl ion concentrations and alter-

Table 6.7 Active alkalis in various cementitious materials

	Total alkali content (% Na_2O_{eq})	Active (available) alkali fraction (% of total alkali content)
Portland cement		
SA	0.3–1.0 or higher	85–100
Ground granulated blastfurnace slag		
SA	0.3–1.0	42–50
UK*	0.4–1.1	0 or 50*
Fly ash		
SA	1.0–3.0	17–40
UK*	3.0–3.8	0 or 20*
CSF		
SA	1.0	30

Sources: Adapted from Oberholster, 2001 (South Africa); Concrete Society, 1999; Hobbs, 1986, 1988 (UK).

Notes
* The Concrete Society document (1999) refers to three classifications of binders, with Group B being the one most commonly used in concrete mixes, particularly when ASR is a problem. Such binders generally have more than 40% GGBS or more than 25% fly ash in the binder, in which case the alkali contribution of the extender can be assumed to be nil. However, if the GGBS proportion of the total binder is between 25 and 40%, or if the fly ash proportion is between 20 and 25%, then 50 or 20% of the mean alkali content of the extender must be taken into account respectively.

At high extender content, the mitigating effect on ASR is large enough to counteract the higher total alkali content. In terms of the South African approach, the total alkali content contributed by the binder increases as the extender content increases (if its total alkali content is more than that of the clinker it replaces) but the active alkali content decreases.

ing diffusion rates of alkali and calcium to reaction sites, thereby resulting in the formation of harmless non-expansive reaction products. Two types of fly ash exist, Class F and Class C (ASTM C618), distinguished by their lime (CaO) contents, with Class C ashes having greater lime content often exceeding 10 per cent or even 20 per cent. For purposes of preventing expansive ASR, the minimum permissible proportion of $(SiO_2 + Al_2O_3 + Fe_2O_3)$ for Class F ash is 70 per cent, and for Class C ash 50 per cent. A further requirement for fly ash is a 1.5 per cent maximum limit on available alkalis as determined using ASTM C311.

In general, provided a sufficient proportion of extender is used, ASR can be effectively prevented. The commonly accepted minimum proportions are:

Ground-granulated blastfurnace slag	40–50%
Fly ash	20–30%
Silica fume	10–15%

However, Swamy (1994) points out that these minimum amounts may not be adequate in all cases and testing should be carried out when it is suspected that the aggregates may be unusually expansive. Figure 6.19 illustrates this for a reactive greywacke aggregate from Cape Town where replacement of 50 per cent of the portland cement $(Na_2O_{eq} = 0.58\%)$ with either a

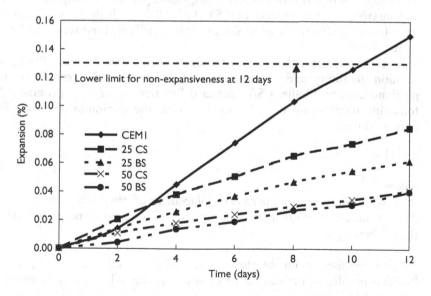

Figure 6.19 Effectiveness of CS and BS in controlling ASR expansion, accelerated mortar-bar test. Numbers refer to percentage replacement of portland cement (CEM I) with slag. CS = Corex slag; BS = Blastfurnace slag; Na_2O_{eq} of portland cement = 0.58% (adapted from Jaufeerally, 2002).

granulated blast furnace slag or a granulated Corex slag gave a result in the standard accelerated mortar-bar test (SANS 6245) indicating innocuous conditions. For the Corex slag, 25 per cent replacement came close to the limit where ASR would be avoided, with blast furnace slag performing somewhat better (Jaufeerally, 2002). Incorporation of cement extenders of the correct type and amount into an ASR-susceptible mix imparts long-term benefits in eliminating or controlling damaging expansions and the generation of concrete tensile strains.

The suitability of fly ash (and GGBS) for controlling ASR can be assessed using ASTM C441. This involves replacing 25 per cent of the cement by mass with an amount of fly ash equal to the *volume* of cement replaced. For GGBS, 50 per cent of the portland cement is replaced by slag equal to the *volume* of cement replaced. Alkali content of the cement should be about 1 per cent. Mixes are made up with highly reactive Pyrex glass as fine aggregate, with companion mortars being made up with and without the extender. Specimens are stored over water in sealed containers at 38 °C and expansion is determined at 14 days. ASTM C618 requires expansion of the test mortar not to exceed 0.02 per cent at 14 days for mortars containing Class F or Class C ash, and requires a minimum reduction in expansion of 75 per cent for Class N (raw or calcined) natural pozzolan. For GGBS, ASTM C989 similarly requires that reduction in expansion be at least 75 per cent. The main criticism of ASTM C441 concerns the use of Pyrex glass as the standard reactive aggregate. The glass is highly reactive, and meeting the test criteria of ASTM C618 is difficult even though the fly ash might effectively suppress expansive ASR in long-term field tests (Stark, 1994).

Research indicates that the efficiency of silica fume in controlling pore solution alkalinity and expansion due to ASR depends strongly on the portland cement alkalis. CSA Standard Practice A23.2-27A proposes the following relationship in order to determine the minimum level of silica fume required:

$$SF = 2.5 \times AL \qquad\qquad (6.6)$$

where
SF = silica fume content (per cent replacement of cement by mass)
AL = total alkali content of concrete due to portland cement component (kg/m^3 Na_2O_{eq}).

The Canadian document also recommends that not less than 7 per cent by mass of silica fume should be used if it is the only cement extender in the concrete.

Avoiding the use of alkali-reactive aggregates   The obvious way to prevent ASR is to avoid the use of alkali-reactive aggregates! Table 6.2 gives details of rocks and minerals that may be potentially deleteriously reactive. Many naturally occurring rock types are potentially reactive. This underlines the importance of testing new or untried sources of aggregates for ASR prior to their use in concrete. Aggregates from generic rock types given in Table 6.2 may, however, not be alkali-reactive or only slightly reactive. Nevertheless, there are occasions when it is not possible to avoid an alkali-reactive aggregate. In these cases, blending with a non-reactive aggregate may reduce the problem. Another measure is to exclude ASR-susceptible constituents by beneficiation (selective quarrying and crushing, heavy media separation, etc.). Alternatively, attention must be paid to reducing or minimising the alkalis in the mix or modifying the environment to eliminate moisture.

Modifying the environment to reduce the moisture content of the concrete If it is impossible to avoid or minimise ASR by the measures given above, then the only alternative is to prevent wetting of the structure or reduce its moisture content. This can be done in a number of ways: shrouding or cladding the structure with a protective cover; providing sufficient fall in flat elements to ensure rapid drainage; paying attention to details of drainage such as expansion joints, embedded gutters and drain pipes, and so on; use of hydrophobic coatings to shed external water while allowing the concrete to 'breathe'. None of these measures will succeed, however, if the structure is partly submerged in water (including groundwater), or has water seeping into it from another face (such as in wing walls and abutments), or is simply in too moist an environment to properly dry (average RH in excess of about 80 per cent).

A word on use of lithium compounds (LiF, LiCl, and $LiCO_3$) to control ASR expansion is appropriate. This is a relatively recent development, despite the first reports appearing in the early 1950s, but has been given new impetus due to a major research project under the US-based Strategic Highway Research Programme (Stark, 1992). Expansion with most aggregates is effectively eliminated provided Li/(Na + K) molar ratios in the range of 0.60–1.00 are used (Blackwell *et al.*, 1997; Lumley, 1997). Certain lithium compounds show a pessimum effect where low dosages may actually increase expansion. This is related to the increased hydroxyl ion concentration in the pore solution when lithium compounds react with calcium hydroxide to produce insoluble calcium salt and lithium hydroxide. The use of lithium nitrate does not cause this effect due to the high solubility of calcium nitrate. The use of lithium as a practical measure to control ASR requires long-term monitoring before it can be unequivocally recommended.

REMEDIATION OF ASR-AFFECTED STRUCTURES

Alkali–aggregate reaction, if unchecked, can continue for decades. Eventually, it will cease when the available alkalis have been exhausted and provided there is no external source of alkalis. Usually, a decision will need to be made before this time as to whether to remediate the structure, 'do nothing', or possibly demolish it. These decisions can be made only on the basis of proper condition assessments and engineering evaluations, which must address the present state of the structure, its use, and the consequences of continuing ASR. Remediation will usually involve modifying the environment of the structure to reduce its moisture content. Newer techniques are now available such as lithium impregnation which can also assist. In some cases, very little can be done about the environment and other measures may be required to manage the effects of continuing ASR. For example, many arch dams around the world suffer from ongoing ASR expansion and it is impossible to modify their moisture conditions or to treat the concrete due to it massive nature. Expansion in these cases usually results in the build-up of stresses and deformations which can be alleviated by techniques such as slot cutting and provision of movement joints (Gilks et al., 2001). For normal AAR-affected structures, it is often possible to undertake remediation to retain the serviceability of the structure depending on the severity of the environment. The Johannesburg Motorway is a case in point, as is a sports stadium in Stellenbosch where cracks stabilised after crack injection with epoxy, application of a silane coating to exposed surfaces, and further protective coating with an elastomeric material. Details of these and other remedial measures are beyond the scope of this book and the reader is referred to specialist texts (ISE, 1992; Concrete Society, 1999; Johnston et al., 2000; Stokes et al., 2000). Other useful information on ASR can be found in CSA A23.1, Appendix B (2000), Hobbs (1987), Swamy (1992), and West (1996).

Internal sulphate attack

Concrete can deteriorate severely from the action of external sulphates which react with hydroxides and with hydrated aluminates of the cement to form expansive ettringite (calcium sulpho-aluminate). Strong sulphates such as ammonium and magnesium sulphate can also attack the hydrated silicates. However, the concern here is not external sulphate attack, but whether such attack can arise from the presence of aggregates embedded in concrete. Two such forms of attack are possible.

SULPHATE ATTACK FROM OXIDATION OF SULPHIDES IN AGGREGATES

Certain aggregates contain sulphide minerals, notably pyrite (FeS_2) and pyrrhotite (FeS), and in the presence of moisture and oxygen these minerals can convert into sulphates. Pyrite reacts as follows:

$$2FeS_2 + 2H_2O + 7O_2 \rightarrow 2FeSO_4 + 2H_2SO_4 \qquad (6.7)$$

The ferrous sulphate produced can further oxidise to ferric sulphate, which in turn can react with the remaining sulphide to set up a self-perpetuating reaction (Oberholster, 1986b). This results in the formation of sulphate and sulphuric acid, both of which are aggressive to concrete. Acids attack the cementitious binder forming calcium sulphate (gypsum) in the case of sulphuric acid, leading to its destruction, while sulphates react with the products of cement hydration to form reactants (such as ettringite) that occupy a larger volume than the original constituents, leading to internal cracking and breakdown of the matrix.

Sulphide-bearing rocks are common in mining operations and aggregates derived from these rocks must be treated with some caution. Pyrite-rich gravels also occur in many parts of the world such as in southeast England as well as in other parts of the UK (BRE, 2001). Two problems exist: first, the aggregates themselves may be contaminated with sulphates and acids from the oxidation of the sulphide minerals, which may be problematic if used in concrete; second, sulphide minerals in the aggregates may oxidise in the concrete leading to internal breakdown. The first problem can be alleviated by washing the aggregates prior to use; the second issue is a major problem only when the concrete is highly porous and permeable allowing ingress of moisture and oxygen, or where the pyrite particles exist close to an exposed surface. Isolated instances of concrete staining due to surface-exposed sulphide minerals in the aggregates have been reported (Oberholster, 1986b). Table 3.13 covered this subject previously.

THAUMASITE ATTACK

Sulphate attack of concrete can occur internally by the formation of thaumasite, a complex calcium carbonate-silicate-sulphate hydrate ($CaCo_3 \cdot CaSio_3 \cdot CaSo_4 \cdot 15H_2o$). This form of sulphate attack reduces the concrete matrix to a mushy paste rather than inducing expansive cracking. The calcium silicate hydrate phases break down in the presence of an available supply of sulphate and carbonate ions, destroying the binding properties of the matrix in the process (Crammond and Halliwell, 1996; Construction Research Communications, 2002). The necessary constituents and conditions for the attack are calcium silicates from the cement, sulphates and/or sulphides in the ground, carbonate-bearing aggregates such as limestones and dolomites, moisture usually in the form of mobile groundwater, and low temperatures (generally below 15 °C). Both thaumasite and ettringite are end products of a continuous series of solid solutions of similar structure and morphology. The formation of thaumasite is usually preceded or accompanied by that of ettringite. In general, good quality impermeable concrete will not suffer deleterious strength reduction from this form of sulphate attack.

The likelihood for thaumasite attack depends on the proportion of carbonate ($CaCO_3$) present in the fine and coarse aggregate fractions. The risk of attack is reduced when the $CaCO_3$ proportion is lowered to a threshold level of between 10 and 30 per cent, depending on whether this is in the fine or coarse aggregate fractions respectively. To reduce further the risk of attack, a second threshold of between 2 and 12 per cent of the total aggregate mass may be adopted. In general, the better the quality of the concrete, the more total carbonate can be tolerated.

The report of the Thaumasite Expert Group (DOE, 1999) and BRE Special Digest 1 (BRE, 2001) recommend classifying carbonate-bearing aggregates into three ranges depending on their carbonate content, with corresponding concrete qualities. Range C has the lowest carbonate content (between 2 and 12 per cent) and concretes containing such aggregates will not generally be susceptible to attack, even in sulphate-rich ground. The lower limit of 2 per cent applies when all the carbonate is in the fine fraction, while the upper limit applies when all the carbonate is in the coarse fraction of the aggregate. Linear interpolation between the limits is done when carbonates exist in both the coarse and the fine fractions. Range B (carbonate content between 10 and 30 per cent) permits a higher carbonate level in aggregates, but additional precautionary measures such as higher quality concrete are needed if sulphate conditions in the ground require such. Range A permits an unlimited proportion of carbonates in aggregate (above the upper bounds of Range B), but stringent additional precautionary measures are needed for most sulphate conditions, such as low water/binder ratios (generally <0.45), and proper attention to surface protection and drainage.

Thaumasite can also occur in other concrete products as well as in conventional concretes. Thaumasite formation was found to be the main reaction product in concrete bricks in South Africa that expanded because of the oxidation of pyrrhotite (FeS) in the aggregate (Oberholster et al., 1984).

Dimensional incompatibility

Dimensional incompatibility between aggregates and matrix can arise from thermal or moisture effects, and can lead to internal breakdown of the concrete. This was covered previously in Chapter 3. Major problems arise under cyclic conditions of temperature or moisture, depending on the nature of the incompatibility. Extreme cases usually involve aggregates with very low coefficients of thermal expansion such as certain limestones which may have coefficients of linear expansion as low as $3.5 \times 10^{-6}/°C$ or aggregates with high shrinking or swelling potential (covered in Chapter 5). In these cases, concretes with aggregates of low modulus of elasticity withstand temperature cycles better than those containing high E aggregates.

Extrinsic mechanisms

Extrinsic mechanisms refer to physical, mechanical, and chemical deterioration arising from external agents acting upon the surface of concrete, or from external environmental conditions. Such mechanisms have been studied extensively with a focus on the nature of the binder or matrix. Nevertheless, aggregates do have a role to play in these effects, and this role will be explored here.

Physical and mechanical actions

FREEZE–THAW

Resistance of concrete to freeze–thaw-induced cracking is important in cold harsh climates. This resistance can be influenced by the freeze–thaw resistance of the aggregates *per se*, dealt with in Chapter 3. Certain aggregates may be susceptible to freeze–thaw damage, in particular those with smaller pore sizes (<4 μm) and those containing deleterious clay minerals. Chapter 3 also discussed tests to assess the susceptibility of aggregates to freeze–thaw damage. Susceptible aggregates may render an otherwise freeze–thaw-resistant concrete liable to deterioration. Important factors governing freeze–thaw-resistance of aggregates are size, pore-structure, permeability, and tensile strength of the aggregate. These factors combine to determine the condition of 'critical saturation', in which an aggregate particle may crack upon being subjected to freeze–thaw cycles. Dense, fine pore structure aggregates attain this condition at smaller sizes than coarse-grained open-pore structure rocks which can allow the dissipation of freezing pressures. However, fine aggregates do not in general suffer from this phenomenon. ACI Committee Report 221 R has the following useful comment:

> Various properties related to the pore structure within the aggregate particles such as absorption, porosity, pore size and distribution, or permeability, may be indicators of potential durability problems for an aggregate used in concrete that will become saturated and freeze in service. Generally, it is the coarse aggregate particles with higher porosity or absorption values, caused principally by medium-sized pore spaces in the range of 0.1–5 μm, which are most easily saturated and contribute to deterioration of concrete. Larger pores usually do not become completely filled with water. Therefore damage does not result from freezing.

> (ACI Committee 221 R, 2001)

Distress due to freeze–thaw action in critically saturated aggregates is usually revealed by general disintegration and pop-outs, or it results in a phenomenon known as D-cracking (see Figure 3.28). D-cracking occurs in ground slabs exposed to freeze–thaw conditions and with continuous moisture on some faces, such as sidewalks, roads, and highway and airfield pavements. The phenomenon results in the development of fine, closely spaced cracks adjacent and parallel to joints and free edges. Distress is caused by delamination of the aggregate within the concrete. ACI 221 R mentions that nearly all occurrences of D-cracking are associated with absorptive sedimentary rocks including limestone, dolomite, shale, and sandstone. In Canada, absorptive limestone aggregates occur in parts of Manitoba, Ontario, and Quebec, while in the USA, several mid-Western states suffer from the problem, for example Ohio's sedimentary carbonate coarse aggregate sources. Pop-outs are often associated with aggregates showing susceptibility to D-cracking. From Canadian experience, a suitable test for such aggregates is a modified version of ASTM C666, using only two cycles of freeze–thaw per day, with length change being the criterion for failure (maximum expansion of 0.035 per cent after 350 cycles). Reduction of aggregate particle size is an effective means of controlling these problems.

Tests for susceptibility to freeze–thaw damage can be done on aggregate particles (e.g. the magnesium or sodium sulphate soundness test of ASTM C88), but it is probably better to test the aggregate embedded in the concrete mixture intended for use. This allows the concrete system to be evaluated rather than unrealistic tests on aggregate particles. Chapter 3 referred to ASTM C666 which evaluates the freeze–thaw resistance of air-entrained concrete that might contain frost-susceptible aggregates, by rapid cycles of freezing and thawing between 4.4 and $-17.8\,°C$. This test uses an unrealistically high rate of freezing (between 4 and $10\,°C/h$) whereas for in-situ concrete, rates rarely exceed $3\,°C/h$. An alternative test is ASTM C682 which is intended to evaluate the frost resistance of coarse aggregates in air-entrained concrete. The test was developed particularly for use with normal weight aggregates not having vesicular, highly porous structures. The test has the advantage of allowing the user to simulate exposure conditions closely to match the expected service conditions, using a continuous soaking period and then a slow cycle of freezing and thawing every two weeks, but has the disadvantage of a long test period. Damage occurs when a dilation or length increase is noted after the normal contraction as the concrete is cooled below freezing. The 'period of frost immunity' is the total number of weeks of test necessary to cause the critical dilation to occur. The test procedure is specified in ASTM C671. The test method of ASTM C682 specifically stresses that great care is required in performing the tests and in interpreting the results, mainly because the test permits variations in simulated exposure conditions that can greatly affect the performance of aggregates.

WETTING AND DRYING

Most dense aggregates have low shrinking and swelling potential and aggregates impart dimensional stability to concrete. However, certain aggregates are exceptions to this general rule and they have been discussed earlier in Chapter 3. Concrete surfaces may be exposed to wetting and drying cycles due to rainfall, rise and fall of water as in tidal conditions, periodic washing and inundation, and so on. These influences often do not penetrate more than a few centimetres into the concrete, but give rise to alternate swelling and shrinking of the concrete surface and may lead ultimately to surface crazing and cracking. This deterioration arises chiefly from the movement of the concrete matrix. Aggregates will influence this process adversely only if they exhibit excessive swelling and/or shrinking tendencies, and such aggregates were covered in Chapter 5. Local pop-outs may be caused by swelling aggregates and by synthetic aggregates such as crushed metallurgical slags which may contain components that expand on exposure to moisture (see Figure 3.26).

Chemical actions

Chemical properties of aggregates *per se* were covered in Chapter 3. In general, the aggregate phase in concrete is more durable and resistant to chemical attack than the paste or matrix component. However, there are rare occasions when the aggregates are either more vulnerable to chemical attack than the matrix, or they modify the resistance of the composite material. Examples of these will be discussed below.

ACID ATTACK

Acids attack concrete but most aggregates are immune to acid attack, particularly siliceous aggregates. Calcareous aggregates on the other hand react readily with acids, and this can be used to good effect in acidic conditions by ensuring that the acid attack on the concrete is spread over both the paste and the aggregate phases. The rate of acid attack on concrete depends on the type of acid, and in some cases a self-protecting skin of corrosion products (e.g. gypsum) can build up on the surface thus limiting the attack.

Acid resistance of concrete sewer pipes   Sulphuric acid generation and attack on concrete sewer pipes is a pervasive problem in hot climates, particularly where sewers are laid at relatively flat grades. The problem occurs in Australia, Cyprus, Egypt, South Africa, and parts of the USA. Design of concretes for such conditions is based on one of two different philosophies, either creating an acid-resistant material that will resist acid attack, or producing an acid-soluble concrete that will help to neutralise the attacking acid. In the former category are attempts to create acid-resistant matrixes by reducing their acid-susceptible components such as calcium hydroxide or by introducing a polymer phase, coupled with acid-resistant aggregates. Concrete with high supplementary cementitious material contents and insoluble aggregates may be used. Other than the practical manufacturing difficulties created by this approach (e.g. low early strengths or need for polymerisation), hydraulic cement-based binders remain susceptible to acid attack at low pHs of around 1 which are typical in aggressive sewer conditions. This is illustrated in Figure 6.20 with reference to concretes with different binders: ordinary portland cement (CEM I), ground granulated slag (50 per cent), and silica fume (10 per cent), at w/b ratios varying from 0.29 to 0.36, immersed in concentrated hydrochloric acid solution of pH 1. Maximum size of coarse aggregates was 13.2 mm in all cases. The aggregates for the OPC and SF mixes were rounded, while those for the slag mix were crushed making them angular and irregular. The improved performance of the slag mix was ascribed to an aggregate interlock effect, hindering removal of corrosion debris

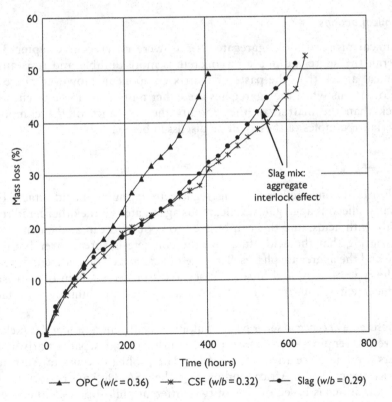

Figure 6.20 Acid attack on sewer pipe concrete mixtures made with different binders and aggregates (based on data from Fourie, 2003).

and slowing attack on the binder. For the SF mix, the better performance was ascribed to the improved aggregate bonding and denser matrix.

For acid-soluble concretes, the aggregates must also be acid-soluble and this limits them to carbonates such as limestone or dolomite. Two inter-related effects come into play. First, the acid generated is spread over a much larger surface area – in effect, all of the concrete becomes acid-soluble and consequently a given amount of acid will result in less removal of surface material. The second effect is related to the first, which is that as the acid is continually neutralised by the entire concrete surface, so the pH at the surface is higher. This reduces the metabolic rate of the acid-loving bacteria, causing less acid to be generated. The overall effect is one of considerable improvement in concrete performance. Table 6.8 shows data from a real sewer used for test purposes in South Africa. Different pipes with various concrete materials were laid sequentially along a particularly aggressive section of the sewer. The 900-mm diameter pipes were kept in service for about 12 years, after which they were removed and inspected for acid attack. The portland cement pipes with siliceous aggregates had lost their entire wall thickness of 80 mm at the sides, and almost all at the crown of the pipe. By contrast, the pipes with dolomitic aggregate had only lost about 40 mm of their walls at the sides and somewhat less at the crown. At the pipe crowns the loss of material was due to acid corrosion. At the sides, it was due to a combination of corrosion and hydraulic erosion.

The approach of creating conditions unfavourable for acid generation can be extended by use of binders such as calcium aluminate cements which act to

Table 6.8 Acid attack on various concrete materials in pipes in the Virginia Experimental Sewer, South Africa

Pipe material	Corrosion rates	
	Predicted* (mm/year)	Measured† (mm/year)
OPC/siliceous	>6	>7.5
OPC/dolomitic	2–3	3.9
CAC/siliceous	1–2	2.5
CAC/dolomitic‡	–	0.6

Source: Goyns, 2003.

Notes
* Predicted values inferred from alkalinity (or neutralisation capacity) of the pipe materials.
† Measured values show some variability, as would be expected. Values were measured at the sides of the pipes where maximum loss of material occurred.
‡ CAC/dolomitic pipes were not used in the experiment. This value was estimated based on other materials and performance of coupons of this material exposed in the pipeline.

'poison' the bacteria and reduce their metabolism. When coupled with chemically similar aggregates such as the synthetic aggregate Alag©, these concretes can give extremely good acid resistance in biologically generated acid environments. Results from CAC pipes used in the South African sewer experiment referred to are also given in Table 6.8, where it can be seen that a CAC binder coupled with a siliceous aggregate gave the best overall performance.

Occasionally, other factors come into play, which affect the acid resistance of carbonate-bearing aggregate concretes. One such effect is where the aggregate is weak and internally fractured. The acid enters the fractures and dissolves the aggregate from within leaving a 'cratered' appearance of the surface. The attack in such cases is usually much greater than if the aggregates are hard and sound. On the other hand, acid can remove the matrix portion around insoluble coarse aggregates, leaving them prone to fall-out. Figure 6.21 shows the surfaces after mineral acid attack of similar concretes but made with (a) strong competent dolomite aggregate, (b) weak fractured limestone aggregate, and (c) strong, acid-insoluble granite aggregate. In the case of the competent dolomite, the concrete surface is relatively smooth indicating that the attack is evenly spread over the entire surface; for the granite aggregate, only the binder has been attacked resulting in rapid degradation and premature aggregate fall-out.

(a)   (b)   (c)

Figure 6.21 Acid-attacked concrete cylinders: (a) Smooth surface – competent dolomite; (b) 'Cratered' surface, fractured limestone; and (c) Very rough surface due to aggregate fall-out. Cylinders had original dimensions of 80-mm diameter by 150 mm (photographs courtesy of Clyde Fourie, UCT).

ALKALI ATTACK

In strong alkalis, siliceous aggregates may become vulnerable – the reason for the alkali–aggregate reactivity problem discussed earlier. The strong alkalis that attack such aggregates usually derive from the cement hydration, and the process is essentially self-sustaining in the presence of sufficient moisture. Obviously, additional alkalis that might arise from external sources will exacerbate the problem. Such sources may be marine or de-icing salt environments, or they may be industrial environments where strong alkalis are generated. In the latter case, the cement matrix may also be attacked. It should be stressed, though, that not all siliceous aggregates are susceptible to alkali attack, and many such aggregates have a long history of successful use in concrete.

What is clear from studies of concretes in aggressive chemical environments is that the binder phase is usually far more susceptible to chemical attack than the embedded aggregates. The vast majority of dense, sound aggregates possess more than adequate resistance to aggressive chemicals. Also, since the aggregates are embedded in concrete, they are largely protected from aggressive external agents.

Closure

Transport properties of concrete govern its durability. A notable exception is the intrinsic mechanism of AAR, but even in this case, internal transport of ions and moisture must occur. While many of the important durability properties of concrete are related primarily to the matrix and binder, aggregates play a key role in several ways. They 'dilute' the binder, thus generally improving the durability, they impart dimensional and chemical stability to the material, but they can also occasionally react with the products of hydration. Through the presence of ITZs, they can allow more permeable paths through the material for aggressive agents. Thus, the concrete engineer should be aware of how aggregates influence concrete durability and select appropriate aggregates for the application in hand.

References

Alexander, M.G., Mackechnie, J.R. and Ballim, Y. (2001) 'Use of durability indexes to achieve durable cover concrete in reinforced concrete structures', in J.P. Skalny and S. Mindess (eds) *Materials Science of Concrete VI*, Westerville, OH: American Ceramic Society, 483–511.

Alpha (Pty) Ltd. (2000) Alpha Product Manual, Technical Information, Roodepoort, Alpha (now Holcim) Corporate Communications, 43pp.

American Concrete Institute Committee 221 (2001) 'Guide for use of normal weight and heavyweight aggregates in concrete', ACI 221R-96 (Re-approved 2001), *American Concrete Institute Manual of Concrete Practice*, Farmington Hills, MI: American Concrete Institute.

Bentz, D.P., Stutzman, P.E. and Garboczi, E.J. (1992) 'Experimental and simulation studies of the interfacial zone in concrete', *Cem. Conc. Res.*, 22: 891–902.

Bentz, D.P., Schlangen, E. and Garboczi, E.J. (1995) 'Computer simulation of interfacial zone microstructure and its effect on the properties of cement-based composites', in S. Mindess and J.P. Skalny (eds) *Materials Science of Concrete IV*, Westerville, OH: American Ceramic Society, 155–200.

Blackwell, B.Q., Thomas, M.D. and Sutherland, A. (1997) 'Use of lithium to control expansion due to alkali–silica reaction in concrete containing UK aggregates', ACI SP-170, Farmington Hills, MI: American Concrete Institute, 649–663.

Blight, G.E., Alexander, M.G., Schutte, W.K. and Ralph, T. (1983) 'The effect of alkali aggregate reaction on the strength and deformation of a reinforced concrete structure', *Proceedings 6th Int. Conf. Alkalis in Concrete*, Copenhagen, Denmark, 401–410.

Blight, G.E., Alexander, M.G., Schutte, W.K. and Ralph, T. (1984) 'The repair of reinforced concrete structures affected by alkali–aggregate reaction', *The Civil Engineer in South Africa*, 26(11): 525–538.

Brandt, M.P. and Oberholster, R.E. (1983) 'Ondersoek van formasie Tygerberg-aggregate vir potensiele alkalireaktiwiteit, implikasies van toetsmetodes en toepassing van bevindings in die praktyk' (in Afrikaans), CSIR Research Report BRR 500, Pretoria: CSIR.

Bretton, D., Ollivier, J.P. and Ballivy, G. (1992) 'Diffusivite des Ions Chlore dans la Zone de Transition entre Pate Ciment et Roche Granite', in J.C. Maso (ed.) *Interfaces in Cementitious Composites*, London: E&FN Spon, 269–278.

Building Research Estabishment (BRE) (2001) *Concrete in Aggressive Ground*, Special Digest 1, 4 Parts, London: Building Research Establishment.

Canadian Standards Association (2000) *Concrete Materials and Methods of Concrete Construction/Methods of Test for Concrete*, CSA A23.1-00/A23.2-00, Toronto: CSA International.

Concrete Society (1999) *Alkali–Silica Reaction: Minimising the Risk of Damage to Concrete*, Technical Report 30 (3rd edn), Slough: Concrete Society.

Construction Research Communications (2002) *Proceedings of the First Int. Conf. Thaumasite in Cementitious Materials*, Garston, UK: Construction Research Communications.

Costa, U., Facoetti, M. and Massazza, F. (1990) 'Permeability of the cement–aggregate interface: Influence of the type of cement, water/cement ratio, and superplasticizer', in E. Vazquez (ed.) *Admixtures for Concrete: Improvement of Properties*, London: Chapman & Hall, 392–401.

Crammond, N.J. and Halliwell, M.A. (1996) 'The Thaumasite form of sulphate attack in concretes containing a source of carbonate ions – a microstructural overview', *Advances in Concrete Technology, Proceedings 2nd CANMET/ACI Int. Symposium*, ACI SP-154, Farmington Hills, MI: American Concrete Institute, 357–379.

Department of the Environment (DOE), Thaumasite Expert Group (1999), *The Thaumasite Form of Sulphate Attack: Risks, Diagnosis, Remedial Works and Guidance on New Construction*, London: Department of the Environment.

Diamond, S. (2003) 'Percolation due to overlapping ITZs in laboratory mortars: A microstructural evaluation', *Cem. Conc. Res.*, 33: 949–955.

Dixon, S. (2001) The effects of bleeding on the durability of concrete, Final Year Thesis, Department of Civil Engineering, University of Natal, Durban, KwaZulu Natal.

Dolar-Mantuani, L. (1983) *Handbook of Concrete Aggregates*, New Jersey: Noyes Publications.

Fourie, C. (2003) Unpublished doctoral research, Department of Civil Engineering, University of Cape Town, Cape Town.

Fulton's Concrete Technology (2001) Eds B.J. Addis and G. Owens, 8th edn, Midrand: Cement and Concrete Institute.

Gilks, P., May, T. and Curtis, D. (2001) 'A review and management of AAR at Mactaquac Generating Station', *Canadian Dam Association Annual Conference*, Fredericton, NB, 10pp.

Goyns, A. (2003) *Annual Report on Virginia Sewer*, CMA Pipe Div., Midrand: Concrete Manufacturer's Association.

Grattan-Bellew, P.E., Cybanski, G., Fournier, B. and Mitchell, L. (2003) 'Proposed universal accelerated tests for alkali–aggregate reaction: The concrete microbar method', *Cement, Concrete and Aggregates*, 25(2): 29–34.

Helmuth, R. and Stark, D. (1992) 'Alkali–silica reactivity mechanisms', in J.P. Skalny (ed.) *Materials Science of Concrete III*, Westerville, OH: American Ceramic Society, 131–208.

Hobbs, D.W. (1986) 'Deleterious expansion of concrete due to alkali–silica reaction: Influence of pfa and slag', *Magazine of Concrete Research*, 38(137): 191–205.

Hobbs, D.W. (1987) 'Some tests on fourteen years old concretes affected by the alkali–silica reaction', in P.E. Grattan-Bellew (ed.) *Proceedings 7th Int. Conf. Concrete Alkali–Aggregate Reactions*, New Jersey: Noyes Publications.

Hobbs, D.W. (1988) *Alkali–Silica Reaction in Concrete*, London: Thomas Telford Ltd.

Hooton, R.D. and Rogers, C.A. (1992) 'Development of the NBRI rapid mortar bar test leading to its use in North America', *Proceedings 9th Int. Conf. Alkali–Aggregate Reaction in Concrete*, London, 461–467.

Hoppe, G.G. (1990) 'Rehabilitation and impregnation of a concrete bridge to inhibit the further effects of alkali–aggregate reaction', *Proceedings G M Idorn Int. Symp. on the Durability of Concrete*, Ontario, Canada, 357–380.

Houst, Y.F., Sadouki, H. and Wittmann, F.H. (1992) 'Influence of aggregate concentration on the diffusion of CO_2 and O_2', in J.C. Maso (ed.) *Interfaces in Cementitious Composites*, London: E&FN Spon, 279–288.

Institution of Structural Engineers (1989) *Load Testing of Structures*, London: The Institution.

Institution of Structural Engineers (1992) *Structural Effects of Alkali–Silica Reaction: Technical Guidance on the Appraisal of Existing Structures*, London: The Institution.

Jaufeerally, H. (2002) Performance and properties of structural concrete made with Corex slag, MSc (Eng) Thesis, University of Cape Town, Cape Town.

Johnston, D.P., Surdahl, R. and Stokes, D.B. (2000) 'A case study of a lithium-based treatment of an ASR-affected pavement', *Proceedings 11th Int. Conf. Alkali–Aggregate Reaction*, Québec, Canada, 1079–1087.

Jones, T.N. and Poole, A.B. (1986) 'Alkali–silica reaction in several UK concrete structures: The effect of temperature and humidity on expansion and the significance of ettringite development', *Proceedings 7th Int. Conf. Alkali–Aggregate Reaction*, Ottawa, Canada, 446–450.

Jones, F.E. and Tarleton, R.D. (1958) *Reactions between Aggregates and Cements, Parts V and VI, Alkali–Aggregate Interaction*, National Building Studies, Research Paper No. 25, London: HMSO.

Landgren, R. and Hadley, D.W. (2002) *Surface Popouts Caused by Alkali–Aggregate Reactions*, Research and Development Bulletin RD121, Skokie: Portland Cement Association.

Lumley, J.S. (1997) 'ASR suppression by lithium compounds', *Cem. Conc. Res.* 27(2): 235–244.

Mindess, S. and Gilley, J.C. (1973) 'The staining of concrete by an alkali–aggregate reaction', *Cem. Conc. Res.*, 3: 821–828.

Mindess, S., Young, J.F. and Darwin, D. (2003) *Concrete*, Upper Saddle River, NJ: Prentice Hall.

Nixon, P. and Sims, I. (1996) 'Testing aggregates for alkali-reactivity', Report of RILEM TC 106, *Materials and Structures*, 29(190): 323–334.

Nyame, B.K. (1985) 'Permeability of normal and lightweight mortars', *Magazine of Concrete Research*, 37(130): 44–48.

Oberholster, R.E. (1983) 'Alkali reactivity of siliceous rock aggregates: Diagnosis of the reaction, testing of cement and aggregate and prescription of preventive measures', *Proceedings 6th Int. Conf. Alkalis in Concrete*, Copenhagen, 419–433.

Oberholster, R.E. (1986a) 'Alkali–aggregate reaction', in B.J. Addis (ed.) *Fulton's Concrete Technology*, 6th edn, Midrand: Cement and Concrete Institute.

Oberholster, R.E. (1986b) *The Oxidation of Sulphides in Aggregates Produced From Witwatersrand Quartzite Gold-mine Waste Dumps and Its Contribution to the Sulphate Attack on Concrete*, National Building Research Institute, South Africa-CSIR, Pretoria, Council for Scientific & Industrial Research.

Oberholster, R.E. (2001) 'Alkali–silica reaction', in B.J. Addis and G. Owens (eds) *Fulton's Concrete Technology*, 8th edn, Midrand: Cement and Concrete Institute.

Oberholster, R.E. and Davies, G. (1986) 'An accelerated method for testing potential alkali reactivity of siliceous aggregates', *Cem. Conc. Res.*, 16: 181–189.

Oberholster, R.E., du Toit, P. and Pretorius, J.L. (1984) 'Deterioration of concrete containing a carbonaceous sulphide-bearing aggregate', *Proceedings 6th Int. Conf. Cement Microscopy*, Albuquerque, New Mexico, 360–373.

Pike, D.C. (1990) *Standards for Aggregates*, Chichester: Ellis Horwood.

Ping, X., Beaudoin, J.J. and Brousseau, R. (1991) 'Flat aggregate–portland cement paste interfaces, II. Transition zone formation', *Cem. Conc. Res.*, 21: 718–726.

Princigallo, A., van Breugel, K. and Levita, G. (2003) 'Influence of the aggregate on the electrical conductivity of portland cement concretes', *Cem. Conc. Res.*, 33: 1755–1763.

RILEM Technical Committee TC 20-TBS (1984) 'General recommendations for statical loading test of load-bearing concrete structures in situ (TBS2)', *Rilem Technical Recommendations for the Testing and Use of Construction Materials*, London: E&FN Spon.

RILEM Technical Committee TC 106-AAR (2000) 'Aggregates for alkali–aggregate reaction. International assessment of aggregates for alkali–aggregate reactivity', Committee Report, *Materials and Structures*, 33: 88–93.

Scrivener, K.L. and Gartner, E.M. (1988) 'Microstructural gradients in cement paste around aggregate particles', in S. Mindess and S. Shah (eds) *Bonding in Cementitious Materials*, Pittsburgh: Materials Research Society, 77–85.

Sims, I. (2000) 'Alkali-reactivity – Solving the problem world-wide', *Concrete*, 34(10): 64–66.

Sims, I. and Nixon, P. (2001) 'Alkali-reactivity – a new international scheme for assessing aggregates', *Concrete*, 35(1): 36–39.

Sims, I. and Nixon, P. (2003a) 'Towards a global system for preventing alkali-reactivity: The continuing work of RILEM TC 191-ARP', *Proceedings 6th CANMET/ACI Int. Conf. Durability of Concrete*, ACI Special Publication SP-212, Farmington Hills, MI: American Concrete Institute, 475–487.

Sims, I. and Nixon, P. (2003b) 'RILEM Recommended Test Method AAR-0: Detection of alkali-reactivity potential in concrete – Outline guide to the use of RILEM methods in assessments of aggregates for potential alkali-reactivity', Report of RILEM TC 191-ARP, *Materials and Structures*, 36: 472–479.

Sims, I. and Nixon, P. (2003c) 'RILEM Recommended Test Method AAR-1: Detection of potential alkali-reactivity of aggregates – Petrographic method', Report of RILEM TC 191-ARP, *Materials and Structures*, 36: 480–496.

Stark, D. (1978) 'Alkali–silica reactivity in the Rocky Mountain region', *Proceedings 4th Int. Conf. Effect of Alkalis in Cement and Concrete*, Purdue University, Indiana, 235–243.

Stark, D.C. (1991) 'How to evaluate the state of alkali–silica reactivity (ASR) in concrete', *Concrete Repair Digest*, August–September, 104–107.

Stark, D.C. (1992) 'Lithium-salt admixtures – an alterative method to prevent expansive alkali–silica reactivity', *Proceedings 9th Int. Conf. Alkali–Aggregate Reaction in Concrete*, Slough: Concrete Society, 1017–1025.

Stark, D. (1994) 'Alkali–silica reactions in concrete', *Significance of Tests and Properties of Concrete and Concrete Making Materials*, ASTM STP 169C, West Conshohocken, PA: American Society for Testing and Materials, 365–371.

Stark, D. and Bhatty, M.S.Y. (1985) 'Alkali–silica reactivity: Effect of alkali in aggregate on expansion', *Alkalis in Concrete*, ASTM STP 930, West Conshohocken, PA: American Society for Testing and Materials, 16–30.

Stokes, D.B., Thomas, M.D.A. and Shashiprakash, S.G. (2000) 'Development of a lithium-based material for decreasing ASR-induced expansion in hardened concrete', *Proceedings 11th Int. Conf. Alkali–Aggregate Reaction*, Québec, Canada, 1089–1098.

Swamy, R.N. (1992) *The Alkali–Silica Reaction in Concrete*, New York: Van Nostrand Reinhold.

Swamy, R.N. (1994) 'Alkali–aggregate reactions in concrete. Material and structural implications', *Advances in Concrete Technology*, Ottawa: CANMET, Natural Resources Canada, 533–581.

Swamy, R.N. (1997) 'Assessment and rehabilitation of AAR-affected structures', *Cem. Conc. Comp.*, 19: 427–440.

Swenson, G.E. and Gillott, J.E. (1964) 'Alkali–carbonate rock reaction', *Highway Research Board Symposium on Alkali–Carbonate rock Reaction*, Highway Research Record No. 45, Washington, DC: Highway Research Board, 21–40.

United States Bureau of Reclamation (USBR) (1975) *Concrete Manual*, 8th edn, 37.

West, G. (1996) *Alkali–Aggregate Reaction in Concrete Roads and Bridges*, London: Thomas Telford.

Winslow, D.N., Cohen, M.D., Bentz, D.P., Snyder, K.A. and Garboczi, E.J. (1994) 'Percolation and pore structure in mortars and concrete', *Cem. Conc. Res.*, 24(1): 25–37.

Yang, C.C. and Su, J.K. (2002) 'Approximate migration coefficient of interfacial transition zone and the effect of aggregate content on the migration coefficient of mortar', *Cem. Conc. Res.*, 32(10): 1559–1565.

Chapter 7

Special aggregates and special concretes

Hitherto, this book has focused primarily on *ordinary* aggregates for use in the normal concretes that make up about 90 per cent of concrete production. However, with the advent of a new generation of concrete admixtures, combined with a much better understanding of the fundamental mechanisms governing concrete rheology, strength, cracking, and durability, we can now largely 'tailor-make' concretes to provide the precise properties required for a particular project. In this chapter, we will examine the aggregate requirements for some of the special concretes listed in Table 7.1, for which the aggregate properties are particularly important.

We will also examine the properties and uses of 'artificial' or synthetic aggregates, either those that are produced purposefully for concrete applications, or those that occur as wastes or by-products of other industrial processes. 'Marginal' aggregates and their beneficiation and use will be mentioned as well.

Low density aggregates

Low density aggregates (also referred to as *lightweight* aggregates – see Chapter 5) are characterized by their highly porous or cellular microstructure. They may be classified either by their origin (natural or artificial) or by their end-use:

1   In structural concrete (ASTM C330)
2   In concrete masonry units (ASTM C331)
3   In insulating concretes (ASTM C332).

In all three cases, the standards state 'the aggregates shall be composed predominantly of lightweight-cellular and granular inorganic material'. This excludes materials such as cork, sawdust and the like, though as we shall see later organic materials may well be used to make certain types of lightweight

Table 7.1 Some special concretes

Abrasion resistant concrete
Coloured aggregate concrete
Exposed aggregate concrete
High density concrete
Low density concrete
Mass concrete
Preplaced aggregate concrete
Radiation shielding concrete
Recycled aggregate concrete
Roller compacted concrete
Self-compacting concrete
Very high strength concrete
White concrete

concrete. The following general types of lightweight aggregates are defined in ASTM C330:

1   Aggregates prepared by expanding, pelletizing, or sintering products such as blast-furnace slag, clay, diatomite, fly ash, shale or slate.
2   Aggregates prepared by processing natural materials, such as pumice, scoria, or tuff.

In addition, for concrete masonry units, ASTM C331 adds to this list 'aggregates consisting of end products of coal or coke combustion', and for insulating concretes, ASTM C332 adds further 'aggregates prepared by expanding products such as perlite or vermiculite'.

The basic properties of some of these low density aggregate types are given in Table 7.2 (Mindess *et al.*, 2003). It should be noted that any particular aggregate type may exhibit a wide range of properties, depending on the details of the aggregate processing employed, and on the exact origin and nature of the parent material. The range of concrete properties that may be obtained is correspondingly very wide, as shown in Figure 7.1 (ACI 213R-87).

Natural lightweight aggregates

Pumice, scoria, and tuff are porous, glassy materials derived from igneous rocks. They differ primarily in their pore structure. Pumice contains a network of interconnected tube-like voids; scoria tends to contain more spherical voids; and tuff (formed from the cementing together of volcanic ash) has an irregular pore structure. These materials may be crushed and screened to obtain the desired gradation. Such materials were, of course, the first used

Table 7.2 Properties of selected lightweight aggregates

Aggregate	Aggregate dry bulk density (kg/m³)	Absorption (wt. %)	Origin[a]
Expanded shale, clay, slate	550–1050	5–15	PN
Foamed slag	500–1000	5–25	S
Sintered fly ash	600–1000	14–24	S
Exfoliated vermiculite	65–250	20–35	PN
Expanded perlite	65–250	10–50	PN
Pumice	500–900	20–30	N
Expanded glass	250–500	5–10	S
Expanded polystyrene beads	30–150	–	S
Brick rubble	~750	19–36	S
Crushed stone[b]	1450–1750	0.5–2.0	N

Notes
a PN processed natural material; N natural material; S synthetic material.
b Natural aggregate listed for comparison.

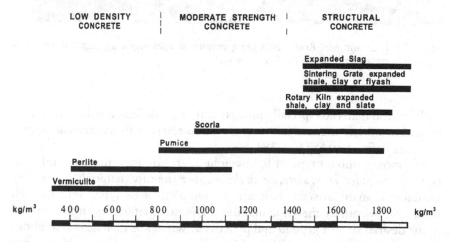

Figure 7.1 Classification of lightweight concretes (adapted from ACI Committee 213, 2003).

to make lightweight concrete. For instance, the Pantheon in Rome, built in the second century, was constructed in part of concrete containing pumice to reduce the weight of the great dome (Figure 7.2). The Pantheon is still in use today, which attests to the potential durability of lightweight concrete.

Synthetic and specially processed aggregates

Because the natural aggregates mentioned above are rather limited in their geographic distribution, most lightweight concretes today are made either

Figure 7.2 The Pantheon, Rome, built using pumice as lightweight aggregate in the concrete of the dome (2nd century AD).

with natural materials specially processed to provide lightweight aggregates, or with synthetic materials. These are often referred to as *artificial* aggregates, as defined in ASTM C294.

The most common type of lightweight aggregate used in structural concrete is *expanded clay, slate,* or *shale* because the raw materials are readily available in many areas. (These are treated as a single type, since they all behave in essentially the same manner.) The raw material is either crushed to the desired size, or ground and pelletized. When it is then heated to about 1000–1200 °C, the material 'bloats' (puffs up to several times its original size, like popcorn). This is due to the rapid generation of gas from the combustion of the small quantities of organic material that either occurs naturally within the material, or may be deliberately added. The composition of the material is such that partial melting occurs at the bloating temperature. The lowered viscosity of the material allows it to expand, and an impervious viscous coating forms on the outside that prevents the gas from escaping too quickly. The resulting material may then be crushed and screened to obtain the desired particle size distribution. Commercial versions of these materials available in the United Kingdom are Aglite, which has angular-shaped particles, and Leca, which has smooth, spherical particles. Sintag is produced from shale which is crushed and then sintered at

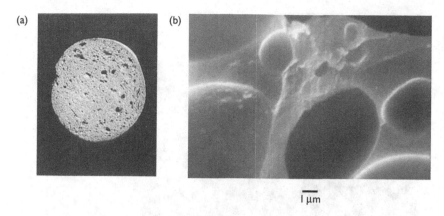

Figure 7.3 Expanded clay aggregate (UK Leca): (a) Typical particle, nominal size 16 mm; (b) SEM micrograph of internal structure (photo courtesy of Prof. J. Newman and Prof. T. Bremner).

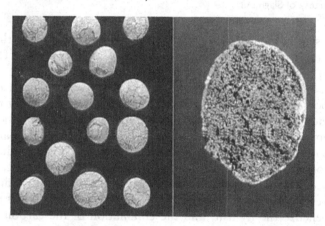

Figure 7.4 Expanded shale aggregate (German Liapor). Typical particles are shown on left (nominal sizes 8–16 mm), and cross section through particle on right. Particles can be of different sizes and densities (photo courtesy of Prof. J. Newman).

about 1200 °C. Commercial expanded clay and shale aggregates are shown in Figures 7.3 and 7.4.

Expanded (or foamed) blast furnace slag is produced by treating the molten slag from blast furnaces, which escapes at a temperature of about 1500 °C, either with a controlled amount of water, or with steam and compressed air, in such a way that the steam is trapped in the molten slag. This imparts a porous structure to the slag, similar to that of pumice – see Figure 7.5. Again, the material is then crushed and screened. The foaming process renders the material relatively inert in a cementitious system.

Figure 7.5 Typical particles of blast furnace slag aggregate (South Africa) (photo courtesy of Slagment).

If the molten slag is bloated with water in a rotating drum, and then thrown through air, the resulting product is *pelletized expanded slag*. This produces smooth particles, with a lower density than foamed slag.

Sintered fly ash is produced by taking the fly ash produced from the combustion of powdered coal (preferably bituminous coal), pelletizing the fly ash by the addition of water, and then firing the mixture at about 1200 °C. This temperature is not high enough to completely melt the particles, but permits the fly ash to coalesce into larger particles. This provides a good quality, economical lightweight aggregate, since the fly ash is already in a finely divided state, and generally contains enough carbon to reduce fuel costs. The resulting cellular structure is due to the evaporation of the water used to pelletize the fly ash, and the elimination of the carbon by combustion. One commercial version of this aggregate is known as Lytag. Figure 7.6 shows a typical sintered fly ash aggregate from the UK.

Perlite is a volcanic glass that contains about 2–6 per cent water within its structure. When it is heated quickly to about 1800 °C, the water dissociates itself from the mineral structure; the steam that results causes a cellular structure to develop within the material. This leads to a very low density. This material is widely used, primarily for insulation purposes, and to protect steel from the effects of fire.

Vermiculite is a clay mineral with a platy, layered structure. When it is heated to about 1000 °C, the interlayer water turns to steam, causing it to expand greatly (or *exfoliate*), up to about 30 times its original volume. Like perlite, it too is used primarily in insulating concretes. Vermiculite in its various forms, including the expanded flakes used for aggregates, is shown in Figure 7.7.

(a)  (b)

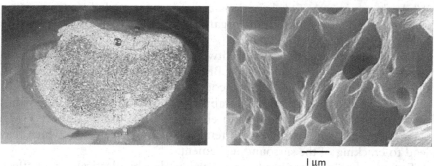

I μm

Figure 7.6 Sintered fly ash aggregate (UK Lytag): (a) Cross section through typical particle nominal size; (b) SEM micrograph of internal structure (photo courtesy of Prof. J. Newman and Prof. T. Bremner).

Figure 7.7 Expanded vermiculite aggregate. Bottom centre is a vermiculite chunk, which is flaked to produce particles on top right, which are then further expanded to produce exfoliated particles at top left. Nominal size of exfoliated particles is 8 mm (photo courtesy of Mandoval Vermiculite, South Africa).

Organic materials

While the various ASTM definitions quoted above exclude organic materials, there are nonetheless some organic materials that are widely used as lightweight aggregates.

Sawdust has been used as a lightweight aggregate for many years, in products such as tiles and blocks. Because raw sawdust contains sugar, tannins, and various other organic compounds that can interfere with the setting, hydration, and resulting durability of portland cement, the sawdust must be pretreated in some way to eliminate these harmful effects. Even then, however, wood particles will tend to swell when wetted, which may lead to cracking of the surrounding cement.

Expanded polystyrene beads (or other low density plastics) are, in effect, foamed resins. While expensive to produce, and difficult to incorporate uniformly into concrete, they too are used in insulating concretes for certain applications. Their maximum size is about 4 mm.

Properties and test methods for lightweight aggregates

Clearly, from the descriptions given above of the various lightweight aggregates that are available, their shapes and structures (both internal and external) are highly variable, depending upon the particular materials and the way in which they were processed. Their engineering properties are thus highly variable as well. Similarly, while many of the ASTM, BS, and other test standards developed for 'normal' aggregates are also applicable to lightweight aggregates, this is not universally true, and care must be taken in characterizing the properties of lightweight aggregates.

Grading, shape and texture

Many lightweight aggregates are angular or highly irregular in shape. This may have detrimental effects on the workability of the fresh concrete, as well as on its finishability and bleeding characteristics. In such cases, the same remedial measures may be taken as with unfavourable normal weight aggregates: more entrained air (up to 10 per cent), the use of mineral admixtures such as fly ash, or substitution of all or part of the fine lightweight aggregate with smoother normal weight sand. Otherwise, the grading characteristics for lightweight aggregate are similar to those for normal aggregates, and are determined in the same way by sieve analysis.

Absorption

Because of their porous structure, lightweight aggregates tend to have high absorption values, as shown in Table 7.2, as well as high *rates* of absorption. This requires a modified approach to concrete mixture design, to account for the volume of mix water absorbed into the aggregate, not all of which will then be available for either workability or hydration. The design process is thus even more empirical than it is for normal concrete. (The details of the mixture design process for lightweight concrete are beyond the scope of this book; they may be found in any number of standard concrete textbooks, e.g. Mindess *et al.*, 2003.)

It is difficult to measure the absorption of lightweight aggregates because their highly irregular and vesicular surfaces make it difficult to achieve either 'surface dry' or 'saturated surface dry' conditions. Thus, in addition to the ASTM tests for specific gravity (relative density) and absorption of coarse (ASTM C127) and fine (ASTM C128) aggregates, a number of other agencies have developed their own tests for these properties. However, none of these tests are considered to be particularly reliable or reproducible. It should also be noted that, in general, the relative density of lightweight aggregates increases with decreasing aggregate size, and so any change in the grading of a particular aggregate type would be expected to change the absorption. As a further complication, if lightweight aggregate particles are fractured during the mixing process, they may expose a more absorptive internal pore system, which will rapidly increase the effective absorption and thus reduce the workability of the concrete.

Strength

The strength of lightweight aggregate cannot be estimated adequately using methods such as the Los Angeles Abrasion test described earlier for normal aggregates (see Chapter 3). The values obtained from these tests cannot be correlated sensibly with the strengths of concretes made using these aggregates. Thus, to evaluate the effects of different lightweight aggregates on concrete strength, it is necessary to do comparisons of otherwise identical mixes.

In other respects, lightweight aggregates can usually be tested and used in the same way as normal weight aggregates. However, it is generally more difficult to produce a homogeneous lightweight concrete mix, due to the greater tendency of the lightweight aggregates to segregate during mixing, handling, and placing of the concrete. The lightest particles will tend to 'float up' towards the surface, interfering with proper finishing. Thus, more attention to detail will be required when producing and placing lightweight aggregate concrete.

High density aggregates

Heavyweight aggregates (also referred to as high density aggregates) are used primarily in concrete for radiation shielding, though they may also occasionally be used in concretes for which a high mass-to-volume ratio is

Table 7.3 Physical properties of some heavyweight aggregates[a]

Material	Chemical composition	Classification[b]	Relative density	Granular bulk density (kg/m^3)
Goethite	$Fe_2O_3 \cdot H_2O$	N	3.5–3.7	2100–2250
Limonite	Impure Fe_2O_3	N	3.4–4.0	2100–2400
Barytes	$BaSO_4$	N	4.0–4.6	2300–2550
Ilmenite	$FeTiO_3$	N	4.3–4.8	2550–2700
Magnetite	Fe_3O_4	N	4.2–5.2	2400–3050
Haematite	Fe_2O_3	N	4.9–5.3	2900–3200
Ferro phosphorus	$Fe_2O_3 \cdot P_2O_3$	N	5.8–6.8	3200–4150
Steel	Fe (scrap iron steel punchings)	S	7.8	3700–4650

Notes
a See ASTM C637 and C638.
b N – naturally occurring; S – synthetic.

desirable (e.g. counterweights). Heavyweight aggregates are, as the name suggests, composed of materials (either natural or synthetic) that have a high relative density; some of these are listed in Table 7.3 (Mindess *et al.*, 2003). (Heavyweight aggregates used specifically for radiation shielding are described in more detail in ASTM C638; their specifications are given in ASTM C637.) Naturally occurring high density aggregates can be used to produce concrete densities of up to about 4000 kg/m^3; beyond that, synthetic aggregates such as steel punchings must be used.

In the design of concrete for *radiation shielding*, two types of radiation must be considered: gamma rays and neutrons. High density materials are good attenuators of gamma rays, and hence the use of heavyweight aggregates is desirable for this purpose. The attenuation of fast neutrons also requires a high density material. However, materials containing hydrogen atoms best absorb moderate and slow neutrons. For ordinary concrete, a hydrogen content of about 0.45 per cent is required for this purpose, corresponding to about 4 per cent by weight of water. However, fully dried concrete will not contain sufficient water, and hence aggregates containing some hydrogen are desirable; boron-containing aggregates are particularly useful for this purpose. Table 7.4 (Mindess *et al.*, 2003) provides a list of aggregates that are recommended for radiation shielding. Figure 7.8 shows a cross section through a concrete cube containing haematite coarse aggregate, but normal density fine aggregate. The matrix was stained a pinkish colour by the oxide dust that adhered to the coarse aggregate particles.

Heavyweight aggregate concretes generally require some special mix design and placement considerations. Heavyweight concrete mixes tend to be harsh, and so a higher fine aggregate content, a lower fineness modulus of the sand and often a higher content of cementitious materials are used to impart sufficient workability and cohesiveness to the mix. Heavyweight

Table 7.4 Aggregates recommended for radiation shielding

Aggregate type	Relative density	Shielding capability
Natural		
Bauxite	~2.0	Fast neutrons (H)[a]
Serpentine	~2.5	Fast neutrons (H)
Goethite	~3.5	Fast neutrons (H)
Limonite	~3.5	Fast neutrons (H)
Borocalcite	~2.5	Neutrons (B)[b]
Colemanite	~2.5	Neutrons (B)
Barytes	~4.2	Gamma rays
Magnetite	~4.5	Gamma rays
Ilmenite	~4.5	Gamma rays
Haematite	~4.5	Gamma rays
Synthetic		
Heavy slags	~5.0	Gamma rays
Ferro phosphorus	~6.0	Gamma rays
Ferrosilicon	~6.7	Gamma rays
Steel punching or shot	~7.5	Gamma rays
Ferroboron	~5.0	Neutrons (B)
Boron carbide	~2.5	Neutrons (B)
Boron frit	~2.5	Neutrons (B)

Notes
a (H) – moderation by hydrogen.
b (B) – moderation by boron.

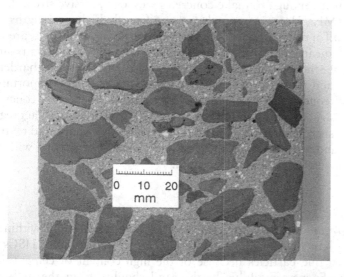

Figure 7.8 Cross section of heavyweight aggregate concrete. Haematite coarse aggregate, normal density fine aggregate. Concrete density approximately 3480 kg/m^3.

mixes also have a tendency to segregate because of the high relative density of the aggregates; to mitigate against this, it is preferable that both the coarse and the fine particles should be of high density. The use of *preplaced aggregate concrete* (see p. 375) can minimize both the segregation and the drying shrinkage of these concretes.

Apart from these qualifications, however, heavyweight concrete mixtures can be proportioned in much the same way as normal concrete mixes (as described in ACI 211.1). The strength properties of these concretes are also similar to those of ordinary concretes, and the familiar w/c (or w/cm) versus strength relationships can be used. It should be noted, however, that some heavyweight aggregates, such as limonite, have relatively low strengths. Also, some high density aggregates may not be very durable when exposed to the environment or to abrasive forces.

As a final note, when iron or steel punchings are used as aggregate, it is recommended that they be allowed to rust somewhat before use (Popovics, 1979), in order to increase the cement/aggregate bond strength.

Aggregate requirements for special concretes

High strength concrete

For normal strength concretes, aggregate properties are not often considered explicitly, except for the usual cleanliness, grading and other requirements needed for mix design. Provided good quality ingredients are available, it is straightforward enough to make concretes with compressive strengths up to about 100 MPa. However, to achieve significantly higher strengths, the properties of *all* of the components of the concrete mixture must be carefully considered: cement, aggregate, mineral fillers, and chemical admixtures. In addition, the interfacial transition zone (ITZ) between the hardened cement paste and the aggregate (see Chapter 5) is of major importance. Here, we will concentrate upon the aggregate requirements for concretes with compressive strengths greater than 100 MPa. The critical aggregate properties for such concretes are the grading, the particle shape and texture, the strength and stiffness, and the possible chemical reactivity with the cement that may affect the bond strength.

Grading

The *fine aggregate* for high strength concrete (HSC) should fall within the same grading limits used for ordinary concretes. However, since HSCs are usually rich in fine particles because of the high content of cementitious materials, the fineness modulus of the sand should be on the high side (2.7–3.0, or even higher), corresponding to coarser sands (Aïtcin, 1998). This has the advantage of slightly reducing the water requirement for a given workability, thus leading to somewhat higher strengths. In order to provide a rougher surface and thus better bond with the cement, it is often recommended that crushed sands rather than natural sands be used. In areas in which natural sands, which tend to have smoother surfaces, are normally used, it has been suggested that about half of this sand should be replaced with crushed sand. However, this is rarely done in practice, largely because of practical considerations in a concrete batch plant, such as the necessity then of maintaining an extra storage silo for a second aggregate source.

The *coarse aggregate* should also adhere to the normal grading requirements. Again, crushed rock is generally preferred because of its better bond with the cement and more angular shape, promoting higher strengths. It is commonly assumed (or asserted) that the maximum size of coarse aggregate (MSA) should be relatively small, in the range of 10–15 mm. However, this cannot be taken as a general guideline. The optimum MSA will depend upon the initial size of the rocks, the extent of damage to the rock during blasting and crushing, the type of crusher used, and the shape and texture of the resulting aggregate particles. While North American practice is mostly to use an MSA of 10–14 mm, work carried out in South Africa has shown the optimal MSA to be in the range of 19–26.5 mm (Addis, 1991, 2001; Addis and Alexander, 1992). There is some evidence, however, that larger aggregate particles tend to be somewhat weaker than smaller aggregate particles of the same rock type (Aïtcin, 1988). Larger particles may also lead to a larger ITZ.

Nature and strength of the rock

For ordinary concretes, the aggregate strength is almost never a consideration, since it is assumed (correctly) that almost all common, normal weight aggregates are stronger than the cement paste matrix. However, for HSCs, the aggregate itself may become the strength-limiting factor, though, at least for compressive strengths up to about 70 MPa, it has been found that the aggregate strength is not the strength controlling factor (Cement and Concrete Association of New Zealand, 1991). Fortunately, however, a number of different aggregate types have been found to be suitable for HSC (limestone, granite, diabase, quartzite, etc.), and so at least in principle suitable aggregates should be fairly widespread in the world. The mechanical properties of some of these are given in Table 7.5. (A much more extensive table of aggregate properties is provided in Table 3.2.) While there is no clear correlation between aggregate strength and concrete strength, particularly when silica fume is used in the mix (Aulia and Deutschmann, 1999), soft or weathered rock should be avoided. In some areas of North America, dolomitic rocks with high compressive strength but relatively low elastic modulus have been found to be best suited for HSC. In general, the elastic modulus of the aggregate will have a greater effect on the elastic modulus of the concrete than the aggregate strength does on the concrete strength.

Table 7.5 Strength and elastic modulus of some common aggregates

Aggregate	Compressive strength (MPa)	Elastic modulus (GPa)
Granite	35–450	15–60
Limestone	90–270	10–80
Sandstone	35–240	5–50
Quartzite	110–470	~80
Marble	50–240	20–65
Gneiss	95–235	25–70
Schist	90–290	~50

Sources: Adapted from Popovics, 1979; Barksdale, 1991.

To somewhat reduce the brittleness inherent in HSC, less brittle aggregates should be used (Wu *et al.*, 2001). In practice, this means that the strength of the aggregate should be less than twice the concrete strength, and the elastic modulus of the aggregate less than 1.5 times the concrete elastic modulus (Aulia and Deutschmann, 1999).

River-run (*fluvial*) gravels generally consist of rounded and smooth siliceous particles; they tend to be strong because the weaker constituents will have been ground away. Aggregates derived from crushing operations are sometimes weaker because of the microcracking induced during the crushing process; this depends, of course, on the type of rock and the type of crusher. Thus, it has been found that for concretes with compressive strengths above about 120 MPa, many crushed rocks appear to be the weakest component in the system, with failure initiated in the aggregate itself. In the Seattle area of the United States, glacial gravels have been used successfully to produce 120 MPa field concretes. These seemed to combine the qualities of fluvial gravels (rounded and strong) and crushed rock (rough surfaces).

The major disadvantage of some fluvial gravels is that they may be too smooth. This surface smoothness results in the formation of a layer of water due to surface tension, leading in turn to the formation of a relatively weak ITZ (Chapter 5). This can result in poor bonding between the cement and the aggregate. Using fluvial gravels, even with silica fume additions, it is difficult to achieve compressive strengths beyond about 100 MPa. Similarly, crushed sand rather than natural sand seems to provide better sand/cement interlocking and somewhat higher strengths, though they require higher superplasticizer dosages to maintain workability (Donza *et al.*, 2002).

Ultra high strength concrete

Using conventional (albeit high quality) aggregates and normal mixture design procedures, followed by the careful application of standard mixing, handling, and placing practices, the maximum compressive strength that can currently be achieved is about 150 MPa. It is, however, possible to produce ultra HSCs, with compressive strengths of over 600 MPa and flexural strengths of up to 100 MPa, by using highly specialized mixtures. Such mixtures might contain, in addition to portland cement and water, some combination of very fine sand, silica fume, ground quartz, precipitated silica, superplasticizers, other admixtures, and steel or synthetic fibres. In some cases, steam curing of the concrete may also be necessary to achieve the best results. Here, we will look only at the aggregate requirements for such concretes. It is beyond the scope of this book to examine the details of their mix design and production processes which are highly specialized.

DSP (Densified with small particles)

One of the first ultra HSCs to be developed, and probably the first one to be commercialized, was that by Bache (1981, 1989) at the Aalborg Cement Company in Denmark; it is marketed under the trade name DENSIT®. To produce DSP, silica fume and superplasticizers are used to reduce the porosity of the material by greatly improving its particle packing characteristics. Bache also noted that for these very high strength materials, it was the coarse aggregate that was the 'weak link' in the system; the smaller the MSA, the higher the strength that could be achieved; for his highest strength materials (compressive strength ≈250 MPa), the MSA was only 4 mm. DSP also required particularly strong aggregates: granite, diabase, and calcined bauxite were used.

RPC (Reactive powder concrete)

The remarkably high strength properties of RPC are based in large part on the very careful control of the particle size distribution of *all* of the solid ingredients in the mix. Most notably, these materials contain no coarse aggregate at all. The RPC first described by Richard and Cheyrezy (1994) had as its 'aggregate' extremely fine sand, with particles in the size range of 150–300 μm. In some cases, to achieve the highest strengths, ground quartz with a particle size of about 4 μm was added. Silica fume and the portland cement itself made up the rest of the solid fine material. The elimination of the coarse aggregate particles made possible the production of a more homogeneous material. Optimization of the particle size distribution of all of the solid materials combined led to a mix approaching optimum density. (Of course, such particle packing also requires large additions of compatible superplasticizers to make the concrete mixtures workable in the plastic state.)

A material of this type has recently been commercialized under the name of DUCTAL®*. This material contains no coarse aggregate; the 'aggregate' is a combination of fine sand, with an MSA of less than 2 mm, supplemented

* A joint venture between Bouygues Construction, Lafarge, and Rhodia.

by crushed quartz. When made with steel fibres, it has compressive strengths in the range of 150–180 MPa, and flexural strengths of about 32 MPa. These values are reduced by about 25 per cent when polypropylene fibres are used instead of steel. To improve the durability and dimensional stability, it is preferable to heat treat these materials with steam at about 90 °C, though they may also be cured at ambient temperature conditions. They are now used primarily in precast operations, but have not yet been adapted for ready-mix or cast-in-place construction.

A somewhat similar material, BSI®-CERACEM concrete was used to construct the toll gate roofs for the new Millau viaduct in the south of France (Thibaux et al., 2004). Another version of this technology, also developed in France, has been patented under the name of CEMTEC$_{multiscale}$® (Parent and Rossi, 2004). It is characterized by much higher cement and fibre contents than the materials just referred to, though the underlying principles are the same. This material can achieve flexural strengths of about 60 MPa. It also has an extremely low permeability. For comparison, typical mix proportions for two of these materials are given in Table 7.6.

There are other ultra HSCs that have either been produced or are under development. Their common features are a very low w/cm ratio, the use of silica fume and superplasticizers, high contents of fibres, limitations on the maximum aggregate size, and careful control of the particle size distribution so as to approach maximum packing densities. They also require very tight quality control both in their production and in their placement. Consequently, these materials are very expensive, but they are finding a market as architects and engineers are beginning to take advantage of the possibilities that these materials offer.

Roller compacted concrete (RCC)

Roller compacted concrete ('rollcrete') is a *no-slump concrete* mixture that is used for both mass concrete and concrete pavements. For mass concrete, the

Table 7.6 Compositions of some commercial RPCs

Material	DUCTAL® (kg/m³)	CEMTEC$_{multiscale}$® (kg/m³)
Portland cement	710	1050
Silica fume	230	268
Crushed quartz	210	–
Sand	1020	514
Water	140	180
Fibres	40–160[a]	858[b]
Superplasticizer	10	44

Notes
a Either steel or polypropylene fibres (13 mm × 0.20 mm).
b A mixture of three different geometries of steel fibres.

RCC is placed using rock fill and earthmoving equipment; for pavements, heavy-duty pavers with tamping and vibrating screeds made especially for RCC construction are used. Sometimes, conventional asphalt plants and asphalt paving equipment may be used.

For mass concrete, the maximum aggregate size is generally limited to 75 mm (Saucier, 1994), as using larger size aggregate can lead to difficulties in batching and to segregation. There is a move, at least in the United States, to go to a 50-mm maximum aggregate size, to reduce further the likelihood of segregation. The use of crushed stone rather than natural aggregates also reduces the amount of segregation. While going to 150 mm can reduce cement requirements by about 15 per cent, the batching and placing complications may not lead to any overall cost savings. In any case, the lift thickness should be at least three times the maximum aggregate size. For pavements, the maximum aggregate size is usually either 19 or 9.5 mm, though this will generally lead to increases in the contents of sand, cement and water (Perrie, 2001).

The grading limits recommended for RCC are essentially the same as those used for ordinary concretes. However, the suggested 3–8 per cent passing the 75-μm sieve is higher than would normally be used for ordinary concretes; the extra fines are needed for RCC to improve the workability and cohesiveness and fill the voids. Most good RCC mixes contain a total content of fine material (including cement, aggregate fines, pozzolanic material, and mineral fillers) of 8–12 per cent by volume or 12–16 per cent by mass, or even higher. For instance, the aggregate gradings shown in Table 7.7 (Schrader, 1997) for various United States projects have total fines in the range of 20 per cent. Otherwise, the general quality of the aggregate is not particularly different from that required for ordinary concretes. Table 7.7 also shows typical ranges of aggregate grading used for several rollcrete dams in South Africa. In comparison with the USA projects, the South African gradings may contain higher proportions of finer materials.

However, because of the relatively higher aggregate content of RCC compared to plain concrete, aggregate quality can have a considerable effect on concrete strength, due to the effect of aggregate on water requirements. While some projects have successfully used marginal aggregates because of their proximity to the site, this is not generally recommended.

Self-compacting concrete (SCC)

Self-compacting concrete is concrete that can flow purely under gravity (i.e. without vibration) to fill forms of virtually any geometry, and to become fully compacted without segregation or bleeding. SCC will typically have free slump values in the range of 250–275 mm. This fluidity is achieved by severely limiting the coarse aggregate content, and by increasing the amount of fine material without changing the water content (i.e. not increasing the

374   Special aggregates and special concretes

Table 7.7 Combined aggregate gradings percentage passing for RCC from various projects in USA and South Africa

Sieve size	Willow Creek	Upper Stillwater	Christian Siegrist	Zintel Canyon	Stage-coach	Elk Creek	Typical SA dams
4 in. (100 mm)	–	–	–	–	–	–	–
3 in. (75 mm)	100	–	–	–	–	100	100
2.5 in. (62 mm)	–	–	–	100	–	96	–
2 in. (50 mm)	90	100	–	98	100	86	–
1.5 in. (37.5 mm)	80	95	100	91	95	76	70–90
1 in. (25 mm)	62	–	99	77	82	64	–
0.75 in. (19 mm)	54	66	91	70	69	58	55–65
3/8 in. (9.5 mm)	42	45	60	50	52	51	40–55
No. 4 (4.75 mm)	30	35	49	39	40	41	35–50
No. 8 (2.36 mm)	23	26	38	25	32	34	30–45
No. 16 (1.18 mm)	17	21	23	18	25	31	25–40
No. 30 (0.60 mm)	13	17	14	15	15	21	17–32
No. 50 (0.30 mm)	9	10	10	12	10	15	10–25
No. 100 (0.15 mm)	7	2	6	11	8	10	3–10
No. 200 (0.075 mm)	5	0	5	9	5	7	1–5
Cement + Pozzolan kg/m^3	48+19	80+173	59+42	74+0	71+77	70+33	
Total fines*	20%	21%	19%	21%	–	21%	
Workability	Poor	Excellent	Excellent	Excellent	Good	Excellent	

Note
* Total fines = all materials in full mixture with particle size smaller than No. 200 sieve.

water/binder ratio), which can be achieved by the use of superplasticizers. The maximum aggregate size is limited to 19 mm. Typical mix proportions are (EFNARC, 2002):

- Water/powder ratio (by volume) of 0.8–1.1
- Total powder content of 400–600 kg/m^3
- Coarse aggregate content of 28–35 per cent of the mix volume
- Water content less than 200 kg/m^3
- Fine aggregate content to make up the rest of the volume.

The fine material added may be any combination of stone powder (e.g. finely crushed limestone or granite <0.125 mm), silica fume (which improves durability), fly ash or ground granulated blast furnace slag (which improves workability), or ground glass (<0.1 mm). Of course, the exact proportions of these materials must be chosen to satisfy the appropriate mechanical properties of the hardened concrete.

Self-compacting concrete must be highly deformable so that it flows easily, and yet viscous enough so that the coarse aggregate does not segregate, but instead 'floats' in the mortar. To achieve this, the total content of solid particles passing the 150-μm sieve should preferably be in the range of 520–560 kg/m³.

Preplaced aggregate concrete

Preplaced aggregate concrete refers to a method of placing concrete in which an appropriately graded coarse aggregate is first packed into the forms, and is then injected with a structural mortar (or grout) to fill the voids. This method is now widely accepted for underwater concreting, and for some types of structural repair. It is particularly suitable for the high density concretes used for radiation shielding; where steel punchings or other particularly heavy aggregates are used, this method minimizes the segregation that would otherwise take place in a plastic mix. Because the aggregates can, in principle, be packed more densely than in ordinary concrete, less cement paste should be needed. Also, because the coarse aggregate particles are in direct contact with each other, the material exhibits significantly less drying shrinkage than ordinary concrete. A schematic illustrating this concept is shown in Figure 7.9.

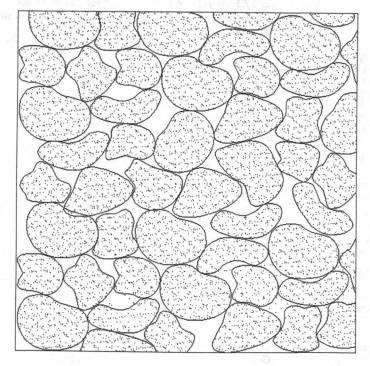

Figure 7.9 Schematic of preplaced aggregate concrete, showing aggregate particles touching at numerous places.

The grading of the *fine aggregate* in preplaced aggregate concrete is more important than that of the coarse aggregate. If the material is too coarse, it may block the void channels in the preplaced coarse aggregate, and prevent proper penetration of the grout. If it is too fine, it will increase the water requirement for fluidity, and thus reduce the strength of the concrete. Generally, the fineness modulus of the fine aggregate is in the range of 1.2–2.0, with most of the material passing the 1.25-mm sieve (Kosmatka *et al.*, 2002). Suitable gradings for both fine and coarse aggregates are given in Table 7.8. Though this table is taken from ASTM C637, *Standard Specification for Aggregates for Radiation-Shielding Concrete*, it is suitable for normal weight concretes as well.

The *coarse aggregate* requirements for grading are less stringent. However, it is essential that the aggregates be clean (free of very fine surface

Table 7.8 Grading requirements for coarse and fine aggregates for preplaced aggregate concrete

Percentage passing

Sieve size	Grading 1 For 37.5 mm ($1^1/_2$ in.) nominal maximum size aggregate	Grading 2 For 25 mm (1 in.) nominal maximum size aggregate
Coarse aggregate		
50 mm (2 in.)	100	–
37.5 mm ($1^1/_2$ in.)	95–100	100
25.0 mm (1 in.)	40–80	95–100
19.0 mm ($^3/_4$ in.)	20–45	40–80
12.5 mm ($^1/_2$ in.)	0–10	0–15
9.5 mm ($^3/_8$ in.)	0–2	0–2
Fine aggregate		
2.36 mm (No. 8)	100	–
1.18 mm (No. 16)	95–100	100
600 μm (No. 30)	55–80	75–95
300 μm (No. 50)	30–55	45–65
150 μm (No. 100)	10–30	20–40
75 μm (No. 200)	0–10	0–10
Fineness modulus	1.30–2.10	1.00–1.60

Grading of aggregate

Relative density of fine aggregate	Coarse aggregate	Fine aggregate
Up to 3.0	Grading 1	Grading 1
Greater than 3.0	Grading 1	Grading 2
Full range	Grading 2	Grading 2

Source: After ASTM C637.

coatings) so that the grout will bond well to the aggregate. They should also be saturated, so that they do not absorb water from the grout and thus reduce its flowability. Of course, the aggregate should also be tough enough, or strong enough, that it does not fracture or break down while it is being placed in the forms, as this will inhibit the grout penetration. In other respects, however, the aggregate should simply meet the same requirements as for ordinary concretes.

Marginal aggregates

As stated previously, it is possible to produce concrete from a very wide range of aggregates, including those that fall well outside the normal ASTM, CSA, BS, SANS or other national specifications in terms of grading or other properties. This will almost always result in lower quality concrete and/or increased costs. However, there may be times when it is necessary to use 'marginal' (or 'borderline') aggregates because of their proximity, or because they may be the only aggregates available in a particular region. Marginal aggregates may be defined as those that are deficient in one or several respects in terms of the applicable specifications. Alternatively, they may be aggregates already known to lead to premature deterioration of concrete. In either case, they may be quite serviceable once their deficiencies are understood, and matched against the particular job requirements. Often, the aggregate may be beneficiated in some way so that neither the strength nor the durability of the resulting concrete are compromised. It must be remembered that marginal aggregates are likely to be suitable at least for some concrete applications. For instance, a weak or porous aggregate may be suitable for relatively low strength concretes, but not for HSC; aggregates that are susceptible to chemical or physical attack may be satisfactory for concrete not exposed to weathering. In any event, a thorough petrographic examination is essential for a proper evaluation of the aggregate being considered. This is a necessary step in assessing the suitability of an aggregate for any particular application. Thus, when contemplating the use of aggregates that are classed as marginal, there are four principal considerations:

1 The desired concrete properties
2 The deficiencies of the aggregates
3 The means of beneficiating the aggregates
4 The use of protective measures for the concrete.

Beneficiation of marginal aggregates

Once the deficiencies of the particular aggregate source are identified, and the desired concrete properties are defined, there are a number of remedies that may be applied in order to make possible the use of these aggregates.

Table 7.9 Beneficiation of aggregates

Treatment	Removal
Crushing	Friable particles
Heavy-media separation	Lightweight particles
Reverse air or water flow	Lightweight particles
Hydraulic jigging	Lightweight particles
Elastic fractionation	Lightweight particles
Washing and scrubbing	Surface coatings, finely divided materials, organics
Selective quarrying, crushing, and blending	Control or removal of deleterious components

(It should be noted that the remedies discussed below are largely the same as those used for even more unsatisfactory aggregates.) Some of the ways in which marginal aggregates may be treated are listed in Table 7.9.

The simplest remedy is, if possible, to *blend* the marginal aggregate with a more suitable one. For example, a fine aggregate with a poor particle size distribution may be blended with a different sand in order to improve the grading; weaker aggregates may be blended with a stronger aggregate; in the case of alkali–aggregate problems, the use of fly ash and/or cement with a lower alkali content may be used; the judicious use of admixtures may help to overcome workability problems; crushing of the coarse aggregate to a finer size may eliminate freeze–thaw problems; and so on.

If this relatively straightforward type of remediation is not possible, then rather more 'aggressive' (and costly) steps may be taken. One common procedure is to use *selective quarrying*. This may involve not using ('stripping') the upper layers of the pit, where the aggregate may be badly weathered or heavily contaminated with organic materials. In other cases, certain strata lower down in the pit may be eliminated.

Once the aggregate has been quarried, there are other treatments that may be applied. *Crushing* (and sometimes *re-crushing*) may eliminate soft, friable particles, and may also improve the particle shape. *Reverse water or air flow* can be used to remove wood or other lightweight materials. *Heavy media separation* can also be used to separate out soft, light particles, which float to the top and are then skimmed off. By using a liquid with a higher specific gravity, this method may be particularly suitable for improving heavyweight aggregate sources. *Elastic fractionation* involves dropping the aggregate on an inclined steel plate; the rebound of each particle depends on its elastic modulus. Hard, dense particles will rebound further than soft, porous particles with a lower elastic modulus, and the appropriate placement of collection bins can provide effective separation. This method is well suited to remove materials such as wood, coal, clay lumps, chert, sandstones, chalk, and the like (see also Chapter 2 on these issues).

Aggregates derived from industrial waste materials

As a society, we produce huge volumes of industrial waste materials. At the same time, the concrete industry is using more and more natural aggregates, and this resource is not only becoming depleted, but also having a considerable adverse environmental impact. There is thus increasing interest in using some of these waste materials as concrete aggregates (Grieve, 2001). As shown in Table 7.10, in the United Kingdom alone, about 150 million tons of waste materials that might make suitable aggregates at least for some applications are produced annually. Here we will look at some of these types of materials.

Recycled concrete aggregate

Recycled concrete may be defined as concrete reclaimed from the demolition of old structures or pavements that has been processed to produce aggregates suitable for use in new concrete. The processing (as with many natural aggregates) generally involves: crushing; removal of contaminant materials such as reinforcing steel, remnants of formwork, gypsum board, and other foreign materials; grading; and washing. The resulting *coarse* aggregate is then suitable for use in concrete. The *fine* aggregate, however, generally contains a considerable amount of old cement paste and mortar. This tends to increase the drying shrinkage and creep properties of the new concrete, as well as leading to problems with mix stability and strength. Therefore, a RILEM report (Hansen, 1994) recommends that any material smaller than 2 mm (passing the No. 8 sieve) should be discarded. BS 8500-2: 2002 (British Standards Association, 2002) goes even further, and does not permit the use of any fine recycled

Table 7.10 Potential quantities of material available for use as aggregate in concrete in the UK

Material	Arisings (per annum)	Uses in concrete
Recycled aggregates	109 million tons	As coarse aggregate Strength classes 10–80 MPa
Glass	>2 million tons	As fine aggregate Strength classes 10–50 MPa Improved freeze/thaw and abrasion resistance
Incinerator ash	1 million tons	As aggregate. Finer fraction has potential pozzolanic properties Strength classes 5–35 MPa
Granulated rubber	>40 million used tyres	Specialist concretes for improved freeze/thaw resistance, thermal insulation, or impact resistance

Source: After Dhir et al., 2004.

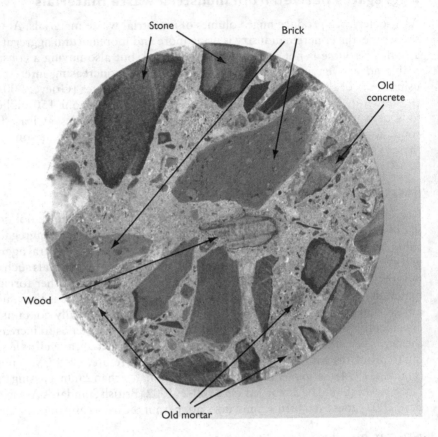

Figure 7.10 Cross section through 68-mm core containing recycled concrete and rubble as coarse aggregate.

concrete in new concrete mixes. Figure 7.10 shows a cross section through a core made with recycled concrete as coarse aggregate. In this case, there are also some contaminants present such as brick and wood particles, since the recycled aggregate is derived from old building rubble.

Inevitably, even properly processed recycled concrete aggregate will have some contaminant materials. While the ASTM standards do not have requirements that are specific to recycled concrete, BS 8500-2: 2002 does have such compositional requirements, as shown in Table 7.11.

In general, the strength of the source of the recycled aggregate has relatively little effect on the strength of the resulting concrete. While the compressive strength of recycled aggregate concrete may be as much as 8 MPa lower than that of natural aggregate concrete at the same w/c ratio, this difference can easily be made up by a small reduction in w/c ratio; other engineering properties are not much affected, as shown in Tables 7.12

Table 7.11 Compositional requirements for coarse recycled concrete aggregate (RCA) and recycled aggregate (RA) in BS 8500-2: 2002

Requirement	RCA	RA
Grading	Same as natural aggregate to BS EN 126020: 2000	
Maximum content masonry (%m/m)	5	100
Maximum fines (%m/m)	5	3
Maximum lightweight material* (%m/m)	0.5	1.0
Maximum asphalt (%m/m)	5.0	10.0
Maximum other foreign materials (%m/m) (e.g. glass, plastics, metals)	1.0	1.0
Maximum acid-soluble sulphates SO_3 (m/m)	1.0	1.0

Note
* Density $<1000\,kg/m^3$.

Table 7.12 Engineering and durability properties of natural aggregate (NA) and recycled concrete aggregate (RCA) concrete

Concrete type				
Property	NA	RCA (% coarse)		
		30	100	100 (equal strength)
Engineering property				
28-day cube strength (MPa)	41.5	40.5	37.0	41.0
Flexural strength (MPa)	4.9	4.8	4.6	4.9
Elastic modulus (GPa)	27.5	28.0	25.5	27.0
Shrinkage ($\mu\varepsilon$) (90 days)	565	570	630	639
Durability property ISA* (ml/m^2/s $\times$ 10^{-2})	29	31	47	35
Air permeability (m$^2 \times$ 10^{-17})	2.7	3.8	143	6.6
Carbonation depth (mm)	13.5	13.5	16.5	12.5
Abrasion depth (mm)	0.61	0.65	1.02	0.72
Freeze/thaw durability factor (%)	97	98	96	97
Coefficient of chloride diffusion (cm^2/s $\times$ 10^{-6})	1.16	1.17	–	1.05

Source: After Dhir et al., 2004.

Note
* ISA = Initial surface absorption.

Table 7.13 Comparison of engineering and durability properties of natural aggregate (NA) and recycled concrete aggregate (RCA) high strength concrete (cube strength = 60 MPa)

Concrete type				
Property	NA	RCA (% coarse)		
		30	50	100
Water/cement ratio	0.36	0.36	0.34	0.32
Engineering property Flexural strength (MPa)	7.0	6.9	7.0	7.2
Elastic modulus (GPa)	30.0	30.0	31.0	30.5
Shrinkage ($\times 10^{-6}$) (90 days)	769	781	792	818
Durability property ISA* ($ml/m^2/s \times 10^{-2}$)	28.0	29.0	33.0	36.7
Air permeability ($m^2 \times 10^{-17}$)	1.1	1.2	4.0	7.7
Coefficient of chloride diffusion ($m^2/s \times 10^{-6}$)	48.1	49.1	46.3	41.4

Note
* ISA = Initial surface absorption.

and 7.13. Indeed, more recent work (Levy and Helene, 2004) has shown that concretes made with recycled aggregates (both coarse and fine) can have the same workability and compressive strengths as concretes made with natural aggregates in the range of 20–40 MPa 28-day strengths. A 20 per cent replacement of natural aggregates with recycled aggregates presented the same, and sometimes better, performance than the reference concrete in terms not only of strength, but also of water absorption, total pore volume, and carbonation. In general, however, it has been found that absorption is higher than for ordinary concretes, while relative density is lower.

Of course, some caution must be exercised when contemplating the use of recycled concrete, particularly if the original source of the material is unknown. A thorough petrographic examination would reveal any susceptibility to alkali–aggregate reactions or other durability problems that might be caused by the contaminants in the aggregate.

Other industrial wastes

The other waste materials mentioned in Table 7.10 are much less commonly used in concrete, though their use may increase in the future as a result of some combination of economic and environmental pressures.

Waste glass cannot be used to replace coarse aggregate, since the glass particles are brittle and flaky, and are likely to break down during the mixing process. However, it may be used as fine aggregate in some circumstances (Jozsa and Nemes, 2004). It has been used in precast concrete to produce a 'glittery' surface. However, it tends to lead to lower compressive strengths, and may be particularly susceptible to alkali–aggregate reactivity when used with high alkali cements.

Granulated rubber has been used in low strength flowable concretes for fill applications. It leads to concretes with more flexibility and better thermal insulating properties. However, it is unlikely to be suitable for ordinary structural concrete.

Incinerator bottom ash is one of the by-products of the incineration of municipal wastes. However, its use as aggregate leads to substantially lower strengths than with natural aggregates, and there have also been problems with the volume stability of such materials. Also, it has been found to be unsuitable for use in reinforced or prestressed concrete, because it tends to have high chloride contents. Thus, its practical use is severely limited.

Porous aggregates for internal curing of concrete

High strength concretes, which are characterized by a low w/c ratio, are very sensitive to cracking at early ages, due to self-desiccation of the cement paste and the associated autogenous shrinkage. One novel way of overcoming this problem is to use low density (porous) fine aggregates, such as pumice or absorbent polymer particles, to replace all or part of the conventional fine aggregate (Jensen and Lura, 2003). If these particles are saturated with water before being incorporated into the concrete, they do not affect the initial w/c ratio, but provide reservoirs of water that is released as the cement hydrates and 'uses up' the initial mix water, thus minimizing the self-desiccation of the cement. This technique, first suggested in the early 1990s, is now beginning to gain acceptance in practice. (Of course, it must be emphasized that this internal curing *cannot* replace conventional external curing; the two techniques must be used together to solve the problem of autogenous shrinkage.)

The lightweight aggregate itself may be a naturally occurring material such as pumice, but is more likely to be an expanded clay or shale (Roberts, 2004), or sintered fly ash. To be suitable for this purpose, the fine aggregate must have a pore structure that makes it relatively easy for water to be absorbed, and then released. (For instance, Lura *et al.* [2004] described work with a pumice aggregate that could only be saturated by immersion in boiling water or by vacuum saturation, neither technique being particularly practical.) As well, it should have approximately the same gradation as the natural sand that it is replacing, and should not have adverse affects on workability because of the particle shape or texture.

Concretes produced using this method of internal curing have been found to have improved compressive strengths compared to normally cured concretes, particularly at early ages. They have also displayed a higher degree of cement hydration, and a significant reduction in shrinkage. It should be pointed out that this technique is really suitable only for relatively low w/c ratio concretes ($w/c < 0.40$). Concretes with higher initial w/c ratios already have sufficient water for hydration, and in any event are permeable enough at early ages for the water supplied by conventional curing to penetrate into the concrete.

Aggregates for refractory concrete

The performance of castable refractory (high temperature) concretes which are designed to be used in furnace or kiln applications is greatly influenced by the aggregates used. In particular, as the aggregates are heated and cooled they undergo volume changes; if the thermal coefficient of expansion of the aggregate is much different from that of the cement paste, high thermal stresses may be generated which will lead to cracking. Also, the aggregates themselves must be chemically and physically stable at high temperatures. A low thermal conductivity to slow down the temperature rise in the concrete may also be helpful.

The types of refractory concrete are shown in Table 7.14 (Mindess et al., 2003). It may be seen that naturally occurring aggregates are suitable only at temperatures below about 1000 °C. Carbonate aggregates break down chemically at temperatures less than 1000 °C, releasing CO_2. Siliceous aggregates remain chemically stable at higher temperatures, but the large volume changes that they undergo lead to a significant thermal

Table 7.14 Types of refractory concrete

Type	Service temperature limit (°C)*	Cement	Aggregate
Structural	300	Portland	Common rocks
Heat resisting	1000	High alumina	Fine-grained basic igneous rocks (basalt, dolerite); heat-treated porous aggregates (bricks, scoria, expanded shale)
Ordinary refractory	1350	High alumina	Firebrick
Super-duty refractory	1450	Calcium aluminate	Firebrick
	1350–1600	High alumina	Refractory (magnesite, silicon carbide, bauxite, fired clay)
	–	Calcium aluminate	Refractory

Note
* $(°C \times \frac{9}{5}) + 32 = °F$.

Table 7.15 Some high temperature aggregates

Aggregate	Temperature limit (°C)
Calcined diatomite; Vermiculite	1000
Expanded clay or shale	1000–1200
Porous fireclay	1350
Fired kaolin (molochite)	1500
Fired kyanite (sillimanite)	1550
Calcined bauxite	
High-alumina firebrick chrome	1650
Chrome-magnesite	
Carborundum (silicon carbide)	1700
Corundum (fused bauxite or fused alumina)	
Brown	1700
White	1800

incompatibility with the cement paste. Thus, beyond 1000 °C, the various synthetic aggregates listed in Table 7.15 must be used with, of course, the appropriate cement. At high temperatures, some of these aggregates are able to form ceramic bonds with the cement, further enhancing the performance of the refractory concrete.

Abrasion-resistant aggregates

Aggregates play an important role in determining the resistance of concrete to surface abrasion and wear. The abrasion resistance of aggregates themselves has been discussed in Chapter 3. However, the tests described there, such as the Los Angeles abrasion test and the Micro-Deval test do not necessarily correlate well with concrete behaviour in the field (Liu, 1981). Relating test data to field performance is further complicated by the fact that there are three different types of wear to be considered:

1 *Abrasion.* This refers to wear by rubbing or friction (*attrition*), such as is caused by traffic on pavements or industrial floors.
2 *Erosion.* This is due to the abrasive action of solid particles suspended in fluids, such as occurs in canals, piers, or spillways.
3 *Cavitation.* This is caused by turbulent high-velocity liquid flow, in particular at spillways and sluiceways in dams and irrigation installations.

For all of these different types of attack, it is clear that the aggregates should be hard, strong, and free of soft or friable particles. As well, a low *w/c* ratio and high compressive strength will improve the concrete performance. However, there are also some different requirements for the three types of wear. For *erosion*, a larger maximum aggregate size seems to be advantageous, while for *cavitation*, a smaller aggregate size (<20 mm) and good cement–aggregate bond provide the best performance.

It is *abrasion* that is most affected by the aggregate properties. For concretes with compressive strengths greater than about 40 MPa, the aggregate properties have relatively little effect on abrasion resistance. However, the coarse aggregate properties become increasingly important as the concrete strength decreases below this value. For instance, the amount of abrasion will increase if the volume of coarse aggregate is reduced, or if very fine sand is used as fine aggregate. For heavy-duty industrial floors, it may be necessary to use especially hard aggregates, such as emery or iron shot in the topping layer.

There is little correlation between the precise mineralogy of an aggregate and the abrasion resistance of concrete made with that aggregate. However, there is some correlation between aggregate composition and *skid resistance*. The gradual polishing of the aggregate due to rubber tyres on pavements may eventually cause a smooth, skid-prone surface to develop. Thus, aggregates that are either more likely to fracture than to polish, or which are made up of minerals with different *rates* of polishing, will be better in this respect. (Crushed aggregates will tend, initially, to provide better skid resistance than rounded aggregates, but this effect will disappear after a period of time in service.) It is often considered that limestone aggregates are more susceptible to polishing than most, and will thus not provide good skid resistance. This is particularly true of limestones with high $CaCO_3$ contents. However, not all limestones are the same, and each aggregate source should be considered on its own merits. Sandstones have shown good anti-skid properties, as have coarse-grained granites and trap rock. Chert and rhyolite, which polish relatively easily, have been shown to have poorer anti-skid properties. It should be noted in this regard, however, that it is the properties of the rubber tyres that have a much larger effect on skid resistance than the properties of the aggregate.

Closure

As may be seen from the above discussion, there are many types of aggregate available to produce 'special' concretes. Many of these are artificial aggregates, such as industrial by-products specifically manufactured for use in concrete. The use of industrial waste materials, and marginal natural aggregates, will continue to increase, due to both environmental and economic pressures. Fortunately, 'ordinary' concrete (compressive strength ~20–30 MPa) is a remarkably forgiving material, and with reasonable care in mix proportioning and the judicious use of both mineral and chemical admixtures, perfectly suitable concretes can be made with aggregates that were once considered to be unsuitable for use in concrete.

References

Addis, B.J. (1991) Properties of high-strength concrete made with South African materials, PhD thesis, University of the Witwatersrand, Johannesburg.

Addis, B.J. (2001) 'High-performance concrete', in B.J. Addis and G. Owens (eds) *Fulton's Concrete Technology*, 8th edn, Midrand: Cement and Concrete Institute, 267–273.

Addis, B.J. and Alexander, M.G. (1992) 'A method of proportioning trial mixes for high strength concrete', *Concrete Beton*, 63: 16–22. (Reprinted from *Proceedings of 2nd Int. Symp. High Strength Concrete*, 1990, ACI SP-121, American Concrete Institute, Farmington Mills, MI 287–308.)

Aïtcin, P.-C. (1988) 'From gigapascals to nanometers', in E. Gartner (ed.), *Engineering Foundation Conference on Advances in Cement Manufacture and Use*, New York: American Society of Civil Engineers, 105–130.

Aïtcin, P.-C. (1998) *High Performance Concrete*, London: E&FN Spon.

American Concrete Institute Committee 211 (1991) 'Standard practice for selecting proportions for normal, heavyweight, and mass concrete', ACI 211.1 (Re-approved 2002), *American Concrete Institute Manual of Concrete Practice*, Farmington Hills, MI: American Concrete Institute.

American Concrete Institute Committee 213 (2003) 'Guide for structural lightweight concrete', ACI 213R-87, *American Concrete Institute Manual of Concrete Practice*, Farmington Hills, MI: American Concrete Institute.

Aulia, T.B. and Deutschmann, K. (1999) 'Effect of mechanical properties of aggregate on the ductility of high performance concrete', *Leipzig Annual Civil Engineering Report*, No. 4, Universitat Leipzig, 133–147.

Bache, H.H. (1981) 'Densified cement/ultrafine particle-based materials', *Proceedings 2nd Int. Conf. Superplasticizers in Concrete*, June, Ottawa, Canada.

Bache, H.H. (1989) 'Fracture mechanics in integrated design of new, ultra-strong materials and structures', in L. Elfgren (ed.) *Fracture Mechanics of Concrete Structures: From Theory to Applications*, RILEM, Technical Committee 90 FMA, London: Chapman & Hall, 382–398.

Barksdale, R.D. (ed.) (1991) *The Aggregate Handbook*, Washington, DC: National Stone Association.

BS EN 8500-2: 2002 (2002) *Concrete – Complementary British Standard to BS EN 206-1: Specification for Constituent Materials and Concrete*, London: British Standards Institution.

Cement and Concrete Association of New Zealand (1991) *The Influence of Aggregates on the Strength Parameters of High Strength Concrete*, GLR 29, December.

Dhir, R., Paine, K., Dyer, T. and Tang, A. (2004) 'Value-added recycling of domestic, industrial and construction waste arising as concrete aggregate', *Concrete Engineering International*, 8(1): 43–48.

Donza, H., Cabrera, O. and Irassar, E.F. (2002) 'High-strength concrete with different fine aggregates', *Cem. Con. Res.*, 32(11): 1755–1761.

EFNARC (2002) *Specification and Guidelines for Self-Compacting Concrete*, Farnham, Surrey: EFNARC.

Grieve, G.R.H. (2001) 'Aggregates for concrete', in B.J. Addis and G. Owens (eds) *Fulton's Concrete Technology*, 8th edn, Midrand: Cement and Concrete Institute.

Hansen, T.C. (ed.) (1994) *Recycling of Demolished Concrete and Masonry*, RILEM Report No. 6, London: Chapman and Hall.

Jensen, O.M. and Lura, P. (2003) 'Techniques for internal water curing of concrete', in D.A. Lange, K.L. Scrivener and J. Marchand (eds) *Advances in Cement and Concrete, Proceedings Conf. Copper Mountain*, Colorado, USA. University of Illinois, Urbana-Champaign, 67–78.

Jozsa, Z. and Nemes, R. (2004) 'An innovative material from recycled glass to lightweight concrete', in A. La Tegola and A. Nanni (eds) *Proceedings 1st Int. Con. Innovative Materials and Technologies for Construction and Restoration*, Liguore Editore, Naples, Italy, Vol. 1, 229–240.

Kosmatka, S.H., Kerchoff, B., Panarese, W.C., Macleod, N.F. and McGrath, R.J. (2002) *Design and Control of Concrete Mixtures*, Engineering Bulletin 101, 7th Canadian edn, Ottawa, ON: Cement Association of Canada.

Levy, S.M. and Helene, P. (2004) 'Durability of recycled aggregate concrete: A safe way to sustainable development', *Cem. Conc. Res.*, 34(11): 1975–1980.

Liu, T.C. (1981) 'Abrasion resistance of concrete', *ACI Journal*, 7B(5): 341–350.

Lura, P., Bentz, D.P., Lange, D.A., Koules, K. and Bentur, A. (2004) 'Pumice aggregate for internal water curing', in K. Kovler, J. Marchand, S. Mindess and J. Weiss (eds), *Concrete Science and Engineering: A Tribute to Arnon Bentur*, RILEM Proceedings PRO 36, Bagneux: RILEM Publications s.a.r.l., 137–151.

Mindess, S., Young, J.F. and Darwin, D. (2003) *Concrete*, 2nd edn, Upper Saddle River, NJ: Prentice Hall.

Parent, E. and Rossi, P. (2004) 'A new multi-scale cement composite for civil engineering and building construction fields', *Advances in Concrete through Science and Engineering*, CD-ROM Paper No. 14, Hybrid-Fiber Session, Bagneux: RILEM Publications s.a.r.l.

Perrie, B. (2001) 'Roller-compacted concrete', in B.J. Addis and G. Owens (eds) *Fulton's Concrete Technology*, 8th edn, Midrand: Cement and Concrete Institute, 293–301.

Popovics, S. (1979) *Concrete-Making Materials*, Hemisphere Publishing Corporation.

Richard, P. and Cheyrezy, M.H. (1994) 'Reactive powder concretes with high ductility and 200–800 MPa compressive strength', in P.K. Mehta (ed.) *Concrete Technology Past, Present and Future, Proceedings of V. Mohan Malhotra Symposium*, S.P – 144, Farmington Hills: American Concrete Institute, 507–518.

Roberts, J. (2004) 'The 2004 practice and potential of internal curing of concrete using lightweight sand', *Advances in Concrete through Science and Engineering*, CD-ROM, Bagneux: RILEM Publications s.a.r.l.

Saucier, K.L. (1994) 'Roller compacted concrete (RCC)', *Significance of Tests and Properties of Concrete and Concrete Making Materials*, ASTM STP 169C, West Conshohocken, PA: American Society for Testing and Materials, 567–576.

Schrader, E.K. (1997) 'Roller-compacted concrete', in E.G. Nawy (ed.) *Concrete Construction Engineering Handbook*, Boca Raton: CRC Press, Chapter 18, 1–60.

Thibaux, T., Hajar, Z., Simon, A. and Chanut, S. (2004) 'Construction of an ultra-high-performance fibre-reinforced concrete thin-shell structure over the Millau viaduct toll gates', in M. di Prisco, R. Felcetti and G.A. Plizzari (eds)

Fibre-reinforced Concretes BEFIB 2004, Bagneux: RILEM Publications s.a.r.l., Vol. 2, 1183–1192.

Wu, K., Chen, B., Yao, W. and Zhang, D. (2001) 'Effect of coarse aggregate type on mechanical properties of high-performance concrete', *Cem. Conc. Res.*, 31(10): 1421–1425.

Chapter 8

Standards for aggregates

Concrete construction relies on codes and specifications that aim to ensure basic standards of good practice and minimum requirements for quality. Standard specifications, test methods, and codes of practice play an important role in ensuring that the concrete industry delivers an acceptable product to the market. Aggregates, being generally natural materials, display wide variation in their properties. Despite this, most aggregates can be used to produce acceptable quality concrete although the aggregate properties will be reflected in the concrete properties to some degree.

Standards in the form of specifications and test methods are necessary and value-adding for the following reasons:

- They represent benchmarks for quality and good practice by laying out necessary requirements and characteristics, and approved procedures.
- They help ensure that clients and owners of a structure receive a 'product' of acceptable standard and fitness-for-purpose, and thus protect both clients and specifiers from unacceptable practice.
- They assist in achieving uniformity of delivered product. This will have a beneficial effect on reducing the variability of the concrete produced.
- They permit construction to be carried out nationally and, where appropriate, internationally to the same minimum levels of quality and performance. In an international context, they are also a vehicle for sharing of knowledge and good practice.
- They provide a common basis for tenderers and constructors to bid for work and for engineers to prepare design and contract documents.

Standards also are required when problems are experienced in industry with a product or process. Thus, writing, specifying, and enforcing standards to avoid construction problems is essential, not least to preserve the viability of the concrete and construction industries.

On the other hand, standards can also have negative consequences such as restricting innovation and being over-prescriptive rather than performance-based. They may take a long time to change, and may impose unsuitable limits or test methods, especially when foreign standards are applied in an inappropriate context.

Designers and specifiers use and enforce standards while aggregate producers must comply with the requirements of standards due to contractual obligations. Consequently, the designer or specifier needs to understand the scope and limitations of standards and be able to apply them intelligently, giving alternatives or exceptions as appropriate. An example is the case where an aggregate falls outside standard grading limits, but no other suitable economical source is available. Provided the concrete mix can be designed to achieve the necessary performance requirements, there is no reason to reject the aggregate. Rigid adherence to standards without proper understanding of the context can be detrimental even to good practice!

Care must be exercised when applying national standards in a foreign context. For example, experience with British Standards applied in the Middle East and in parts of Africa has not always been positive (Fookes and Collis, 1975a,b). A mechanically strong aggregate in the dry state may comply with a specification but turn out to be unsound and impair the durability of concrete in other environments. Unnecessarily restrictive requirements for the application of standards or codes are also frequently a problem, particularly where local practice may demand a different approach. Unfortunately, the risk of claims and litigation frequently stifles innovation. When specifications are prepared for areas where local standards are unavailable, it is essential to acquire detailed and up-to-date knowledge about available aggregate resources and their mineralogical, physical, and chemical properties. Thus, a good aggregate specification must be relevant to the available resources, to the characteristics required of the concrete in which the aggregates will be used, and to the environmental exposure of the structure.

Standards for aggregates are of two types. The first is standard specifications for aggregates, and the second is standard test methods. Specifications relate to the required properties and characteristics of aggregates in order to ensure acceptable quality of concrete made with them, in the short and long term. Specifications give limiting values for various properties, and therefore they are inextricably linked with standard tests for aggregates in which procedures are provided to obtain reliable values of the aggregate properties. Both these topics will be dealt with here by referring to relevant publications and by providing a summary review of the important standards, attempting to synthesise them where possible. The aim is to give a general overview that may be useful when dealing with different standards or when wishing to cross-reference between them.

There are several publications which cover standards for aggregates directly or indirectly. ASTM publishes a *Manual of Aggregate and Concrete Testing* (ASTM R0030) which, while not a Standard, is intended to supplement but not to supersede ASTM Standard Test Methods for sampling and testing aggregates and fresh and hardened portland cement concrete. First published in 1965, it is an informative and interpretive treatment of test methods. Since many specifications are based on ASTM Test Methods, the manual stresses the importance of strict adherence to test method requirements in order to obviate inaccurate data and incorrect conclusions being drawn. The manual deals with many of the factors that might affect the results of aggregate tests and covers the full range of tests, including unit weight and voids content, surface moisture, sieve analysis, specific gravity and absorption, evaporable moisture content, and sampling. The manual serves as a useful reference for laboratory staff and could be used to train staff in performing the standard tests competently. Other useful texts were mentioned in Chapter 1.

Standard specifications for aggregates

Most countries possess national standard specification for concrete aggregates as well as standard test methods. In some cases, these may be adopted from other national standards. Standard specifications may contain sections dealing with aggregates for other uses such as bituminous mixtures or unbound materials. The focus here is on concrete aggregates and the review will be restricted to this use.

Table 8.1 reviews several standard specifications for concrete aggregates: USA, ASTM C33 (Standard Specification for Concrete Aggregates); Canada, CSA A23.1 (Concrete Materials and Methods of Concrete Construction); UK, formerly BS 882 (Specification for aggregates from natural sources for concrete), and now BS EN 12620: 2002 (Aggregates for concrete), and South Africa, SANS 1083 (Aggregates from natural sources – Aggregates for concrete). The review covers scope and limitations, grading requirements, deleterious substances, alkali–aggregate reaction (AAR), soundness, and sampling and testing. It is presented in a synoptic fashion, allowing the reader to compare across the standards.

The ASTM document defines the requirements for grading and quality of fine and coarse aggregate. It is intended to cover most requirements for aggregates in concrete but excludes lightweight and heavyweight aggregates. Due to very harsh weathering conditions in large parts of the continental USA, the standard defines classes of coarse aggregates based on weathering severity, abrasion, and other factors of exposure.

The British Standard for aggregates, BS 882: 1992, covers sampling and testing, quality requirements in terms of particle shape, shell content, mechanical properties for different types of concrete, and grading requirements. Crushed and uncrushed aggregates are permitted as well as all-in aggregates which consist of a mixture of coarse aggregates and sand. Coarse aggregates may be used as graded aggregates in several grading ranges or

Table 8.1 Review of national aggregate specifications

Topic	ASTM C33	CSA A23.1	BS 882	SANS 1083	Comments and notes
1 Scope and Limitations	Defines requirements for grading and quality of fine and coarse aggregates for use in concrete. Excludes lightweight and heavyweight aggregates. Adequate to ensure satisfactory materials for most concrete, but may be restrictive in some cases.	Covers requirements for materials for cast-in-place concrete and concrete precast in the field, and by default covers requirements for aggregates in concrete in these applications. Excludes structural low-density or high-density aggregates. Mentions use of recycled concrete as aggregate.	Specifies the quality and grading requirements for aggregates obtained by processing natural materials for use in concrete. Excludes lightweight aggregates (covered in BS 3797). To be replaced by BS EN 12620, the European standard for aggregates.	Specifies requirements for fine and coarse aggregates from natural sources (i.e. those formed by the natural disintegration of rock, or produced by mechanical crushing or milling of rock) for use in concrete. Properties not covered include presence of sulphate or sulphides, shrinkage, soundness, shape and surface texture, and ASR.	Although BS 882 is in the process of being superseded by an EN specification, it reflects UK aggregate usage in concrete.
2 No. of Parts or Clauses	13 Clauses plus Appendix.	Aggregates chapter has 9 sections, plus an Appendix on AAR.	5 Parts and 3 Appendices.	6 Clauses and 4 Annexes.	
3 Grading: Fine aggregates	Grading envelope specified, FM between 2.3 and 3.1.	Grading envelopes specified for two fine aggregate classes: FA1 with FM between 2.3 and 3.1; FA2 with FM between 3.3 and 4.0. Generally, FA2 to be used blended with FA1. Grading uniformity specified.	Grading envelopes specified, both overall and for 3 classes: C (Coarse), M (Medium) and F (Fine). All-in aggregate gradings for maximum sizes of 40–5 mm also specified.	Very wide grading envelope permitted; FM between 1.2 and 3.5; dust content ($<75\ \mu m$) up to 5% for natural sands, and 10% for crushed sands, and even higher on occasions.	ASTM: Fine aggregates outside specified gradings generally allowed provided they have shown an acceptable performance record.

Table 8.1 (Continued)

Topic	ASTM C33	CSA A23.1	BS 882	SANS 1083	Comments and notes
Coarse aggregates	Gradings specified for various 'Size Numbers', each representing a coarse aggregate size range, e.g. 19–9.5 mm.	Gradings specified for 'Group I' aggregates, representing commonly used combined aggregate gradings, or 'Group II' which provides for special requirements such as gap-grading, pumping, or for blending. Particle shape may be specified.	Gradings specified for either graded aggregates (e.g. 20–5 mm), or single-sized aggregates (from 40 mm down to 5 mm). Limits given for fines content. FI <50 – uncrushed gravel, <40 – crushed rock or gravel.	Gradings given for nominally single-sized aggregates between maximum size of 6.7 and 75 mm. Dust content (<75 μm) limited to 2% and flakiness index to 35.	North American practice generally allows for continuously graded aggregates, while South African practice involves use of nominally single-sized coarse aggregates, i.e. gap-grading.
4 Deleterious substances: Fine aggregates	Limits given for: Clay, −75-μm fraction; Coal and lignite; Organic impurities.	Fine aggregate to be free from organic impurities, as determined colorimetrically, or by a relative strength test, and should not entrain air excessively. Limits also specified for fine and coarse aggregates for clay, low density materials, fines, sulphate soundness, and abrasion loss. Deleterious aggregates containing sulphides or sulphates, or free lime or free magnesia, to be avoided.	Limits given for: • chloride content of aggregates (coarse and fine). • shell content – 20% maximum.	Limits for clay content, organic impurities, chloride content, presence of sugar, and soluble deleterious impurities (on the basis of a strength reduction test).	

Coarse aggregates		Limits given for: clay or friable particles, chert, fines <75 μm, coal and lignite, abrasion loss. Depends on severe, moderate, or negligible weathering region.	Caution on use of staining aggregates (e.g. pyrite) for exposed finishes.	No particular requirements except for dust content; resistance to weathering and mechanical strength requirements given in terms of ACV and 10% FACT values.	ASTM: Coarse aggregates failing the requirements are still acceptable if they have a proven service record when exposed to similar weathering conditions.
5 AAR: Fine aggregates	No AAR-susceptible aggregates permitted in wet or humid conditions, unless low alkali cement or cement extenders used.	AAR-susceptible aggregates that will result in excessive expansion not to be used in concrete, without preventive measures. An Appendix provides guidance on evaluating AAR and applying preventive measures. No distinction is drawn between fine or coarse aggregate. Specification recommends petrographic examination when necessary.	No particular requirements for AAR specified. Appendix B draws attention to other BS, Conc. Soc., and BRE documents for guidance.	Specification does not cover requirements to avoid AAR, either for fine or coarse aggregates. However, Annexure C recommends use of petrographic analysis to check on suitability of aggregates.	ASTM: Appendix provides information on methods for evaluating potential reactivity of an aggregate. SANS: No AAR provisions are given, despite the fairly wide incidence of AAR in South Africa.
Coarse aggregates	Same requirements as for fine aggregates.				
6 Soundness: Fine aggregates	Loss in soundness test <10% (sodium sulphate).	Magnesium sulphate soundness loss not to exceed 16% for fine aggregate.	Information on acid soluble sulphate content and carbonate content of aggregates may be requested (the latter requirement relates to Thaumasite sulphate attack).	No provisions given, although Annexure C indicates this may need to be checked.	ASTM: Fine aggregates failing soundness test may be used if they show satisfactory service record under similar conditions of weathering.

Table 8.1 (Continued)

Topic	ASTM C33	CSA A23.1	BS 882	SANS 1083	Comments and notes
Coarse aggregates:	Loss in soundness test <18% (magnesium sulphate) for relevant classes.	Magnesium sulphate soundness loss <12 or 18% for coarse aggregate, depending on exposure classification.			
7 Sampling and testing	Relevant sampling procedures and test methods specified by reference to other ASTM standards (i.e. ASTM D75).	Aggregates to be sampled in accordance with A23.2-1A. Guidance given on frequency of testing.	Sampling and testing of aggregates to be carried out in accordance with appropriate parts of BS 812.	Reference is made to the SANS Standard Method for sampling aggregates (SANS 5827).	

Notes

1 Regarding lightweight aggregates, ASTM C330 (Standard Specification for Lightweight Aggregates for Structural Concrete) covers the necessary requirements. In addition to clauses covering requirements shown in the table above, the standard also includes requirements for staining, loss on ignition and bulk density for the aggregates, and compressive strength, splitting tensile strength, drying shrinkage, pop-outs, and resistance to freezing and thawing for concrete made with the aggregates.

2 Aggregate specifications generally do not contain requirements for shape or surface texture of aggregates. This is because these factors are either inherent to the aggregate, or are controlled during manufacture, and if these factors were unduly adverse, they would result in unacceptable values for water requirement.

3 The Cement and Concrete Institute of South Africa publishes a 'Commentary on SANS 1083'. See main text for discussion.

4 BS 882 has a requirement for mechanical properties in terms of limiting 10% Fines Value (e.g. >150 kN for heavy-duty floors) or limiting Aggregate Impact Value. In addition, there is a requirement for Flakiness Index of coarse aggregate, FI <50 – uncrushed gravel, <40 – crushed gravel.

5 The equivalent European Standard Specification for Aggregate is EN 12620: 2002. This also covers the specification for air-cooled blast furnace slag aggregate for use in concrete. A further European specification is EN 13055-1: 1997, dealing with lightweight aggregates. EN 12620 has requirements for physical properties such as density (particle and bulk), water absorption, flakiness index, shape index, and mechanical properties such as Los Angeles Abrasion Value (or Impact Value), Micro-Deval coefficient (resistance to wear of coarse aggregate), Polished Stone Value, Aggregate Abrasion Value, Nordic Abrasion Value (for resistance to abrasion from studded tyres), and durability properties such as freeze–thaw resistance of coarse aggregates, magnesium sulphate soundness, volume stability (drying shrinkage), alkali–silica reactivity, and chemical properties such as limiting chloride and sulphate or sulphur contents.

as single-sized aggregates, while fine aggregates are divided into Coarse, Medium, or Fine, depending on their grading. Appendices give guidance on deleterious materials in aggregates and assessment of durability properties such as freeze–thaw, ASR, thaumasite attack, and chloride content. At the time of writing (2004) the standard was still current but in the process of being replaced by European Standard BS EN 12620: 2004.

The Canadian specification is wide in scope covering not only aggregates but also cements and supplementary cementing materials, water, admixtures, reinforcement, formwork, and so on. It is a comprehensive document on all aspects of concrete materials as well as methods of concrete construction.

In South Africa, the standard specification for aggregates is complemented by a commentary published by the Cement and Concrete Institute (1995). The commentary covers requirements for properties of aggregates that are not given in the standard specification and which may be important to concrete properties. It therefore acts as a technical guide to the specification to assist in verifying the suitability of an aggregate for a particular application. The commentary consists of a number of short sections, each covering an aggregate property and assisting in the interpretation of the specification. In particular, it contains a lengthy section on alkali–silica reaction (ASR) which is very important in view of the fairly widespread occurrence of ASR in South Africa.

Standard tests for aggregates

In contrast to a standard specification which is generally a single document (even if in several parts), test methods often reside in multiple documents. Previous chapters referred to many of these tests, and there is no point in dealing with them in detail here. Tests are constantly under revision and improvement, and newer tests are incorporated as standards change from time to time. A full review of all available test methods for aggregates is thus not appropriate. Test methods are important in ensuring that unacceptable aggregates are not permitted in construction, and for providing assurance that quality will not be compromised. Problems arise when unsuitable tests are specified or where no latitude is given in interpretation in non-critical cases. In other cases, the development of test methods may still be in preliminary stages and they may be overly conservative or lenient due to a lack of knowledge. Nevertheless, test methods represent a legacy of generally proven technology, which if used properly impart the benefits of many years of development.

Table 8.2 gives a synoptic summary of test methods covering American, British, Canadian, and South African tests, for ease of comparison and cross-referencing. When referring to the methods, the latest versions should be used since standards are in a continuous state of revision. The vast

Table 8.2 Synoptic table of test methods for aggregates

Test methods	USA (ASTM)	UK (BS)	Canada (CSA)	SA (SANS)
AAR/ASR – Potential expansivity of aggregates (procedure for length change in concrete prisms)		812-123: 1999	A23.2-14A	
Abrasion resistance by LA machine	C13 C535		A23.2-16A; A23.2-17A	5846
Absorption of water	C127 C128		A23.2-6A; A23.2-12A	5843
Acid insolubility				6242
Acid-soluble material in fine aggregate		812-119: 1985		5850
Acid-soluble sulphates of fines		812-118: 1988		
Aggregate abrasion value		812-113: 1990		6239
Aggregate impact value (AIV)		812-112: 1990		6245
Alkali–aggregate reactivity methods: Rock cylinder method, chemical method, and mortar bar method	C586 C289 C227/C126		A23.2-25A	
ASR – Effectiveness of mineral admixtures or GGBS in preventing excessive expansion of concrete	C441		A23.2-28A	
Bulk density (or unit weight)	C29	812-2: 1975	A23.2-10A	5845
Bulk density, voids	C29			5845
Bulking of fine aggregates	C1252			5856
Chloride content of aggregates		812-117: 1988		202
Chlorides, presence of clay content				5831 6244
Clay size particles in aggregate	C117	812-103.1: 1985		201
Clay, friable particles and fine silt (incl. by sieving)	C142	812-111: 1990		5841
Crushing tests of coarse aggregates, including ACV and 10% FACT (TFV)		812-112: 1990	A23.2-3A	5842

Subject				
Descriptive nomenclature for constituents of natural mineral aggregates	C294			
Electrical conductivity of fine aggregates				6240 201
Fineness modulus	C136			5847
Flakiness index of coarse aggregates		812-105.1: 1989	A23.2-13A	
Flat and elongated particles in coarse aggregates			A23.2-24A	
Freezing and thawing tests for concrete	C666			
Frost resistance of coarse aggregates in air entrained concrete by critical dilation procedures; frost heave of aggregates	C682	812-124: 1999		
Grading – see 'Sieve Analysis'				
Impurities in fine aggregate				5834
Low density materials in aggregates	C123		A23.2-4A	5837
Material finer than 0.075 (0.080) mm (No. 200 sieve) in aggregate	C117	812-103.1: 1985	A23.2-5A	
Methylene blue absorption				6243
Moisture content of aggregate	C70 C566	812-109: 1990		5855
Mortar-strength properties of fine aggregates				
Organic impurities in fine aggregate – effect on strength of mortar	C87		A23.2-8A	
Organic impurities; Organic matter in fine aggregate	C40			5832
Organic impurities other than sugar	C87		A23.2-7A	

Table 8.2 (Continued)

Test methods	USA (ASTM)	UK (BS)	Canada (CSA)	SA (SANS)
Particle density (see 'Specific gravity and absorption')				
Particle shape and texture	D3398 D4791	812-105.2: 1990		6244
Particle size distribution: see 'Sieve Analysis'				
Particle size distribution of fine aggregate (pipette method)				
Petrographic examination of aggregates	C295	812-104: 1994		
Polished stone value		812-114: 1989		
Resistance of fine aggregates to degradation by abrasion in the Micro-Deval apparatus			A23.2-23A A23.2-29A	
Sampling and testing of aggregates	D75 C1077 C702	812-101: 1984 812-102: 1989	A23.2-1A	5827
Sand equivalence of fine aggregate				5838
Shell content of coarse aggregates		812-106: 1985		5840
Shell content of fine aggregates		812-106: 1985		5836
Shrinkage and expansion of cement: aggregate mixes; drying shrinkage of aggregates in concrete	C342	812-120: 1989		
Sieve analysis of fine and coarse aggregates (incl. fines content, dust content, etc.)	C136	812-103.1: 1985 812-103.2: 1989	A23.2-2A	201 6241 5834
Soluble deleterious impurities				5839
Soundness of aggregates by use of sodium sulphate or magnesium sulphate	C88	812-121: 1989	A23.2-9A	
Specific gravity (relative density) and absorption	C127 C128	812-2: 1975		5844

Sugar in fine aggregates				5833
Surface moisture in fine aggregate	C70			
Thermal diffusivity of rocks	D4612			
Water-soluble sulphates content		812-118: 1988	A23.2-11A	5850-1
Specifications for aggregates				
Coarse and fine aggregates from natural sources:				
Concrete aggregates	C33	882: 1992	A23.1	1083
Lightweight concrete aggregates	C330 C331 C332	3797: 1990		794
Aggregates for radiation-shielding concrete	C637	4975: 1973		
Standard terminology				
Standard terminology relating to concrete and concrete aggregates	C125			
Standard descriptive nomenclature of constituents of aggregates for radiation-shielding concrete	C638			

Note
See Appendix A for the dual SABS/SANS notation and numbering.

majority of tests are empirical. Even when a test may measure a fundamental parameter such as porosity, the test result is a function of the test method and may not be absolute. Therefore, tests need to be meticulously carried out according to specified procedures, and results treated critically and with due regard to proper interpretation.

Lists of ASTM, CSA, BS, and SANS standards for aggregates (both test methods and specifications) are given in Appendix A. These are relevant for all the chapters.

Developments in aggregate standards

Construction standards are constantly being reviewed and developed, and standards for aggregates are no exception. This is necessary and desirable: necessary because technology and practice are not static and standards to ensure quality and good practice must keep up with technological advances; desirable because well-written standards based on sound science and proven practice can themselves help to advance the science and technology of a field by ensuring their acceptance in practice. Increasingly, environmental concerns are bringing pressure to bear upon available sources of raw materials, thus requiring drafting of innovative standards to cover new requirements. The tendency to litigation in many construction disputes demands that standards be up to date, concise, and clear.

Standards also play an important role in internationalisation of construction. As construction becomes increasingly globalised, so the pressure to draft standards that can be applied everywhere without prejudice to local practice will be considerable. This is in fact a very difficult task, and it is imperative that allowance be made for local conditions. There are many examples of standards drawn up for one set of local circumstances being applied to another set of very different circumstances, often with serious consequences. Aggregates are natural and geological materials and so differ greatly from place to place, and over-rigid specifications are inappropriate. At the same time, principles for desirable aggregate properties that are widely applicable can be established. The way forward is to adopt performance-based specifications, with performance-related tests that will exclude unsuitable aggregates.

Development in aggregate standards in the UK, Europe, and South Africa

In the mid-1980s, an effort was undertaken to re-draft the British Standard for Aggregates: BS 882 (1983) 'Specifications for Aggregates from Natural Sources for Concrete' and BS 812 'Test Methods for Aggregates', which in the 1975 version had existed in four parts. By the early 1990s, BS 812 had been re-drafted in a multitude of parts, some of the parts themselves having different sections. The various parts cover virtually the entire range of aggregate tests that were relevant at that time. The revisions involved their complete overhaul, resulting from the need to include recent developments in standard tests, to incorporate additional non-standard tests, and to obtain reliable precision data. The specification document, BS 882, was also revised and re-issued in 1992, the revision involving improvements towards products being 'fit for purpose'.

However, by the 1990s there was also a strong move towards creating European Standards or 'EN' documents. This was to open up the 'internal market' within Europe for greater trade and the freer flow of goods and services. Pike (1990) wrote that there had been indications from about 1980 that British standards for aggregates would one day be replaced by European standards. Understandably, this brought some concern, primarily because of the risk that unsuitable limits or test methods could be imposed. The obvious answer was that all responsible parties should take an active part in the drafting of standards. By 1988, the CEN (Committee for European Standardisation) had in place a Working Group to write European standards for concrete aggregates. This was shortly replaced by a full Technical Committee on aggregates (CEN TC 154), tasked with setting up a framework for specifying aggregates, and with drafting performance requirements and test methods, with a view to developing standards afterwards. It was recognised that little progress towards full harmonisation of standards could be achieved before test methods that were acceptable to all member states were in place – and there were about 40 such tests to be produced!

The introduction of Europe-wide standards for aggregates is somewhat problematic due to the specific types and usage of aggregates in different parts of Europe. Thus, the European documents cover general requirements for these materials leaving specific issues to member countries or regions. Inevitably, the introduction of these standards will bring changes in the industry, not least because materials may move over greater distances where this was impossible before. Consequently, users and specifiers will need to become more proficient in a wider range of materials for concrete construction.

Work on the introduction of European standards meant that, in effect, work on BS 812 was suspended. Rather, considerable effort was expended not only on producing European standard test methods, but also on resolving policy issues such as appropriate limits, a hierarchy of test methods, and the requirements of quality assurance and certification. By the turn of the millennium, several European test methods were in place. Appendix B contains a table giving the BS EN standards relevant to aggregates as of 2004. Changes in the field of European standardisation are fairly rapid and information quickly becomes out of date. The situation regarding BS 812 is that its many parts are still mostly current (2004), or have been superseded by the relevant EN document (in which case the UK version is given the notation BS EN ...). In some cases, the existing BS document and the replacing EN document are both current, a situation that will fall away in time. Regarding a European specification for aggregates, a document, EN 12620: 2004 has been issued, titled 'Aggregates for Concrete'. This standard specifies quality and grading requirements for aggregates obtained by processing natural material. A further guidance document on the use of BS EN 12620 has also been issued (PD 6682-1: 2003).

A new European code for concrete design and construction EN 206-1 was introduced in the early 1990s in draft form. It was issued as a full BS standard in 2000. This document had the effect of requiring other national standards in Europe to be revised or re-drafted to bring them into line with the European standards. This included changes to well-known test methods such as particle size distribution, particle shape, density and water absorption, durability properties, and so on, as well as new methods not previously in use (e.g. resistance to freeze–thaw, petrographic description, and resistance to fragmentation). Thus, users had to adapt to significant changes in test procedures, and become proficient in new test methods and apparatus (Arens, 2002).

The latest version of the South African standard specification for aggregates, SANS 1083, dates from 1994, and has an accompanying commentary published in 1995, mentioned earlier. The document is generally much sparser than comparable specifications elsewhere, due to a desire not to over-specify or preclude the use of aggregates that may work perfectly well in concrete, despite having, for example, unusual gradings. The absence of clear guidelines particularly for ASR is a shortcoming that needs correcting. South African standards generally are based on BS documents, modified for local conditions. South Africa adopted the European cement specification (EN 197) in February 2002, and will probably continue to adopt relevant European standards, while making allowances for local conditions and materials.

A word is appropriate here on the role of the ISO. The drafters of the EN documents adopted the principle of using ISO documents either in whole or modified form wherever possible, and only to draft new standards where there were no such ISO standards, or where they were severely flawed. While efforts to draw together the activities of ISO and CEN were made, progress towards European standards has been far greater.

The problem with trying to create 'universally' applicable standards among many countries is well illustrated in the production of the new European standard for concrete aggregates. The drafters of this document found that many countries had established grading limits appropriate to their own local materials. Thus, grading criteria were redefined to accommodate whatever material was available. This was done by describing a pair of sieve sizes in mm (d and D) between which most of the particle size distribution should lie, and excluding percentages of oversized and undersized material. The important criteria were established as

- Limits for oversize and undersize fractions
- Consistency in grading
- Overall bandwidth tolerances for defined mid-size sieves in graded aggregates.

The document was drafted to cover the essential principles involved in aggregate gradings, but inevitably, local conditions and materials will continue to dominate practice in any particular country or region.

All standards are revised and modified from time to time, resulting in a continuous state of flux. Notwithstanding this, previous versions of standards often have useful insights that may still be valid. For instance, although the present BS are being replaced by European EN documents, the BS documents are an accurate reflection of the state of concrete manufacture and use in the UK, and are therefore still valuable documents.

Development in aggregate standards in North America

In North America, there has been no recent wholesale revision of the aggregate standards similar to the changes within the European Community. Rather the aggregate standards simply continue to evolve slowly over time, as has always been the custom there. While the most widely used standards are those developed by ASTM, both the Canadian Standards Association and the US Department of Transportation have developed standards and guidelines, as have other federal and state or provincial jurisdictions. Currently, there is more attention being paid to the use of industrial waste products, but changes to the current standards will be slow in coming.

Closure

Aggregate standards, whether in the form of specifications or test methods, have an important role to play in ensuring that concrete with acceptable properties, 'fit for purpose', is produced for construction. The challenge for standards in the future will increasingly be on two fronts: first, that performance-based specifications related to actual service conditions are used, and second that increasing allowance is made for use of marginal and recycled aggregates that may not fit current specifications. The first challenge stems from the need to not over-specify unnecessarily and to permit innovation and flexibility in aggregate use; the latter is a non-negotiable route into the future. The increasing shortages of good quality virgin materials, coupled with environmental restrictions on exploitation of natural resources, will bring strong pressure for use of recycled aggregates and other materials at present considered marginal for concrete construction. For aggregate and concrete practitioners, researchers, specifiers, and producers, these issues will ensure that aggregate science and technology will remain a challenging field in the future.

References

American Society for Testing and Materials R0030 (2003) *Manual of Aggregate and Concrete Testing*, West Conshohocken, PA: American Society for Testing and Materials.

Arens, P. (2002) 'Part 3/4: New test standards for concrete and aggregates – Changes to aggregates testing', *Betonwerk + Fertigteil Technik*, 10(68): 24–34.

Cement and Concrete Institute (1995) *Commentary on SANS 1083:1994, Aggregates from Natural Sources – Aggregates for Concrete*, Midrand: Cement and Concrete Institute.

Fookes, P.G. and Collis, L. (1975a) 'Problems in the Middle East', *Concrete*, 9(7): 12–17.

Fookes, P.G. and Collis, L. (1975b) 'Aggregates and the Middle East', *Concrete*, 9(11): 14–19.

Pike, D.C. (ed.) (1990) *Standards for Aggregates*, Chichester: Ellis Horwood.

Lists of standards
ASTM, CSA, BS and SABS

The lists generally cover relevant:

Test Methods – Aggregates
Specifications – Aggregates
Test Methods – Concrete

Comprising:

ASTM Standards
CSA Standards
BS Standards
SABS (SANS) Standards

Notes

1 The lists of test methods for concrete is restricted to tests that may be relevant to aggregates, such as concrete consistency tests, or that are referred to in the text such as, for example, durability tests.
2 Standards taken as a whole are in a continuous state of revision, and therefore it is important that only the latest versions are used.
3 Virtually all tests in the standards are empirical, although some may measure more fundamental parameters, e.g. porosity. However, even in these cases, the test result is a function of the method of test, and may not be absolute. This also reinforces the need for tests to be meticulously carried out according to the specified procedures. The results also need to be treated with due circumspection and with regard to proper interpretation.

List of relevant ASTM Standards

Standards of the American Society for Testing and Materials (ASTM, West Conshohocken, PA, USA).

Standards referred to are the latest version. Dates are therefore not indicated.

Test methods – Aggregates

C29   Standard Test Method for Bulk Density ('Unit Weight') and Voids in Aggregate.

C40   Standard Test Method for Organic Impurities in Fine Aggregates for Concrete.

C70   Standard Test Method for Surface Moisture in Fine Aggregate.

C87   Standard Test Method for Effect of Organic Impurities in Fine Aggregate on Strength of Mortar.

C88   Standard Test Method for Soundness of Aggregate by use of Sodium Sulphate or Magnesium Sulphate.

C117   Test Method for Materials Finer than 75-μm (No. 200) Sieve in Mineral Aggregates by Washing.

C123   Standard Test Method for Lightweight Particles in Aggregate.

C125   Standard Terminology Relating to Concrete and Concrete Aggregates.

C127   Standard Test Method for Specific Gravity and Absorption of Coarse Aggregate.

C128   Standard Test Method for Specific Gravity and Absorption of Fine Aggregate.

C131   Standard Test Method for Resistance to Degradation of Small-Size Coarse Aggregate by Abrasion and Impact in the Los Angeles Machine.

C136   Standard Test Method for Sieve Analysis of Fine and Coarse Aggregate.

C142   Standard Test Method for Clay Lumps and Friable Particles in Aggregates.

C227   Standard Test Method for Potential Alkali Reactivity of Cement Aggregate Combinations (Mortar-Bar Method).

C289   Standard Test Method for Potential Alkali–Silica Reactivity of Aggregates (Chemical Method).

C294   Standard Descriptive Nomenclature for Constituents of Concrete Aggregates.

C295   Standard Guide for Petrographic Examination of Aggregates for Concrete.

C342   Standard Test Method for Potential Volume Change of Cement–Aggregate Combinations.

C535   Standard Test Method for Resistance to Degradation of Large-Size Coarse Aggregate by Abrasion and Impact in the Los Angeles Machine.

C566   Standard Test Method for Total Evaporable Moisture Content of Aggregate by Drying.

C586 Standard Test Method for Potential Alkali Reactivity of Carbonate Rocks for Concrete Aggregates (Rock Cylinder Method).

C638 Standard Descriptive Nomenclature of Constituents of Aggregates for Radiation-Shielding Concrete.

C641 Standard Test Method for Staining Materials in Lightweight Concrete Aggregates.

C682 Standard Practice for Evaluation of Frost Resistance of Coarse Aggregates in Air-Entrained Concrete by Critical Dilation Procedures.

C702 Standard Practice for Reducing Samples of Aggregate to Testing Size.

C1077 Standard Practice for Laboratories Testing Concrete and Concrete Aggregates for Use in Construction and Criteria for Laboratory Evaluation.

C1137 Standard Test Method for Degradation of Fine Aggregate Due to Attrition.

C1252 Standard Test Method for Uncompacted Void Content of Fine Aggregate (as Influenced by Particle Shape, Surface Texture, and Grading).

C1260 Standard Test Method For Detection of Alkali–Silica Reactive Aggregate by Accelerated Expansion of Mortar Bars.

D75 Standard Practice for Sampling Aggregates.

D2419 Sand Equivalent Value of Soils and Fine Aggregate.

D3398 Standard Test Method for Index of Aggregate Particle Shape and Texture.

D3665 Standard Practice for Random Sampling of Construction Materials.

D3744 Standard Test Method for Aggregate Durability Index.

D4791 Flat and Elongated Pieces in Coarse Aggregate.

Specifications – Aggregates

C33 Standard Specification for Concrete Aggregates.

C330 Standard Specification for Lightweight Aggregates for Structural Concrete.

C331 Standard Specification for Lightweight Aggregates for Concrete Masonry Units.

C332 Standard Specification for Lightweight Aggregates for Insulating Concrete.

C637 Standard Specification for Aggregates for Radiation-Shielding Concrete.

D4612 Standard Practice for Calculating Thermal Diffusivity of Rocks.

Test methods – Concrete and rock

C143   Standard Test Method for Slump of Hydraulic-cement Concrete.

C418   Standard Test Method for Abrasion Resistance of Concrete by Sand-blasting.

C441   Standard Test Method for Effectiveness of Mineral Admixtures or Ground Blast-Furnace Slag in Preventing Excessive Expansion of Concrete due to the Alkali–Silica Reaction.

C469   Standard Test Method for Static Modulus of Elasticity and Poisson's Ratio of Concrete in Compression.

C666   Test Method for Resistance of Concrete to Rapid Freezing and Thawing.

C671   Standard Test Method for Critical Dilation of Concrete Specimens Subjected to Freezing.

C779   Standard Test Method for Abrasion Resistance of Horizontal Concrete Surfaces.

C856   Standard Practice for Petrographic Examination of Hardened Concrete.

C995   Standard Test Method for Time of Flow of Fibre Reinforced Concrete through Inverted Slump Cone.

C1105   Standard Test Method for Length Change of Concrete Due to Alkali–Carbonate Rock Reaction.

C1138   Standard Test Method for Abrasion Resistance of Concrete (Underwater Method).

C1293   Standard Test Method for Determination of Length Change of Concrete due to Alkali–Silica Reaction.

C1362   Standard Test Method for Flow of Freshly Mixed Hydraulic Cement Concrete.

D2936   Direct Tensile Strength of Intact Rock Core Specimens.

D2938   Unconfined Compressive Strength of Intact Rock Core Specimens.

List of relevant BS Standards

Standards of the British Standards Institution (BSI, Chiswick High Road, London, UK).

BS Standards are being replaced by European Standards (BS EN). However, BS standards are referred to in the chapters. A list of BS EN standards is provided separately in Appendix B.

Test methods – Aggregates

BS 812: Parts 1, 2, 3: 1975 and 4: 1976

Part 1: 1975   Testing Aggregates. Methods for Determination of Particle Size and Shape.

Part 2: 1975   Testing Aggregates. Methods for Determination of Physical Properties.

Part 3: 1975   Testing Aggregates. Methods for Determination of Mechanical Properties.

Part 4: 1976   Testing Aggregates. Methods for Determination of Chemical Properties.

BS 812: Part 100: 1990   Testing Aggregates. General Requirements for Apparatus and Calibration.

BS 812: Part 101: 1984   Testing Aggregates. Guide to Sampling and Testing Aggregates.

BS 812: Part 102: 1989   Testing Aggregates. Methods for Sampling.

BS 812: Part 103.1: 1985 (2000)   Testing Aggregates. Methods for Determination of Particle Size Distribution: Sieve Tests.

BS 812: Part 103.2: 1989   Testing Aggregates. Methods for Determination of Particle Size Distribution: Sedimentation Test.

BS 812: Part 104: 1994 (2000)   Testing Aggregates. Method for Qualitative and Quantitative Petrographic Examination of Aggregates.

BS 812: Part 105.1: 1989 (2000)   Testing Aggregates. Methods for Determination of Particle Shape: Flakiness Index.

BS 812: Part 105.2: 1990   Testing Aggregates. Methods for Determination of Particle Shape: Elongation Index of Coarse Aggregate.

BS 812: Part 106: 1985 (2000)   Testing Aggregates. Method for Determination of Shell Content in Coarse Aggregate.

BS 812: Part 109: 1990 (2000)   Testing Aggregates. Methods for Determination of Moisture Content.

BS 812: Part 110: 1990 (2000)   Testing Aggregates. Methods for Determination of Aggregate Crushing Value (ACV).

BS 812: Part 111: 1990 (2000)   Testing Aggregates. Methods for Determination of 10 Per cent Fines Value (TFV).

BS 812: Part 112: 1990 (1995)   Testing Aggregates. Method for Determination of Aggregate Impact Value (AIV).

BS 812: Part 113: 1990 (1995)   Testing Aggregates. Method for Determination of Aggregate Abrasion Value (AAV).

BS 812: Part 114: 1989 (2000)   Testing Aggregates. Method for Determination of the Polished Stone Value.

BS 812: Part 117: 1988   Testing Aggregates. Method for Determination of Water-soluble Chloride Salts.

BS 812: Part 118: 1988 (2000)   Testing Aggregates. Methods for Determination of Sulphate Content.

BS 812: Part 119: 1985   Testing Aggregates. Method for Determination of Acid-soluble Material in Fine Aggregate.

BS 812: Part 120: 1989 Testing Aggregates. Method for Testing and Classifying Drying Shrinkage of Aggregates in Concrete.

BS 812: Part 121: 1989 (2000) Testing Aggregates. Method for Determination of Soundness.

BS 812: Part 123: 1999 Method for Determination of Alkali–Silica Reactivity: Concrete Prism Method.

BS 812: Part 124: 1989 Testing Aggregates. Method for Determination of Frost Heave.

Specifications – Aggregates

BS 882: 1992 Specification for Aggregates from Natural Sources for Concrete.

BS 3797: 1990 Specification for Lightweight Aggregates for Masonry Units and Structural Concrete.

Test methods – Concrete

BS 1881: Part 102: 1983 Testing Concrete. Method for Determination of Slump.

BS 1881: Part 103: 1983 Testing Concrete. Method for Determination of Compacting Factor.

BS 1881: Part 105: 1983 Testing Concrete. Method for Determination of Flow.

BS 1881: Part 121: 1983 Testing Concrete. Method for Determination of Static Modulus of Elasticity in Compression.

BS DD 249. Draft for Development, 1999 Testing Aggregates. Methods for the Assessment of Alkali–Silica Reactivity Potential – Accelerated Mortar-bar Method.

BS 7943: 1999 Guide to the Interpretation of Petrographical Examination for Alkali–Silica Reactivity.

List of relevant CSA Standards

Standards of the Canadian Standards Association (CSA, Mississauga, Ontario, Canada).

Note on applicability of ASTM and CSA Standards
In most cases, the ASTM Standards have been quoted in the text for North American usage. However, CSA standards are also referred to in the text where appropriate. CSA Test Methods sometimes correspond with ASTM or other US Standards.

Standards referred to are the latest version. Dates are therefore not indicated.

Taken from
Canadian Standards Association. Concrete Materials and Methods of Concrete Construction/Methods of Test for Concrete, CSA Standard A23.1-00/A23.2-00.

Test methods – Aggregates

A23.2-1A   Sampling Aggregates for Use in Concrete.
A23.2-2A   Sieve Analysis of Fine and Coarse Aggregate.
A23.2-3A   Clay Lumps in Natural Aggregates.
A23.2-4A   Low-Density Granular Material in Aggregate.
A23.2-5A   Amount of Material Finer than 80 μm in Aggregate.
A23.2-6A   Relative Density and Absorption of Fine Aggregate.
A23.2-7A   Test for Organic Impurities in Fine Aggregates for Concrete.
A23.2-8A   Measuring Mortar-Strength Properties of Fine Aggregate.
A23.2-9A   Soundness of Aggregate by Use of Magnesium Sulphate.
A23.2-10A   Density of Aggregate.
A23.2-11A   Surface Moisture in Fine Aggregate.
A23.2-12A   Relative Density and Absorption of Coarse Aggregate.
A23.2-13A   Flat and Elongated Particles in Coarse Aggregate.
A23.2-14A   Potential Expansivity of Aggregates (Procedure for Length Change due to Alkali–Aggregate Reaction in Concrete Prisms).
A23.2-16A   Resistance to Degradation of Small-Size Coarse Aggregate by Abrasion and Impact in the Los Angeles Machine.
A23.2-17A   Resistance to Degradation of Large-Size Coarse Aggregate by Abrasion and Impact in the Los Angeles Machine.
A23.2-23A   Test Method for the Resistance of Fine Aggregate to Degradation by Abrasion in the Micro-Deval Apparatus.
A23.2-24A   Test Method for Resistance of Unconfined Coarse Aggregate to Freezing and Thawing.
A23.2-25A   Test Method for Detection of Alkali–Silica Reactive Aggregate by Accelerated Expansion of Mortar Bars.
A23.2-26A   Determination of Potential Alkali–Carbonate Reactivity of Quarried Carbonate Rocks by Chemical Composition.
A23.2-27A   Standard Practice to Identify Degree of Alkali-Reactivity of Aggregates and to Identify Measures to Avoid Deleterious Expansion in Concrete.
A23.2-28A   Standard Practice for Laboratory Testing to Demonstrate the Effectiveness of Supplementary Cementing Materials and Chemical Admixtures to Prevent Alkali–Silica Reaction in Concrete.
A23.2-29A   Test Method for the Resistance of Coarse Aggregate to Degradation by Abrasion in the Micro-Deval Apparatus.

Specifications – Aggregates

A23.1   Concrete Materials and Methods of Concrete Construction. Contains a section on Aggregates (Ch. 5), covering inter alia sampling, grading, impurities, deleterious reactions, physical properties.

Test methods – Concrete

A23.2   Methods of Test for Concrete. Contains sections that cover tests on concrete constituents, particularly aggregates (see above), as well as tests for concrete *per se*.

Although CSA has a number of concrete tests, the relevant ASTM standards can also be referred to.

List of relevant SABS/SANS Standards

Standards of the South African Bureau of Standards (SABS, Pretoria, SA).

Standards referred to are assumed to be the latest version. Dates are therefore not indicated.
The dual system SABS/SANS arises from the re-configuration of South African Standard Test Methods (2004).

Test methods – Aggregates

SABS Method 827/SANS 5827   Sampling of Aggregates.
SABS Method 828/SANS 197.   Preparation of Test Samples of Aggregates.
SABS Method 829/SANS 201   Fines Content, Dust Content, and Sieve Analysis of Aggregates.
SABS Method 830/SANS 202   Chloride Content of Aggregates.
SABS Method 831/SANS 5831   Presence of Chlorides in Aggregates.
SABS Method 832/SANS 5832   Organic Impurities in Fine Aggregates.
SABS Method 833/SANS 5833   Detection of Sugar in Fine Aggregates.
SABS Method 834/SANS 5834   Soluble Deleterious Materials in Fine Aggregates (limit test).
SABS Method 835/SANS 5835   Estimation of the Effect of Fine Aggregate on the Water Requirement of Concrete.
SABS Method 836/SANS 5836   Effect of Aggregates on the Shrinkage and Expansion of Mortar.
SABS Method 837/SANS 5837   Low Density Materials Content of Aggregate.
SABS Method 838/SANS 5838   Sand Equivalent Value of Fine Aggregates.
SABS Method 839/SANS 5839   Soundness of Aggregates (Magnesium Sulphate Method).

SABS Method 840/SANS 5840 Shell Content of Fine Aggregate.

SABS Method 841/SANS 5841 Aggregate Crushing Value of Coarse Aggregates.

SABS Method 842/SANS 5842 FACT Value (10% Fines Aggregate Crushing Value) of Coarse Aggregates.

SABS Method 843/SANS 5843 Water Absorption of Aggregates.

SABS Method 844/SANS 5844 Particle and Relative Densities of Aggregates.

SABS Method 845/SANS 5845 Consolidated Bulk Density and Voids Content of Aggregate.

SABS Method 846/SANS 5846 Abrasion Resistance of Coarse Aggregates (Los Angeles Machine Method).

SABS Method 847/SANS 5847 Flakiness Index of Coarse Aggregates.

SABS Method 850-1/SANS 5850-1 Sulphates Content of Fines in Aggregates Part 1: Water-soluble Sulphates in Fines in Aggregates.

SABS Method 850-2/SANS 5850-2 Sulphates Content of Fines in Aggregates Part 2: Acid-soluble Sulphates in Fines in Aggregates.

SABS Method 855/SANS 5855 Free Water Content of Aggregates.

SABS Method 856/SANS 5856 Bulking of Fine Aggregates.

SABS Method 1239/SANS 6239 Aggregate Impact Value of Coarse Aggregate.

SABS Method 1240/SANS 6240 Electric Conductivity of Fine Aggregate.

SABS Method 1241/SANS 6241 Particle Size Distribution of Material of Diameter Smaller than 75 μm in Fine Aggregate (Hydrometer Method).

SABS Method 1242/SANS 6242 Acid Insolubility of Aggregates.

SABS Method 1243/SANS 6243 Deleterious Clay Content of the Fines in Aggregate (Methylene Blue Adsorption Indicator Test).

SABS Method 1244/SANS 6244 Particles of Diameter 20 and 5 μm and Smaller, respectively, in Fine Aggregate (Pipette Method).

SABS Method 1245/SANS 6245 Potential Reactivity of Aggregates with Alkalis (Accelerated Mortar Prism Method).

Specifications – Aggregates

SABS 794/SANS 794 Aggregates of Low Density.

SABS 1083/SANS 1083 Specification for Aggregates from Natural Sources.

SABS 1090/SANS 1090 Aggregates from Natural Sources – Fine Aggregates for Plaster and Mortar.

Test methods – Concrete

SABS Method 862-1/SANS 5862-1 Concrete Tests – Consistency of Freshly Mixed Concrete – Slump Test.

SABS Method 862-2/SANS 5862-2 Concrete Tests – Consistency of Freshly Mixed Concrete – Flow Test.

SABS Method 862-3/SANS 5862-3 Concrete Tests – Consistency of Freshly Mixed Concrete – Vebe Test.

SABS Method 862-4/SANS 5862-4 Concrete Tests – Consistency of Freshly Mixed Concrete – Compacting Factor and Compaction Index.

Appendix B

BS EN Standards

Note
The list of BS EN Standards that follows was current as at mid-2004. New standards are added from time to time, and therefore the list may not be accurate as time progresses.

List of relevant BS EN Standards

BS Standards are being replaced by European Standards (BS EN). However, BS standards are referred to in the chapters.

Test methods – Aggregates

BS EN 932 Parts 1, 2, 3, 5, and 6   Tests for General Properties of Aggregates.

Part 1: 1997   Methods for Sampling.
Part 2: 1999   Methods for Reducing Laboratory Samples.
Part 3: 1997   Procedure and Terminology for Simplified Petrographic Description.
Part 5: 2000   Common Equipment and Calibration.
Part 6: 1999   Definitions of Repeatability and Reproducibility.

BS EN 933 Parts 1 to 10   Tests for Geometrical Properties of Aggregates.

Part 1: 1997   Determination of Particle Size Distribution. Sieving Method.
Part 2: 1996   Determination of Particle Size Distribution. Test Sieves, Nominal Size of Apertures.
Part 3: 1997   Determination of Particle Shape. Flakiness Index.
Part 4: 2000   Determination of Particle Shape. Shape Index.

Part 5: 1998   Determination of Percentage of Crushed and Broken Surfaces in Coarse Aggregate Particles.

Part 6: 2001   Assessment of Surface Characteristics. Flow Coefficient of Aggregate.

Part 7: 1998   Determination of Shell Content. Percentage of Shells in Coarse Aggregate.

Part 8: 1999   Assessment of Fines. Sand Equivalent Test.

Part 9: 1999   Assessment of Fines. Methyl Blue Test.

Part 10: 2001   Assessment of Fines. Grading of Filler (Air-jet Sieving).

BS EN 1097 Parts 1 to 10   Tests for Mechanical and Physical Properties of Aggregates.

Part 1: 1996   Determination of the Resistance to Wear (Micro-Deval).

Part 2: 1998   Methods for Determination of Resistance to Fragmentation.

Part 3: 1998   Determination of Loose Bulk Density and Voids.

Part 4: 1999   Determination of Voids of Dry Compacted Filler.

Part 5: 1999   Determination of the Water Content by Drying in a Ventilated Oven.

Part 6: 2000   Determination of Particle Density and Water Absorption.

Part 7: 1999   Determination of the Particle Density of Filler.

Part 8: 2000   Determination of Polished Stone Value.

Part 9: 1998   Determination of the Resistance to Wear by Abrasion from Studded Tyres. Nordic Test.

Part 10: 2002   Determination of Water Suction Height.

BS EN 1367 Parts 1 to 5   Tests for Thermal and Weathering Properties of Aggregates.

Part 1: 2000   Determination of Resistance to Freezing and Thawing.

Part 2: 1998   Magnesium Sulphate Test.

Part 3: 2001   Boiling Test for Sonnenbrand Basalt.

Part 4: 1998   Determination of Drying Shrinkage.

Part 5: 2002   Determination of Resistance to Thermal Shock.

BS EN 1744 Parts 1 and 3   Tests for Chemical Properties of Aggregates.

Part 1: 1998   Chemical Analysis.

Part 3: 2002   Preparation of Eluates by Leaching of Aggregates.

BS EN 12407: 2000   Natural Stone Test Methods. Petrographic Examination.

Specifications – Aggregates

BS EN 12620: 2002   Aggregates for Concrete.
BS EN 13055 Part 1   Lightweight Aggregates.
Part 1: 2002   Lightweight Aggregates for Concrete Mortar and Grout.

Glossary

This glossary has been gleaned from several sources, most importantly ASTM C125: *Standard Terminology Relating to Concrete and Concrete Aggregates*. Other sources consulted were American Concrete Institute Education Bulletin E1-99 *Aggregates for Concrete*, ACI 116R-00 *Cement and Concrete Terminology*, and Canadian Standards Association CSA A23.1 *Concrete Materials and Methods of Concrete Construction*. The terms were selected primarily for their application to aggregates and their use in concrete.

Abrasion resistance Ability of a surface to resist being worn away by rubbing and friction.

Absorption The process by which a liquid is drawn into and tends to fill permeable pores in a porous solid body; also, the increase in mass of a porous solid body resulting from the penetration of a liquid into its permeable pores.

Aggregate Granular material, such as sand, gravel, crushed stone, or iron blast-furnace slag, used with a cementing medium to form hydraulic-cement concrete or mortar.

Coarse aggregate (1) Aggregate predominantly retained on the 4.75-mm (No. 4) sieve; or (2) that portion of an aggregate retained on the 4.75-mm (No. 4) sieve.

Fine aggregate (1) Aggregate passing the 3/8-in. (9.5-mm) sieve and almost entirely passing the 4.75-mm (No. 4) sieve and predominantly retained on the 75-μm (No. 200) sieve; or (2) that portion of an aggregate passing the 4.75-mm (No. 4) sieve and retained on the 75-μm (No. 200) sieve.

Note: The definitions for *coarse aggregate* and *fine aggregate* are alternatives to be applied under differing circumstances. Definition (1) is applied to an entire aggregate either in a natural condition or after processing. Definition (2) is applied to a portion of an aggregate. Requirements for properties and grading should be stated in the specifications.

Heavyweight or high density aggregate   Aggregate of high density, such as barytes, magnetite, limonite, ilmenite, iron, or steel.

Lightweight or low density aggregate   Aggregate of low density used to produce lightweight concrete, including: pumice, scoria, volcanic cinders, tuff, and diatomite; expanded or sintered clay, shale, slate, diatomaceous shale, perlite, vermiculite, or slag; and end products of coal or coke combustion.

Normal density aggregate   Natural sand, manufactured sand, gravel, crushed gravel, crushed stone, air-cooled iron blast-furnace slag, or any other suitable aggregate from which normal density concrete can be produced.

Alkali–aggregate reaction   Chemical reaction in mortar or concrete between alkalis from portland cement or other sources and certain constituents of some aggregates; under certain conditions, harmful expansion of the concrete or mortar may result.

Bleeding   The autogenous flow of mixing water within, or its emergence from, newly placed concrete or mortar caused by the settlement of the solid materials within the mass, also called water gain.

Bulk density *of aggregate*   The mass of a unit volume of bulk aggregate material (the unit volume includes the volume of the individual particles and the volume of the voids between the particles).

Note: This term is preferred to the term 'unit weight'.

Bulk specific gravity (*saturated surface dry*)   The ratio of the mass of a volume of a material including the mass of water within the pores in the material (but excluding the voids between particles) at a stated temperature to the mass of an equal volume of distilled water at a stated temperature.

Cementitious material (*hydraulic*)   An inorganic material or a mixture of inorganic materials that sets and develops strength by chemical reaction with water by formation of hydrates and is capable of doing so under water.

Cementitious mixture   A mixture (mortar, concrete, or grout) containing hydraulic cement.

Cement, portland   The product obtained by pulverizing clinker consisting essentially of hydraulic calcium silicates with calcium sulphate as an interground addition; when mixed with water it forms the binder in portland cement concrete.

Colorimetric test   A procedure used to indicate the amount of organic impurities present in fine aggregate.

Concrete   A composite material that consists essentially of a binding medium within which are embedded particles or fragments of aggregate;

in hydraulic-cement concrete, the binder is formed from a mixture of hydraulic cement and water.

Consistency *of fresh concrete, mortar, or grout*   The relative mobility or ability to flow.

Crushed gravel   The product resulting from the artificial crushing of gravel with substantially all fragments having at least one face resulting from fracture.

Crushed stone   The product resulting from the artificial crushing of rocks, boulders, or large cobblestones, substantially all faces of which have resulted from the crushing operation.

D-cracking *in concrete*   A series of cracks near to and roughly parallel to features such as joints, edges, and structural cracks.

Density   Mass per unit volume (preferred over the term 'unit weight').

Elongated piece (*of aggregate*)   A particle of aggregate for which the ratio of the length to width of its circumscribing rectangular prism is greater than a specified value (see also **flat piece** (*of aggregate*)).

Extender   A finely divided mineral material, other than cement, that is used in a concrete mix as a supplementary cementitious material; usually comprises latent hydraulic materials such as ground granulated blast-furnace slag, or pozzolanic materials such as fly ash or microsilica.

Fineness modulus   A factor obtained by adding the percentages of material in the sample that is coarser than each of the following sieves (cumulative percentages retained), and dividing the sum by 100: 150 μm (No. 100), 300 μm (No. 50), 600 μm (No. 30), 1.18 mm (No. 16), 2.36 mm (No. 8), 4.75 mm (No. 4), 9.5 mm (3/8 in.), 19.0 mm (3/4 in.), 37.5 mm ($1^1/_2$ in.), 75 mm (3 in.), 150 mm (6 in.).

Flat piece (*of aggregate*)   A particle of aggregate for which the ratio of the width to thickness of its circumscribing rectangular prism is greater than a specified value (see also **elongated piece** (*of aggregate*)).

Free moisture   Moisture not retained or absorbed by aggregate. Also called surface moisture.

Fresh concrete   Concrete which possesses enough of its original workability so that it can be placed and consolidated by the intended methods.

Gradation, grading   The distribution of particles of aggregate among various sizes; usually expressed in terms of total percentages larger or smaller than each of a series of sieve openings or the percentages between certain ranges of sieve openings.

Gravel   Coarse aggregate resulting from natural disintegration and abrasion of rock or processing of weakly bound conglomerate.

Harsh mixture   A concrete mixture that lacks desired workability and consistency due to a deficiency of mortar or aggregate fines.

Igneous rocks Rocks that have solidified from a molten solution.

Los Angeles abrasion test A procedure used to measure the abrasion resistance of aggregates.

Manufactured sand Fine aggregate produced by crushing rock, gravel, iron blast furnace slag, or hydraulic-cement concrete.

Maximum size (*of aggregate*) In specifications for, or description of, aggregate the smallest sieve opening through which the entire amount of aggregate is required to pass.

Metamorphic rocks Rocks altered and changed from their original igneous or sedimentary form by heat, pressure, or a combination of both.

Mortar bar test A procedure used to determine whether an aggregate will expand excessively due to the alkali–aggregate reaction when used in concrete.

Nominal maximum size (*of aggregate*) In specifications for, or description of, aggregate the smallest sieve opening through which the entire amount of the aggregate is permitted to pass.

Note: Specifications on aggregates usually stipulate a sieve opening through which all of the aggregate may, but need not, pass so that a stated maximum proportion of the aggregate may be retained on that sieve. A sieve opening so designated is the *nominal maximum size* of the aggregate.

Pop-out The breaking away of small portions of a concrete surface due to internal pressure which leaves a shallow, typically conical, depression.

Relative density The ratio of mass of a volume of a material at a stated temperature to the mass of the same volume of distilled water at a stated temperature; also called specific gravity.

Reactive aggregate Aggregate containing substances capable of reacting chemically with the products of solution or hydration of the portland cement in concrete or mortar under ordinary conditions of exposure, sometimes resulting in harmful expansion, cracking, or staining.

Roundness A term referring to the relative sharpness or angularity of aggregate particle edges or corners.

Sand Fine aggregate resulting from natural disintegration and abrasion of rock or processing of completely friable sandstone.

Saturated surface dry Condition of an aggregate particle when the permeable voids are filled with water and no water is on the exposed surfaces.

Sedimentary rock Rocks formed by the deposition of plant and animal remains, and of materials formed by the chemical decomposition and physical disintegration of igneous, sedimentary, or metamorphic rocks.

Segregation   The unintentional separation of the constituents of concrete or particles of an aggregate, causing a lack of uniformity in their distribution.

Sieve analysis   Determination of the proportions of particles lying within certain size ranges in a granular material by separation on sieves of different size openings.

Slump   A measure of consistency of freshly mixed concrete obtained by placing the concrete in a truncated cone of standard dimensions, removing the cone and measuring the subsidence of the concrete to the nearest 6 mm (1/4 in.).

Soundness   For aggregate, the ability to withstand the aggressive action to which concrete containing it might be exposed, particularly that due to weather.

Specific gravity   See relative density.

Sphericity   A property of aggregate relating to the ratio of surface area to volume, spherical or cubical particles have a higher degree of sphericity than flat or elongated particles.

Surface moisture   See free moisture.

Surface texture   Degree of roughness or irregularity of the exterior surfaces of aggregate particles or hardened concrete.

Unit weight (*of aggregate*)   Mass per unit volume. Use the preferred term 'bulk density'.

Water-cement ratio   The ratio of the mass of water, exclusive only of that absorbed by the aggregates, to the mass of portland cement in concrete, mortar, or grout, stated as a decimal.

Workability of concrete   That property determining the effort required to manipulate a freshly mixed quantity of concrete with minimum loss of homogeneity.

Notations

a/c	Aggregate/cement ratio (by mass)
ACV	Aggregate Crushing Value
AN	Angularity Number
c/w	Cement to water ratio (by mass) (Inverse of w/c)
C	Concentration of liquid
C'	Yield value
C''	Mobility
CBD	Consolidated bulk density
CV	Coefficient of variation
D_{min}	Minimum aggregate size
D_{max}, d_{max}	Maximum aggregate size
D, D_f	Diffusion coefficient
e	Void content
E	Static modulus of elasticity
E_a	Aggregate (rock) elastic modulus
E_c	Concrete elastic modulus
E_m	Mortar elastic modulus
F	Faraday constant
f'_c	Concrete compressive strength
f'_r	Concrete flexural strength
f_{cu}	Characteristic concrete (cube) strength
FM	Fineness Modulus
g	Gravitational acceleration
HPC	High performance concrete
HSC	High strength concrete
J	Flux per unit cross-sectional area
k	Coefficient of permeability; thermal conductivity
K_0	Aggregate stiffness factor
K_c	Fracture toughness
LBD	Loose bulk density
M	Mass
M_{AD}	Mass of air-dry aggregate
M_D	Mass of (oven dry) solids
M_{SSD}	Mass of SSD aggregate
M_s	Mass of solid
M_T	Total mass

M_{wet}	Mass of wet aggregate (having free surface moisture)
M_w	Mass of evaporable water
N	Angular speed
p	Porosity
P	Pressure
R	Gas constant
R_c	Fracture energy (or specific work of fracture)
RA	Recycled Aggregate
RCA	Recycled Concrete Aggregate
RD	Relative density
S	Sorptivity
S_c, S_p	Shrinkage of concrete and paste respectively
SSD	Saturated-surface-dry
SWR	Standard water requirement
T	Torque
UCS	Unconfined compressive strength
V_a	Volume fraction of aggregate
V_p	Pore volume
V_s	Volume of solids
V_T	Total volume
V_V	Void volume
w	Moisture content
w/c	Water to cement ratio (by mass)
T	Absolute temperature (K)
ΔT	Change in temperature
Z	Electrical charge

Greek symbols used

α	Coefficient of linear thermal expansion and contraction
β	Aggregate–matrix interaction coefficient
ϕ	Packing density
γ	Unit weight
ε	Strain (linear)
$\varepsilon_c, \varepsilon_p$	Creep of concrete and paste respectively
σ	Stress, or strength
ω	Water content
ρ	Density
ρ_{bulk}	Bulk Density

Abbreviations

CSF	Condensed silica fume (microsilica)
FA	Fly ash (Pulverised fuel ash)
GGBS	Ground-granulated blast furnace slag
OPC	Ordinary portland cement